Science and the Garden

Science
and the
Garden

The Scientific Basis of Horticultural Practice

Edited by

David S. Ingram
Daphne Vince-Prue
Peter J. Gregory

Published for the
Royal Horticultural Society
By Blackwell Science

Blackwell Science Ltd, a Blackwell Publishing company
Editorial Offices:
Blackwell Science Ltd, 9600 Garsington Road, Oxford OX4 2DQ, UK
Tel: +44 (0) 1865 776868
Blackwell Publishing Professional, 2121 State Avenue, Ames, Iowa, 50014-8300, USA
Tel: +1 515 292 0140
Blackwell Science Asia Pty, 550 Swanston Street, Carlton, Victoria 3053, Australia
Tel: +61 (0)3 8359 1011

First published 2002 by Blackwell Science Ltd

6 2006

ISBN-10: 0-632-05308-9
ISBN-13: 978-0-632-05308-7

Library of Congress Cataloging-in-Publication Data is available

A catalogue record for this title is available from the British Library

Set in 10/11.5pt Times
by DP Photosetting, Aylesbury, Bucks
Printed and bound in Great Britain
by TJ International Ltd, Padstow, Cornwall

The publisher's policy is to use permanent paper from mills that operate a sustainable forestry policy, and which has been manufactured from pulp processed using acid-free and elementary chlorine-free practices. Furthermore, the publisher ensures that the text paper and cover board used have met acceptable environmental accreditation standards.

For further information on
Blackwell Publishing, visit our website:
www.blackwellpublishing.com

Contents

Foreword

Since its establishment in 1804, the Royal Horticultural Society has set out to foster and encourage the advancement of horticultural science and to stimulate a wider understanding of both the principles and practices involved. This new and important book had its origins in the RHS Science and Horticultural Advice Committee, and the balance of authors involved demonstrates the vitality of the relations between the RHS scientific staff and the wider science community involved with the RHS.

The book progresses logically from consideration of the basic structures and functions of garden plants, through nomenclature and genetics to the environmental factors affecting growth, to methods of propagation and production, to pest and disease control, and finally to post-harvest management and storage. It is principally a book of '*Why*' with strong emphasis on the underlying science but it also, where appropriate, deals with '*How*' and gives the rationale behind practical advice. Although written with the student in mind it will also appeal to gardeners, growers and scientists who will appreciate the width of expertise deployed by the authors in covering the subjects and in bringing an objective perspective to the impact of biotechnology on horticulture.

It is humbling to consider the wonder of plants and to appreciate their structure and function which combine engineering and chemical manufacture beyond the dreams of man: a highly efficient, fully co-ordinated and multi-functional organism with no moving parts and fuelled by natural resources. It is said that environmental factors drive evolutionary responses and this is clearly demonstrated by the infinite variations found in plants to exploit environmental niches. Gardeners need to appreciate that this adaptation to environment is not for decoration and their delectation but for the plant's function and survival. Understanding the science underpinning the differences between form, function and survival strategies of different plant groups will greatly aid their cultivation and the enjoyment of gardening. The relationship between plants and the environment in which they grow is dynamic. The plant can obviously respond immediately to short-term stress caused by factors such as variation in water supply and temperature, but it also must often be sensitive to regular seasonal changes to trigger major physiological processes such as the change from vegetative growth to flowering. Much is already known about these processes and is incorporated into horticultural practice, but the further understanding of these mechanisms, and of the trigger signals that initiate them, is an exciting area of science that will have great implications for the gardener and the commercial grower.

Gardening, fortunately, is not solely the application of science. Indeed gardening is a combination of practical, aesthetic and philosophic ideals. The chapters on 'Designing Plants' and 'Shape and Colour' tackle these wider issues as well as considering the underlying science, while the chapter on 'Controlling the Undesirables' starts off with an interesting perspective on bio-diversity.

> 'In the imagined Arcadian wilderness before gardening was invented there were no undesirables, only a rich bio-diversity. Today's gardeners find this richness excessive and re-label some of it pests, disease and weeds. Remembering, therefore, that a pest, disease or weed is simply bio-diversity being over-assertive, it may be necessary to take some corrective action'.

These rich veins of common sense run throughout the book and temper the hard science with an awareness of the needs, desires and hopes of gardeners. The book is the culmination of much hard work in writing down and then editing the collective expert scientific knowledge of the authors, who were brought together by their interest in and involvement with the Royal Horticultural Society. It is a substantial achievement and will give real benefit and pleasure to all interested in horticulture and gardening.

John MacLeod
RHS Professor of Horticulture

Biographical details

John MacLeod was, for many years, involved in applied agronomic research on the Agricultural Development and Advisory Service Experimental Farms before becoming Director of the Experimental Centres and then, in 1990, Director of the National Institute of Agricultural Botany in Cambridge.

A member of the Royal Horticultural Society Science and Horticultural Advice Committee since 1993 he became Chairman in 2000. He is now the Royal Horticultural Society Professor of Horticulture.

Preface

Science and the Garden has been written primarily for students of horticulture, but we expect that it will also be of interest to amateur gardeners and professional growers who would like to understand more about the science that underlies horticultural practices.

Most conventional gardening books concentrate on how and when to carry out horticultural tasks such as pruning, seed sowing and taking cuttings. In contrast, the aim of the present book is to explain in straightforward terms some of the science that underlies these practices. We address such diverse questions as: why are plants green? Why should one cut beneath a leaf node when taking cuttings? Why do plants need so much water? Why is light so important and what effect does it have on plant growth? How do plants detect drying soils and how is growth modified to improve their survival chances? Why are plants more resistant to freezing in the autumn than in spring? How do plants detect seasonal changes in their environment? Why do chrysanthemums flower in the autumn and onions produce their bulbs in the summer?

The first part of the book is concerned with some fundamental principles. Chapter 1 describes the structural features of the plant, and introduces biochemical and physiological processes such as photosynthesis and water and solute transport, which are expanded on in later chapters in relation to particular aspects of horticultural practice. Chapter 2 introduces the often difficult question of how plants are named. Plant names are a problem for many gardeners and this chapter explains the structure of plant nomenclature in simple terms; it outlines the rules for naming plants, discusses why names sometimes change and, most importantly for the gardener, what is being done to achieve stability in plant nomenclature.

It is often thought that genetic modification (GM) is the 'new' thing in horticulture, but the fact is that most plants grown in gardens (except weeds) have been genetically manipulated in the sense that their genes differ from those of their wild relatives. 'Designing Plants' (Chapter 3) explains how new plants have been developed through cross-breeding and selection processes that have been going on for centuries. The chapter concludes with a look towards the future by showing how new plants can be 'designed' by introducing specific genes using GM technology.

The remainder of the book is more immediately concerned with the practices of horticulture. With the exception of a few aquatics, gardening depends on the soil and Chapter 4 describes the different types of soil, explains how to recognise them and introduces the science underlying soil management practices. Water conservation is an important consideration in many gardens and may well become more important with climate change. The selection of suitable plants is itself a form of water conservation by the gardener and Chapter 5, 'Choosing a Site', describes how certain plants are adapted to grow in dry conditions. All gardens have shady areas and Chapter 5 also explains how plants are able to detect shade from trees and neighbouring buildings and modify their growth accordingly. It ends with advice on how to choose plants for particular situations using scientific principles.

'Raising Plants from Seed' (Chapter 6) and 'Vegetative Propagation' (Chapter 7) are basic horticultural practices. These two chapters discuss the science underlying embryo development, seed maturation and ripening, dormancy and how it may be broken, and the storage of seeds. They also look at vegetative propagation, such as taking cuttings, layering and micropropagation, with special emphasis on the physiological processes underlying these practices, most notably the hormonal control of growth and development. The science of grafting is also considered.

Once plants have been propagated and the site has been selected with due consideration for soil and aspect, the choice of a particular plant for that situation is usually determined by factors such as colour, size and shape, topics that are covered in Chapter 8. The choice of suitable plants also depends on factors such as the time of flowering and, for edible crops, the yields of storage organs such as potato tubers and onion bulbs. The time at which plants enter dormancy and increase their resistance to freezing conditions often determines their ability to grow and even to survive in a particular locality. These processes are largely governed by seasonal factors, although they may be modified by local conditions. Chapter 9, 'Seasons and Weather', focuses on such seasonal factors as day-length and temperature and explains how these are sensed by the plant and how the information is translated into the observed displays. Gardening in the greenhouse is a specialised form of gardening, requiring knowledge of how the physical conditions of the greenhouse interact with the physiology of the plant if optimum yields are to be achieved. Such interactions and their implications for successful greenhouse management are discussed in detail in Chapter 10.

As all gardeners will be aware, no matter how great their horticultural skills, pests, diseases and weeds are a constant problem and can often cause disaster. Chapter 11, 'Controlling the Undesirables', describes how such organisms can be recognised, how they affect plant growth and what strategies are available for combating them. Throughout this chapter the emphasis is on integrated management of pests, diseases and weeds and the use of methods that are environmentally friendly.

Harvesting the flowers, fruit and vegetables that are the product of many hours' labour can be the most satisfying of tasks for the successful gardener. Fittingly then the final chapter in the book considers the physiological basis of the maturation process, and discusses the best ways of harvesting and storing flowers, fruit and vegetables to ensure maximum quality, storage life and flavour.

The book has been edited and written by past and present members of the Royal Horticultural Society's Scientific and Horticultural Advice Committee, past and present members of the scientific staff of the Society, and other specialists. The contents reflect the particular interests of the authors, and their judgement as to the scientific information that is likely to be of greatest importance to gardeners and horticulturists.

We would like to thank all those who have contributed to the volume by writing particular chapters, by commenting on draft chapters and by giving their general support. We also wish to thank Mrs Joyce Stewart, Royal Horticultural Society Director of Horticulture, for allowing her staff to participate in the project and for arranging a grant from the Royal Horticultural Society towards the cost of printing the colour plates. DSI also thanks Napier University for its support during his tenure of a Visiting Professorship there, Mrs Janet Prescott for managing the project and contributing significantly to the editing process, and Mrs Jane Stevens, of St Catharine's College, Cambridge, for her assistance during the final stages of editing the volume.

David S. Ingram
(Cambridge)
Daphne Vince-Prue
(Goring-on-Thames)
Peter J. Gregory
(Reading)

List of Contributors

Guy Barter has worked in commercial horticulture and was superintendent of the Trials Department at the Royal Horticultural Society's Wisley Garden before becoming a horticultural journalist with the Consumer Association publication, *Gardening Which?* He is now Senior Horticultural Advisor to the RHS.
G. Barter, The Royal Horticultural Society, Wisley, Woking, Surrey GU23 6QB.

Anna Dourado was an International Plant Genetic Resources Institute (IPGRI) Intern at the Asian Vegetable Research and Development Center (AVRDC) Taiwan and a visiting Lecturer at Massey University, New Zealand before becoming a Lecturer in Vegetable Production at the University of Bath. She has been Senior Lecturer in Crop Production at the Royal Agricultural College, Cirencester, and Head of Horticultural Science, Advice and Trials at the Royal Horticultural Society. She is now a part-time lecturer in horticulture and a freelance writer.
Dr. A.M. Dourado, Malvern, London Road East, Amersham HP7 9DL.

Peter J. Gregory was Demonstrator then Research Fellow in Physical Chemistry and Soil Science at the University of Nottingham. Next he became Lecturer, then Reader, in Soil Science at the University of Reading. He was then Principal Scientist at CSIRO, Australia and is now Professor of Soil Science and Pro-Vice-Chancellor (Research) at the University of Reading.
Professor P.J. Gregory The University of Reading, Whiteknights, PO Box 233, Reading RG6 6DW.

Andrew Halstead is Senior Entomologist at the Royal Horticultural Society's Garden, Wisley and has 30 years experience of diagnosing and giving advice on garden pest problems. He is a Fellow of the Royal Entomological Society and a past President of the British Entomological and Natural History Society.
A. Halstead, The Royal Horticultural Society, Wisley, Woking, Surrey GU23 6QB.

David S. Ingram OBE was Research Fellow in Plant Pathology at the Universities of Glasgow and Cambridge before becoming Lecturer then Reader in Plant Pathology at the University of Cambridge. Next he was Regius Keeper of the Royal Botanic Garden Edinburgh and Royal Horticultural Society Professor of Horticulture. He was also Chairman of the Science and Horticultural Advice Committee of the Royal Horticultural Society until 2000. He is now Master of St Catharine's College, Cambridge, and Visiting Professor at Edinburgh, Glasgow and Napier Universities.
Professor D.S. Ingram, The Master's Lodge, St Catharine's College, Cambridge CB2 1RL.

David S. Johnson has spent his career at Horticulture Research International, East Malling (formerly East Malling Research Station) and currently leads the team working in post-harvest research on fresh horticultural crops within the Crop Science Department.
Dr. D.S. Johnson, Horticulture Research International, East Malling, Kent ME19 6BJ.

Stephen L. Jury is the Herbarium Curator and Senior Research Fellow in the School of Plant Sciences at the University of Reading. He has a great interest in the taxonomy of cultivated plants and has served on the Advisory Panel on Nomenclature and Taxonomy of the Royal

Horticultural Society, and is currently also a member of its Science and Horticultural Advice Committee.
Dr. S.L. Jury, School of Plant Sciences, University of Reading, Whiteknights, Reading RG6 2AS.

Ray Mathias is Head of Science, Communication and Education Department at the John Innes Centre, where he is responsible for external and media liaison, and science and society issues. He is a research scientist who has been involved in plant biotechnology for over 20 years, firstly at the Plant Breeding Institute, Cambridge and then at the John Innes Centre, Norwich. He has worked on the application of biotechnology in cereal and brassica crops and on the development of alternative oilseed crops.
Dr. R.J. Mathias, John Innes Centre, Norwich Research Park, Colney, Norwich NR4 7UH.

Jon Pickering is a horticultural graduate and received his doctorate from the University of Reading following research on the horticultural uses of green waste compost. He was formerly a horticultural scientist at the Royal Horticultural Society's Garden, Wisley where he provided advice on topics related to soils, plant nutrition, growing media and composting. He currently works for a consultancy firm involved with large scale composting operations and sustainable waste management technologies.
Dr. J. Pickering, Organic Resource Agency Limited, Elm Farm Research Centre, Hamstead Marshall, Newbury, Berks RG20 0HR.

Chris Prior began his scientific career as a research fellow in the University of Papua New Guinea, working on diseases of cocoa. He then worked on biological control of a range of tropical pests with CAB International. He joined the Royal Horticultural Society in 1996 and is currently Head of Plant Pest and Disease Science.
Dr. C. Prior, The Royal Horticultural Society, Wisley, Woking, Surrey GU23 6QB.

Michael Saynor graduated from the University of Leeds in 1965. He then joined the Agricultural Development and Advisory Service, formerly the National Agricultural Advisory Service, based in Cambridge. After postings to the West Country and the South East, he left ADAS in 1991 to become an independent consultant.
Dr. M. Saynor, 169 Reading Road, Wokingham, Berks RG41 1LJ.

Simon Thornton-Wood was a Tropical Botanist at the Natural History Museum, London, before becoming Horticultural Taxonomist for the National Trust. He later became Head of Botany, and then Head of Science, Advice and Libraries, at the Royal Horticultural Society's Garden, Wisley.
Dr. S. Thorton-Wood, The Royal Horticultural Society, Wisley, Woking, Surrey GU23 6QB.

Daphne Vince-Prue was Lecturer in Horticulture and Reader in Botany at the University of Reading before becoming a Scientific Advisor to the Agriculture and Food Research Council. She later became Head of the Physiology and Chemistry Division at the Glasshouse Crops Research Institute. Since her retirement she has been for many years a member of the Science and Horticultural Advice Committee of the Royal Horticultural Society.
Dr. D. Vince-Prue, 2 Maple Court, Goring-on-Thames, Reading RG7 9BQ.

Timothy Walker has spent most of his professional life at the University of Oxford Botanic Garden where he is currently *Horti Praefectus*. He is particularly interested in the practice and theory of plant conservation. He is the Ernest Cook College Lecturer in Plant Conservation at Somerville College and a Research Lecturer in the Department of Plant Sciences at the University of Oxford.
T. Walker, The Botanic Garden, Rose Lane, Oxford OX1 4AX.

1

Know your Plant

- Harvesting the energy of sunlight in photosynthesis and using it in respiration and metabolism
- Cell structure and the water relations of cells
- The structure of the leaf
- Vascular tissues and the transport of water, minerals and carbohydrates to and from the leaf
- The structures of the stem
- Growth and differentiation at the apex
- The role of hormones
- Secondary thickening and bark
- The structure and growth of roots
- Nitrogen fixing nodules; mycorrhizas and the uptake of phosphates
- Structural and physiological modifications of leaves, stems and roots for special functions
- Reproduction: the structure of flowers, fruits and seeds; alternation of generations
- The diversity of plants: algae, liverworts and mosses, ferns and their relatives, conifers and their relatives and flowering plants

GREEN IS BEAUTIFUL: CHLOROPHYLL AND PHOTOSYNTHESIS

The most remarkable thing about plants is that they are green (Fig. 1.1). This property makes it possible for plants to generate the energy required to sustain almost the entire living world.

To appreciate the significance of this statement it is necessary first to consider what happens to the average motor car if, like the one in Fig. 1.2, it is neglected for long enough: the bodywork begins to rust and the non-ferrous components begin to disintegrate and decay. Indeed, it is the usual experience that all inanimate things, left to themselves like this, eventually reach a state of disorder: buildings crumble, books turn to dust and motor cars rust away. This general tendency is expressed in the second law of thermodynamics which states, in essence, that in an isolated system the degree of disorder and chaos (the *entropy*) can only increase.

Fig. 1.1 'The most remarkable thing about plants is that they are green.' Giant redwoods growing in Younger Botanic Garden, Argyll, a Specialist Garden of the Royal Botanic Garden Edinburgh. Photograph by David S. Ingram.

Creating order out of disorder

When one thinks about living things it is immediately apparent that, in contrast, they are able to create order out of disorder, assembling atoms and molecules to form tissues and bodies of great complexity and sophistication. Now, living things must obey the laws of physics and chemistry, just as our motor car must, so how is this creation of order out of disorder possible, thermodynamically? The answer is that the cells of living things are not isolated systems in a thermodynamic sense, as an abandoned motor car is, for they are constantly deriving energy from another external source, the sun. It is necessary to go back in time to find out how this came about.

The earth first condensed from dust and ashes about 4600 million years ago, and life must have appeared some time during the first thousand million years of the planet's existence. The earliest life forms would have derived their organic molecules (those containing carbon) from their surroundings, a legacy from the pre-biotic 'soup' of chemicals that was left on the cooling earth after its genesis. These would have provided them with energy and the building blocks for making cells. But as these natural resources were exhausted, a key event in the further evolution of life was the development of the process called *photosynthesis*, whereby sunlight is harnessed to provide an alternative external source of energy.

The study of very old rock formations in Australia has suggested that this event must have occurred more than 3600 million years ago, for by that time there were present on the planet simple organisms consisting of single cells or chains of cells that resemble the blue-green *Cyanobacteria* ('blue green algae') that grow in shallow, stagnant water or as a greenish slime on the surface of

Fig. 1.2 It is the usual experience that inanimate things, left to themselves, like this VW Beetle, eventually reach a state of disorder. In contrast, living things, like the plants of oilseed rape, are able to create order out of disorder, assembling atoms and molecules to form tissues and bodies of great complexity and sophistication. Photograph by David S. Ingram.

marshy soils and wet lawns even today (Plate 1). These primitive organisms were so successful that they have remained virtually unchanged throughout almost the entire course of evolutionary time. The Cyanobacteria possess the ability to capture the electromagnetic radiation of the sun and incorporate it into a chemical energy source. This is made possible by the presence of light-absorbing pigments which give the blue-greens their characteristic colour, the most significant being the green pigment *chlorophyll* **a** (Fig. 1.3). The great diversity of plants alive today sprang from these humble beginnings. Their chlorophyll (both the **a** and **b** forms of the pigment [Fig. 1.3]) is contained in microscopic disc-like structures, the *chloroplasts*, embedded in the cells of the leaves and other organs (Plate 2). These probably evolved originally by the 'incorporation' of free-living organisms similar to the Cyanobacteria into the cells of the first primitive plants.

Energy flow: photosynthesis and respiration

Photosynthesis (Fig. 1.4) begins when a molecule of chlorophyll is excited by a quantum of light (a photon) and an electron is thereby moved from one molecular orbit to another of higher energy. Chlorophyll absorbs light in the blue and red regions of the spectrum and reflects green light, which is why plants appear green. This happens deep in the chloroplast (Fig. 1.5), where the electrons energised by sunlight are captured by *photosystems*, which are complexes of proteins and pigments located in the thylakoid membranes. This physical energy is then used to split molecules of water (H_2O). Electrons are transferred from the water to an electron acceptor called a quinone and oxygen is released into the atmosphere as a by-product. The energy is then incorporated into energy rich chemical molecules called ATP and NADPH2.

Finally, this chemical energy is used in the main body (the *stroma*) of the chloroplasts to convert carbon dioxide from the atmosphere into sugars. The process is a cyclical one, sometimes called the Calvin cycle after Melvin Calvin who, together with co-workers, first discovered it. During the cycle, which takes place in the dark, the carbon dioxide combines with ribulose-bis-phosphate to form an unstable molecule containing six carbon atoms. It immediately breaks down to form two molecules of the three carbon compound glycerate-3-phosphate, which is then converted to glyceraldehyde-3-phosphate. Finally, this compound is used to regenerate ribulose-bis-

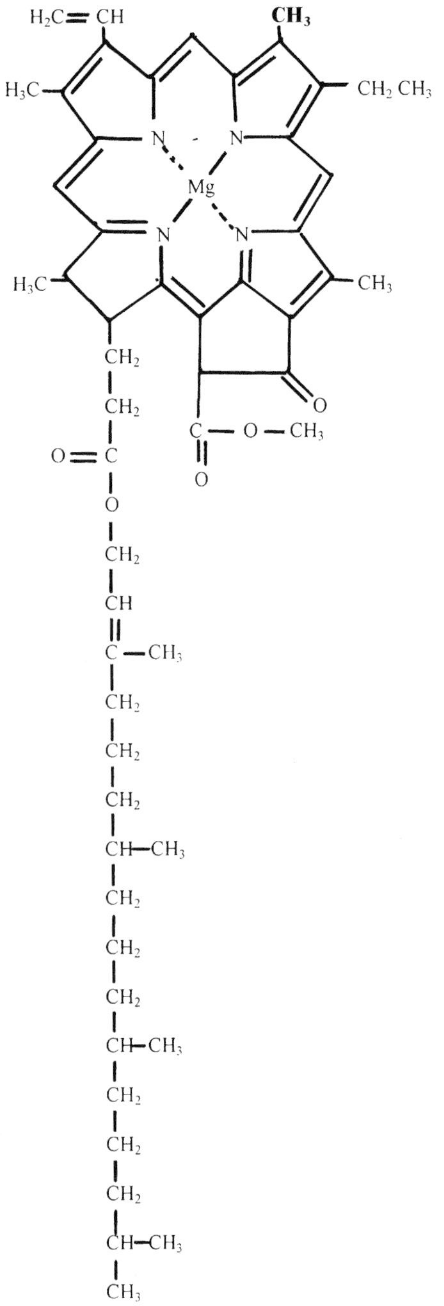

Fig. 1.3 A molecule of the photosynthetic pigment *chlorophyll*. Chlorophyll is a complex molecule comprising a magnesium atom held within a ring of carbon atoms and possessing a long carbon-hydrogen tail. The latter anchors the molecule to the chloroplast membranes (see Fig. 1.5). Chlorophyll exists in various forms which differ from one another in structure and in light absorbing properties. Chlorophyll **a**, illustrated here, occurs in the Cyanobacteria, and in all green plants, where it is the principal light absorbing pigment. It absorbs light in the blue and to a lesser extent, the red parts of the spectrum. Plants also contain chlorophyll **b**, which differs from chlorophyll **a** in having a $-CHO$ group instead of the $-CH_3$ printed in bold on the diagram. It absorbs light in slightly different parts of the spectrum from chlorophyll **a** and passes the energy so captured to chlorophyll **a** itself. It is thus an *accessory pigment* that serves to broaden the range of wavelengths of light that can be used in photosynthesis. A third form of the pigment, chlorophyll **c**, is found only in certain groups of algae.

Another group of light absorbing accessory pigments important in green plants are the red, orange or yellow *carotenoids* (carotenes and xanthophylls), which also occur in the chloroplasts. In most green plants, however, the colour of these pigments is masked by the chlorophylls. Although the carotenoids can also absorb light energy, albeit in different regions of the spectrum from chlorophylls **a** and **b**, the energy must be passed to chlorophyll **a** before it can be used in photosynthesis. One of the carotenoids, beta-carotene, is the main source of vitamin A for animals and humans.

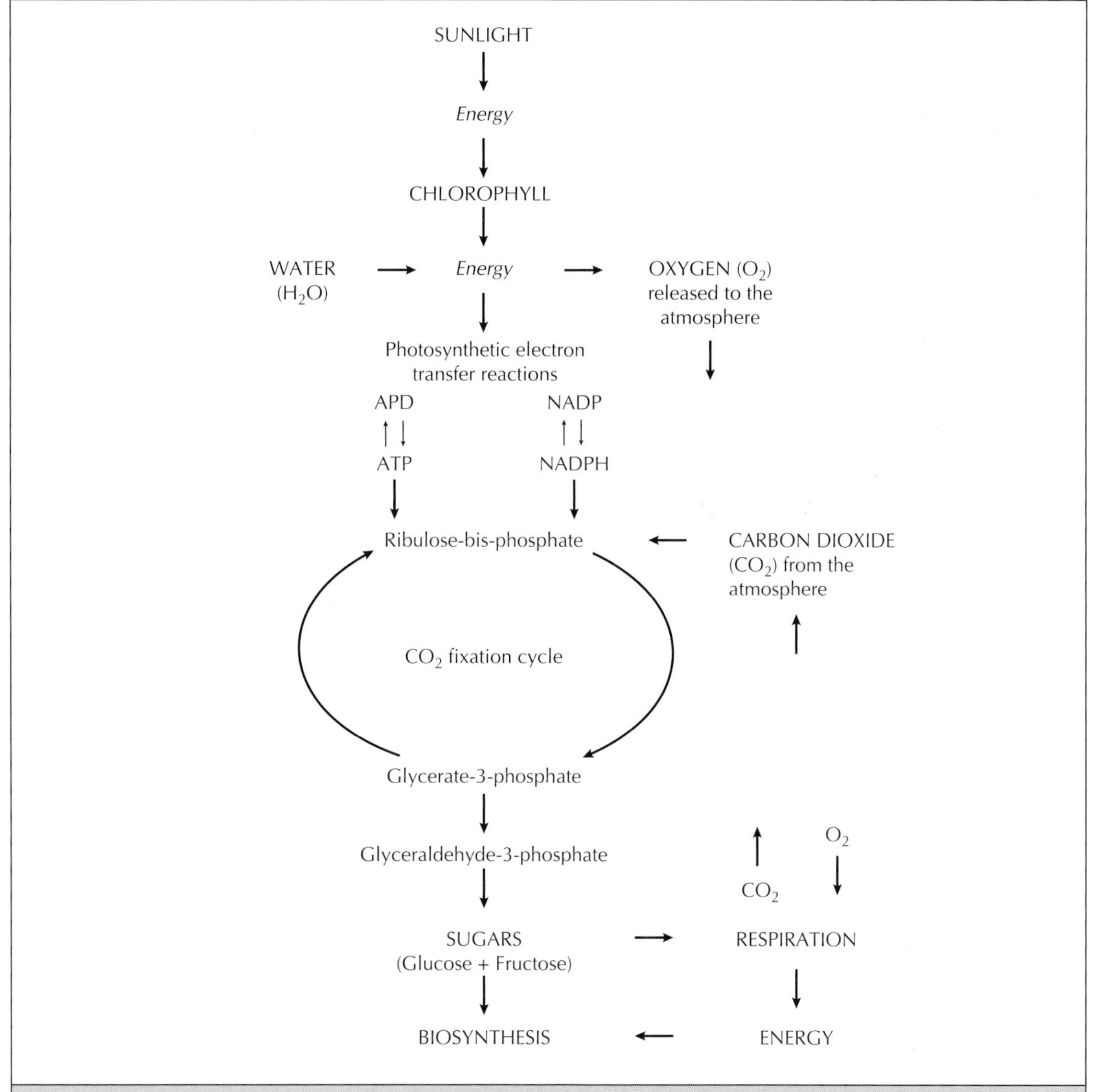

Fig. 1.4 Flow diagram to illustrate the principle stages of photosynthesis. Abbreviations: ADP = adenosine diphosphate; ATP = adenosine triphosphate; NADP = nicotinamide adenine dinucleotide phosphate; NADPH = the reduced form of NADP, formed by the addition of one hydrogen atom and two electrons to NADP.

phosphate and to produce the sugars glucose and fructose.

The Calvin cycle, like all metabolic pathways, is driven by *enzymes*, proteins that act in minute amounts in biological systems to promote chemical changes without being changed themselves. The key enzyme in the Calvin cycle, ribulose-bis-phosphate carboxylase, may be the most abundant protein on earth.

The sugars produced during photosynthesis may be combined with oxygen during *respiration* to release the stored energy. This takes place in

Fig. 1.5 An ultra-thin section of a plant *chloroplast*. Clearly visible are the *thylakoid membranes*, stacked at intervals to form more complex structures called *grana*. Electron microscope photograph by Patrick Echlin, Multi-imaging Centre, University of Cambridge.

organelles called *mitochondria* which occur in all cells of the plant body (Fig. 1.6). It is basically the same process as is used to release energy from carbohydrates in animals. The energy released can be used to drive the biosynthetic processes that lead to the formation of cellulose, fats, amino acids, proteins and the other molecules required by the plant for growth and reproduction.

Some 200 billion tonnes of carbon (0.7×1014 kg) from atmospheric carbon dioxide are fixed by photosynthesis each year. Approximately half of this occurs on land, where the familiar leafy green plants predominate and approximately half is fixed from carbon dioxide in solution in the oceans, lakes and rivers, where the microscopic green plants called *algae* abound as part of the *plankton*. So, through photosynthesis, plants maintain the atmosphere we breathe by consuming carbon dioxide and releasing oxygen; and produce the energy that not only builds the plant body itself but also provides (directly, or indirectly in the case of carnivores) a food source for almost all other living things (Table 1.1).

Table 1.1 Numbers of described and undescribed species of living organisms dependent upon the energy harnessed by photosynthesis in plants.

Group	Species described	Possible number of species remaining to be described
Viruses	5 000	500 000
Bacteria	4 000	up to 3 million
Fungi	7 000	1–1.5 million
Protoza	40 000	200 000
Algae	40 000	up to 10 million
Plants	250 000	500 000
Vertebrates	45 000	50 000
Roundworms	15 000	1 million
Molluscs	70 000	200 000
Crustaceans	40 000	150 000
Spiders, mites	75 000	1 million
Insects	950 000	up to 100 million

Data from *Global Biodiversity* (1992) edited by B. Groombridge and published by Chapman & Hall, London.

INTO THE LABYRINTH: THE LEAF

The manufacturing centre

Although photosynthesis occurs in all the green parts of a plant, the principal site for this process is the leaf. Leaves come in many different shapes and sizes, but most are immediately identifiable as leaves because they have evolved with certain basic features in common, not least the extent of their external and internal surfaces. The external surfaces maximise the collection of sunlight and the internal ones facilitate the movement of the gases carbon dioxide (CO_2) and oxygen (O_2) between the air and the chloroplasts.

Cells

Leaves are made up of *cells* (Fig. 1.6) and although these are usually differentiated to perform particular functions, the majority possess one or more large, membrane-bound, fluid-filled sacs called *vacuoles*, structures not normally found in animal cells. The two most important functions of vacuoles probably relate to photosynthesis, ensuring the most effective deployment of *cytoplasm*, the metabolically active material in the cell which contains the chloroplasts and the other organelles. Firstly, the vacuoles force the cytoplasm towards the edges of the cells with the result that the diffusion paths for gases such as carbon dioxide into and out of the chloroplasts and mitochondria are as short as possible. This is a very important consideration since gases diffuse only very slowly through water, the medium in which the contents of the cytoplasm are dissolved or suspended. Secondly, by restricting the cytoplasm to the edges of the cells, where it is most needed, the vacuoles ensure that the plant is able to achieve a large surface area to volume ratio. As we have seen this is a prerequisite for efficient photosynthesis, with a minimum investment of energy in making cytoplasm. Other, ancillary, functions of vacuoles may be to act as reservoirs for the storage of enzymes and metabolic products not immediately required by the cells, and to act as convenient stores for waste products. The latter are discarded when old leaves are shed, as in the autumn.

Layout of the leaf

The most obvious feature of the leaf (Fig. 1.7) is the relatively large, thin plate of tissue called the *lamina*, with its upper surface usually orientated towards the sun. The lamina is often attached to the stem by means of a stalk, the *petiole*, and this extends into the lamina as the *midrib*, branching to produce a network of *veins*. At the base of the petiole there may be further leafy outgrowths, called *stipules* (as in pea, *Pisum sativum*). In the angle between the petiole and the stem, called the *axil*, is a bud, the *axillary bud*, which may ultimately produce a shoot or a flower. There are many variations on this basic structure (Fig. 1.7), which we shall come to later.

To appreciate the complex arrangement of the different cells and tissues within the leaf it is necessary to examine thin sections with a microscope, and to build up a three-dimensional picture (Fig. 1.8).

Controlling gas and water exchange

An unprotected living structure with a large surface area relative to its volume, constantly exposed to the sun and the wind, will inevitably lose water very rapidly. This does not occur with leaves, because the entire surface is covered with a thin, waterproof layer, the *cuticle*, largely composed of *cutin*, a hydrophobic, fatty molecule called a polyester. Although there may be some microscopic cracks and fissures in the cuticle, by and large it provides a very effective means of restricting the loss of water from the leaf. However, these waterproofing properties also create problems, for the leaf must be able to take up the gas carbon dioxide (CO_2), for photosynthesis, and to lose the excess oxygen (O_2) not required for respiration. It must also be able to take up oxygen for respiration when photosynthesis is not taking place, such as during the night or when the level of sunlight is low. As we shall see shortly, the leaf must also be able to lose a certain amount of water vapour as part of the *transpiration* process in which the movement of water from the soil, through the plant and into the atmosphere provides a means of supplying every cell with water molecules, mineral ions, hormones and other essentials of life. Evaporation of water from the

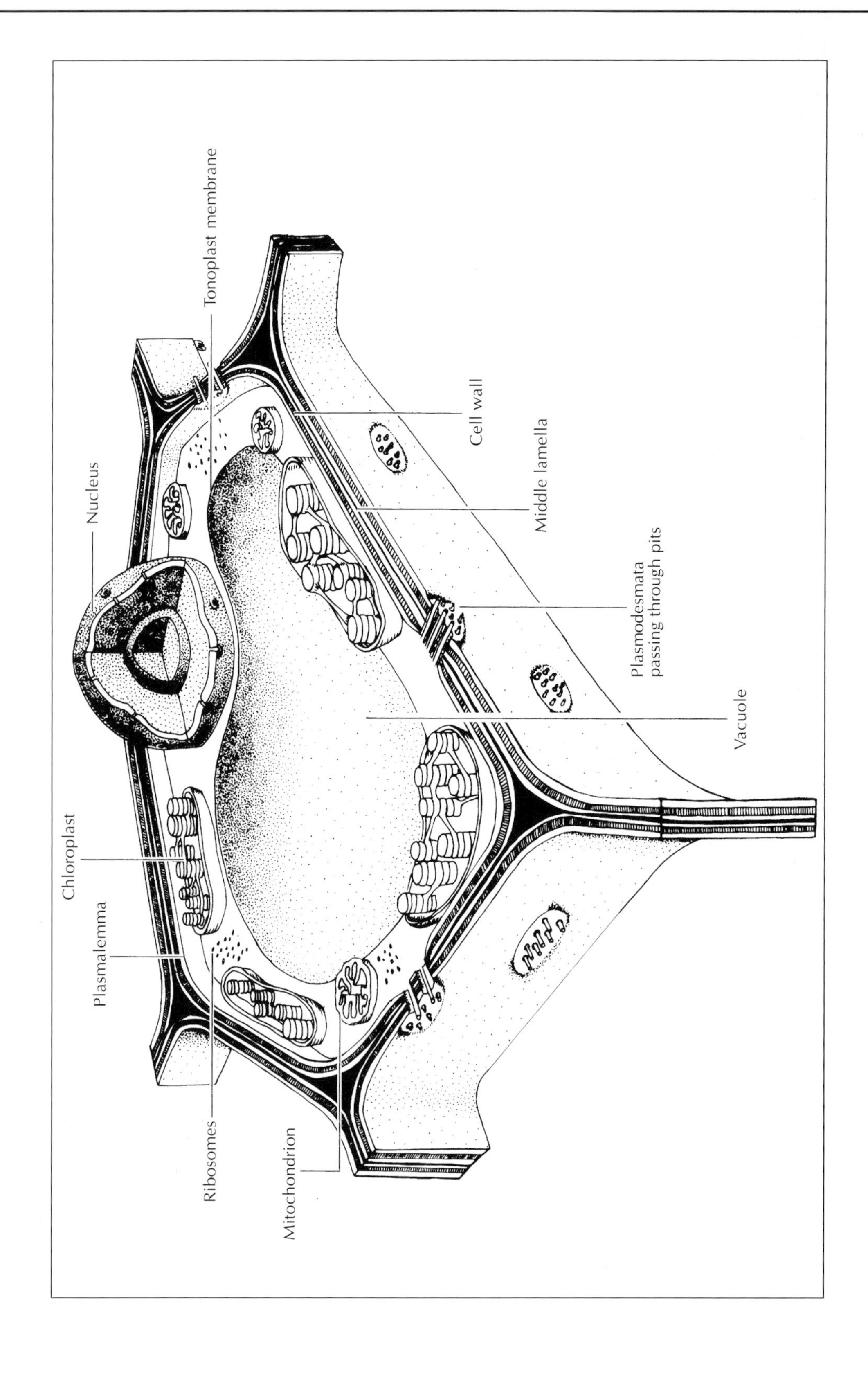
Tonoplast membrane
Nucleus
Cell wall
Middle lamella
Plasmodesmata passing through pits
Vacuole
Chloroplast
Plasmalemma
Ribosomes
Mitochondrion

Fig. 1.6 Diagram of a plant cell seen in section. Clearly visible within the *cytoplasm* are the following organelles: the *vacuole*, bounded by the *tonoplast membrane*; the *nucleus*, which contains the *chromosomes*; *chloroplasts*, the sites of photosynthesis (see Plate 2 and Figures 1.4 and 1.5); *mitochondria*, with their folded inner membranes, the sites of respiration; and *ribosomes*, the sites of protein synthesis. The cytoplasm is bounded by a membrane, the *plasmalemma*, and the whole cell is enclosed by the *cell wall*. The cell is linked to its neighbours by strands of cytoplasm, the *plasmodesmata*, which occur in groups and pass through holes in the cell wall called *pits*.

The two major cell membranes, the plasmalemma and the tonoplast, like all plant membranes, are made up of two layers of lipid (fatty) molecules with globular proteins that straddle the two layers occurring at intervals. These membranes are not impenetrable barriers: they block the movement of most *solutes* (dissolved substances), but allow the movement of water by simple *diffusion*. They are therefore said to be *differentially permeable membranes* (they are sometimes referred to as being *semi-permeable*) and the movement of water through them is called *osmosis*. Osmosis involves the flow of water, through a differentially permeable membrane, from a region of low solute concentration (and, therefore, high water concentration or *high water potential*) to a region of high solute concentration (and, therefore, low water concentration or *low water potential*). Water continues to move across the membrane until equilibrium is reached or until sufficient pressure is applied to the solution with low water potential to balance the pressure created by movement of water from the solution with high water potential.

In the case of the plant cell, water moves across the plasmalemma, into the cytoplasm and vacuoles, until the hydrostatic pressure exerted by the plasmalemma and the rigid cell wall prevents further movement. The cell is then said to be *turgid* and is stiff and hard, as a bicycle tyre is when the inner-tube is pumped full of air. When the cells are not turgid, because water is in short supply, the plant wilts. The hydrostatic pressure in a plant cell that is turgid is often referred to as the *turgor pressure*. The opposing, inwardly directed pressure exerted by the cell, that balances the turgor pressure, is called the *wall pressure*.

Hydrophobic molecules, such as oxygen, and small uncharged molecules, such as carbon dioxide, can also move across the cell membranes by simple diffusion. Other molecules such as nutrients, hormones, waste products and toxins that the cell must take up, secrete or excrete, as appropriate, are carried across the membranes by *transport proteins*. The transport process consumes energy, mainly from the energy-rich molecules of *ATP*.

The cell wall, which surrounds the cell, consists of a mesh or 'basketwork' of rigid *cellulose microfibrils* which have great physical strength. This microfibril mesh is permeated by a *matrix* of softer, somewhat sticky *hemicellulose* and *pectic polymers* and the whole structure is bound together by long sugar-protein molecules called *glycoproteins*. The walls of older cells may be impregnated with the aromatic polymer *lignin*, the principal component of woody tissue, or the fatty polymer *suberin*, the principal component of *cork*.

Cells are joined to one another by a *middle lamella*, a thin layer of pectic substances which are continuous with the pectic substances of the wall matrix.

Drawing by Chris Went.

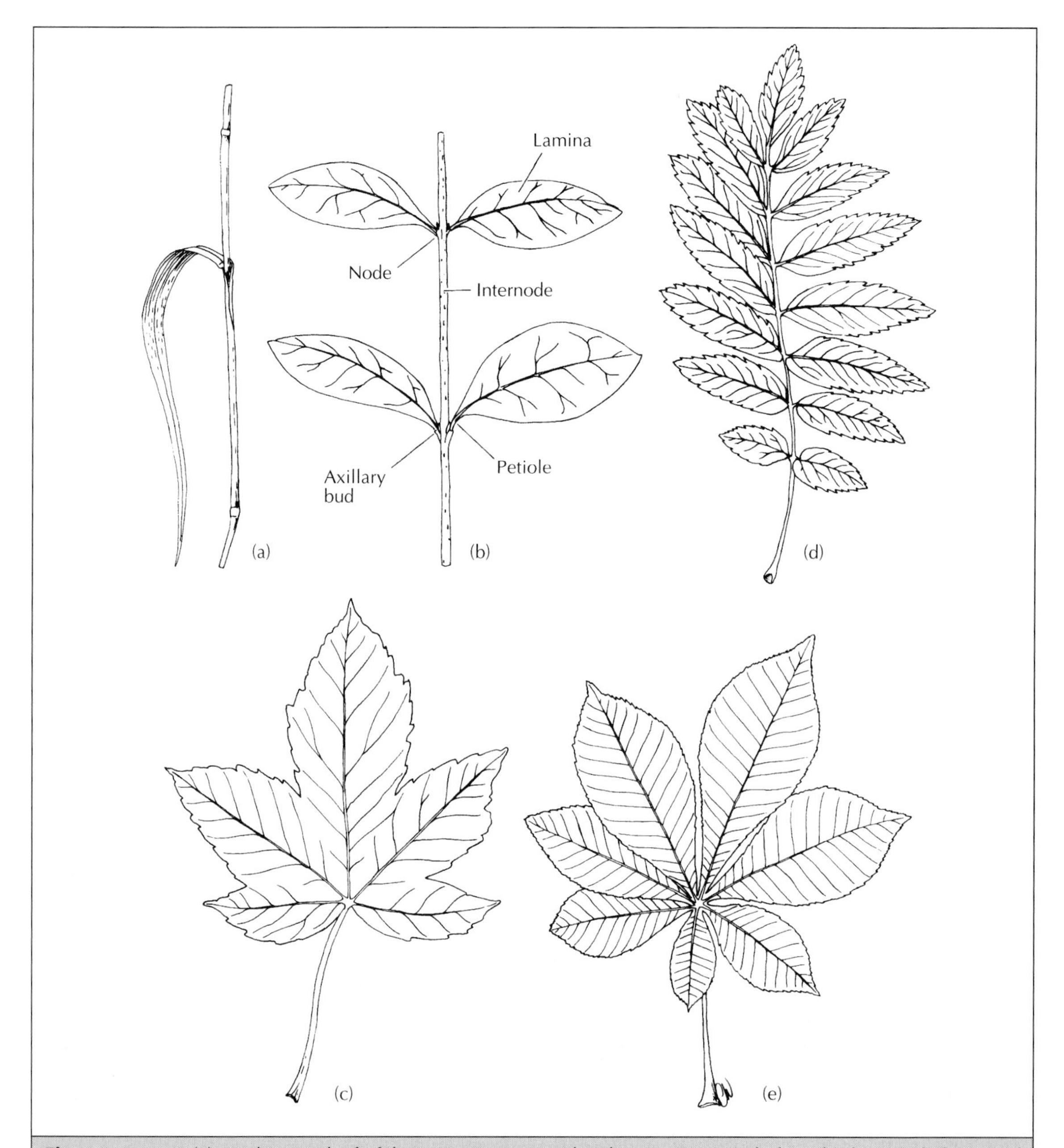

Fig. 1.7 Leaves: (a) an elongate leaf of the grass *Agrostis stolonifera*, a Monocotyledon; (b) the simple leaves of privet (*Ligustrum vulgare*); (c) a divided leaf of sycamore (*Acer pseudoplatanus*); (d) a pinnate, compound leaf of rowan (*Sorbus aucuparia*); (e) a palmate compound leaf of horsechestnut (*Aesculus hippocastanum*). Note that (b), (c), (d) and (e) are all Dicotyledons. Drawing by Chris Went.

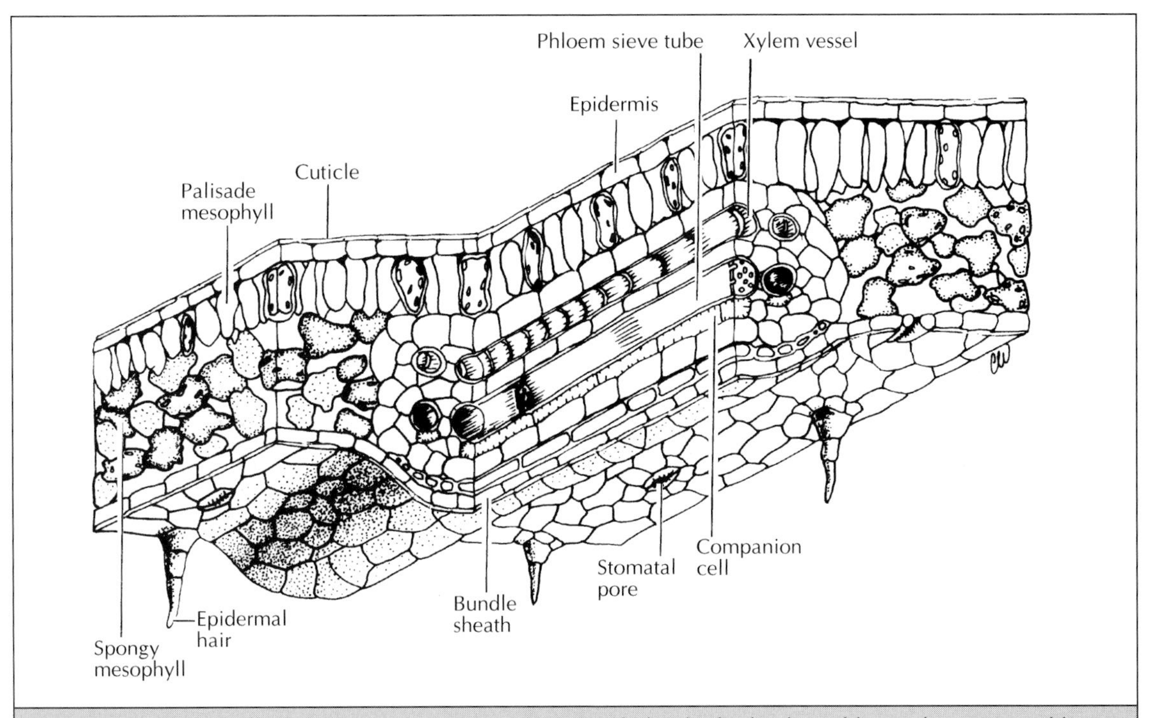

Fig. 1.8 Diagram showing the internal structure of a Dicotyledon leaf. Clearly visible are the upper and lower *epidermises,* covered with a continuous *cuticle,* the *palisade mesophyll* cells, the *spongy mesophyll* cells; *stomata;* a minor *vein* with *xylem* and *phloem* cells and *bundle sheath* cells; and *epidermal hairs* (*trichomes*). Drawing by Chris Went.

leaf surface, a process which consumes heat energy, also helps to cool the leaf – which is why a lawn feels cool when you walk on it in bare feet. This is an essential requirement since sunlight contains large amounts of infra-red radiation, and the exposed leaf is therefore subject to a considerable heat load. The solution to these problems has been the evolution of specialised pores in the cuticle and *epidermis*, the *stomata*, with variable apertures so that gas and water movement can be controlled.

The epidermis surrounds the entire leaf, beneath the cuticle (Fig. 1.8). It comprises a single layer of cells, most of which are relatively unspecialised, without chloroplasts. Some, however, contain chloroplasts and are differentiated as pairs of banana-shaped *guard cells*, which surround the *stomatal pores* (Fig. 1.8). The walls of the guard cells are thickened in such a way that the cells change shape as they take up or lose water, opening or closing the pores. The more water the guard cells take up and the more turgid they become, the wider the pores open. Conversely, as the guard cells lose water, becoming less turgid, the more tightly the pores close. These changes are achieved by the vacuoles of the guard cells taking up or secreting potassium ions very rapidly, in response to hormones activated by environmental stimuli such as light level or water stress (see Chapter 5 and Table 1.2). The more potassium ions the vacuoles acquire and the more water is taken up to balance it, the more turgid the cells become, and vice versa (see also the caption to Fig. 1.6).

Usually most of a plant's stomata are on the lower surface of the leaf, where they are prevented from becoming clogged with rain droplets and where there is relative protection from the drying effects of the wind and sun. Water plants with floating leaves are exceptions to this generalisation; these have stomata on the upper surfaces of the leaves.

The spacing of stomata on the leaf surface is very precise, being ten times the maximum pore diameter. There are good reasons for this. Diffusion of any gas through a perforated surface is assisted by sideways diffusion towards the rims of the pores. Provided that the pores are not so close together that the areas of collection overlap, there is, for each gas, an optimum spacing of pores such that the presence of the barrier between them has a negligible effect on diffusion of the gas. A spacing of ten times the maximum diameter is optimum for CO_2 diffusion. So, although the area for diffusion of CO_2 into a leaf is reduced a hundred-fold by the presence of the cuticle, the spacing of the stomata ensures that when they are fully open the diffusion of this gas occurs almost as readily as if no cuticle were present.

Water vapour diffuses more rapidly than CO_2, so the spacing of stomata in leaves is less optimal for water loss, which occurs only at a rate roughly proportional to the area of the pores themselves. The result is that as stomata begin to close, water vapour loss is restricted far more than is CO_2 gain. This difference is very important for plants with a limited water supply, for water can be conserved by partial closure of the stomata without reducing the supply of CO_2 for photosynthesis by too great an amount.

Protecting against harmful radiation

Another important feature of the epidermis is that in most plants its cells contain pigments called flavanoids which filter out potentially damaging longer wavelength ultra-violet light (320–400 nm). The cuticle of exposed desert plants, like *Yucca* and *Agave* spp., for example, also contains chemicals that filter out shorter wavelength ultra-violet light (280–320 nm). Other cells of the epidermis may be differentiated as hairs (*trichomes*), giving some leaves a downy appearance (e.g. deadnettles, *Stachys* spp.), or may secrete waxes, deposited on the leaf surface as crystals, giving leaves a characteristic bloom (e.g. cabbage, *Brassica oleracea*), or aromatic oils (e.g. rosemary, *Rosmarinus officinalis*), giving plants a characteristic smell. Hairs, waxes and oils may all have specialised functions in adapting plants to particular extreme environments, as we shall see.

Palisade tissues: the sites of photosynthesis

Beneath the upper epidermis is the *palisade tissue* of the leaf (Fig. 1.8), consisting of one or more layers of elongate cells containing numerous chloroplasts. This is where the energy of sunlight is collected and most of the photosynthetic processes occur. Plants that live in the shade, where light levels are low, usually have only one layer of palisade cells, but plants that grow in strong sunlight, where light penetrates deeper into the leaf, invest in two or even three layers of palisade cells to maximise the energy gain. The palisade cells are not attached to one another, but hang from the upper epidermis as sausages hang from the ceiling of an Italian shop. The moist walls of the cells thus have a very large surface area exposed to the internal air spaces of the leaf for the diffusion in and out of carbon dioxide and oxygen, respectively.

Beneath the palisade layer, providing support for the palisade cells, is the *spongy mesophyll* (Fig. 1.8). The cells are large and irregularly shaped, with relatively few chloroplasts. There are large air spaces between these cells, especially around the stomatal pores, where they form the substomatal cavities.

Carbon dioxide represents only about 0.03% of the earth's atmosphere (although this figure is now rising). Since plants have no moving parts to pump the gas into the leaf, it must diffuse along a very weak concentration gradient from its highest concentration in the outside air to the chloroplast molecules of the palisade cells, each stomatal pore connecting with about 100 μm^3 of leaf tissue. The leaf's geometry, just described, is such as to maximise the steepness of this gradient and consequently the rate of flow of the gas. We have already seen that the spacing of the stomata is such that CO_2 enters the leaf at the maximum rate possible, within the constraints imposed by the need to conserve water. It then diffuses freely through the branching labyrinth of air spaces between the spongy and palisade mesophyll cells before entering the chloroplasts of the palisade by diffusing through the liquid phase of the cell walls, cytoplasm and chloroplast membranes (Fig. 1.8). The diffusion of CO_2 in liquid is about 1000 times slower than in air, which means that this last stage is the slowest part of the journey for each CO_2 molecule. However, the distance each molecule has to travel in liquid is very short, being about 1000 times less than the rest of the journey through the leaf. Moreover, the surface area of palisade cells available for absorption is some 15 to 40 times the area of the lower surface of the leaf. The net result is that when the stomata are open, CO_2 reaches the chloroplast at the incredibly high level of about 50–80% of its concentration in the outside atmosphere, and the rate of uptake of the gas by the chloroplasts roughly balances the rate at which it enters the stomata.

The transport system

The leaf, like any good manufacturing unit, requires a reliable and efficient system of transport to bring in raw materials and to take away the manufactured products. This function is served by the *vascular bundles*, which are visible as the veins of the leaf, widest towards the leaf base, branching and narrowing into the body and tip of the lamina so that the entire structure is permeated. As a rough generalisation, the main veins of monocotyledonous plants (Table 1.3) such as grasses and spring flowering bulbs, run roughly parallel with one another along the main axis of the leaf (Fig. 1.7). In contrast, the veins of dicotyledonous plants (Table 1.3) such as *Rosa* and *Acer* species, are branched to a greater or lesser extent. Each vascular bundle includes: *xylem tissues*, comprising *vessels* and *fibres*; *phloem* tissues, comprising *sieve tubes* and *companion cells*, and *cambium*, and is surrounded by a *bundle sheath* (Figs 1.8 and 1.9).

Movement of water, minerals and hormones in the xylem

The xylem cells are elongate, have lost their living contents and have walls thickened with *cellulose* and spirals, rings or a network of the tough aromatic polymer *lignin*, the main component of wood. Together, these cells provide a rigid support for the thin plates of cells that constitute the rest of the leaf lamina. The xylem vessels have a second, equally important, function, which is to carry water to the leaf. This is essential to maintain the turgidity of the leaf cells, to provide an aqueous environment for the metabolic processes (especially photosynthesis and respiration) going on within those cells, and to carry dissolved mineral nutrients such as nitrates and phosphates, taken up from the soil by the roots. Some *plant hormones* (Table 1.2) manufactured in the roots, such as *cytokinins*, may also be carried in the xylem. The water in the xylem also contains dissolved citrate, an agent which helps to keep some mineral ions in solution. The pH of xylem fluid is slightly acid, at pH 5.5–5.7, which helps to prevent precipitation of dissolved minerals.

Xylem vessels, which may be up to 0.2 mm in diameter, lose their end walls during development. Linked end to end they therefore form continuous open tubes up to several metres in length. The xylem of the leaf is in continuity with that of the stem and root and therefore provides an unbroken channel of communication from root to leaf (see Chapter 4). Moreover, the individual xylem vessels are linked to one another laterally through pores called *pits*, so sideways movement of water is also possible.

The conifers and their relatives (Gymnosperms – Table 1.3) unlike the flowering plants (Angiosperms – Table 1.3), do not form vessels like those just described. Instead they possess less efficient water transporting cells called *tracheids*. These too have lost their living contents, and are braced with cellulose and lignin. They are, however, spindle-shaped, with closed ends, and linked to one another only laterally through areas of unlignified wall, the *pits*. They are also only a few millimetres long and very narrow, being only one tenth as wide as the Angiosperm vessels. They thus form a less efficient water transport system than that formed by vessels.

The liquid in the xylem moves upward through the plant to the leaves because it is under tension. This results from the evaporation of water from the stomata at the leaf surface, the process called *transpiration*, generating a pull force that is transmitted to the roots through the unbroken stream or column of water within the vessels or tracheids. Since water molecules, which are 'sticky', are bonded to one another more strongly than to the walls of the vessels and tracheids, this tension literally results in the water being pulled up through the plant and into the leaf by physical force. Indeed, the tensile strength of the water

Table 1.2 The main groups of plant hormones.

Hormones are usually defined as organic compounds that cause a physiological response at very low concentrations. In most cases, they are synthesised in one part of a plant and transported to a distant site of action. The mode of transport is, therefore, important. They may also act at the site of synthesis.

Only the five most important groups are considered here. Other naturally occurring compounds (e.g. brassins, polyamines, jasmonic acid) have growth regulatory activities at low concentrations, but are less well understood.

Name	Main sites of synthesis	Transport	Some effects	Some practical uses of growth regulators in horticulture
Auxins (e.g. indole-3-acetic acid) chapters 7 and 8	Leaf primordia, young leaves, developing seeds	Primarily by unidirectional polar movement in living cells Also bi-directional in phloem	Promote apical dominance, vascular tissue differentiation and fruit growth Induce formation of adventitious roots Stimulate ethylene synthesis Delay abscission	Promoting root initiation in cuttings and in micro-propagation systems (chapter 7) Killing weeds (chapters 7 and 11)
Cytokinins (e.g. zeatin) chapters 7 and 8	Root tips	In xylem from roots to shoots	Promote cell division and lateral bud growth Induce shoot formation Delay senescence	Promoting shoot initiation in micropropagation systems (chapter 7)
Gibberellins (e.g. gibberellic acid) chapters 3, 6, 7, 8 and 9	Young tissues of shoot, developing seeds and probably roots	In phloem and xylem	Promote cell division, cell elongation and seed germination Promote flowering and bolting in long-day plants Cause reversion to juvenile form Prevent tuber formation in potato	Producing dwarfed plants with growth retardants ('anti-gibberellins') (chapter 8) Promoting malting (chapter 6)

Table 1.2 (cont.)

Name	Main sites of synthesis	Transport	Some effects	Some practical uses of growth regulators in horticulture
Ethylene chapters 9 and 12	Most tissues in response to stress and in ripening or senescing tissues	Ethylene is a gas and moves by diffusion in and around the plant May also be transported as precursor solutes	Promotes ripening in fruits and senescence in leaves and flowers Induces abscission of leaves and fruits Promotes flowering in bromeliads	Promoting flowering in bromeliads (especially pineapple) by ethylene-releasing compounds (chapter 9) Promoting flowering in bulbs (chapter 9) Retarding senescence in fruits and flowers by 'anti-ethylene' compounds (chapter 12) Inhibiting flowering in sugar cane (chapter 9)
Abscisic acid chapters 5, 6 and 9	In root tips and leaves, especially in response to stress Seeds	Mainly in phloem, but also in xylem	Closes stomata Maintains dormancy in seed and possibly in buds Increases tolerance to salinity, low temperature and drought Controls deposition of reserve proteins during seed development	None at present but important in connection with water management (chapter 5)

Table 1.3 The main groups of photosythetic organisms mentioned in Chapter 1.

Superkingdom	PROKARYOTAE (cells without a clearly defined nucleus)	
Kingdom	**Monera**	
	Phylum	Cyanobacteria ('blue greens')
Superkingdom	EUKARYOTAE (cells with a membrane-bound nucleus)	
Kingdom	**Protoctista** Phytoplankton green, red and brown 'algae'	
	Plantae	
	Bryophytes	
	Phylum	Hepatophyta ('liverworts') Bryophyta ('mosses')
	Vascular plants without seeds	
	Phylum	*Sphenophyta ('horsetails') *Filicinophyta ('ferns')
	Vascular plants with seeds	
	Phylum	**Cycadophyta ('cycads') **Ginkgophyta ('ginkgo') **Coniferophyta ('conifers') Anthophyta ('angiosperms' – flowering plants): Class Dicotyledons (two cotyledons in the seed) Monocotyledons (a single cotyledon in the seed)

* often referred to together as the 'Pteridophytes'
** often referred to together as the 'Gymnosperms'

column is, amazingly, only ten times less than that of copper wire.

However it is about a million times more difficult to pull water through living cells than through the dead xylem vessels. To facilitate the transport of water in the leaf, therefore, the distance it must travel in the mesophyll is reduced to a minimum by permeating this tissue with many fine veins so that water molecules never have to pass through more than six living cells after leaving a vein. To give some idea of the complexity of this branching system, it has been calculated that a single beech (*Fagus sylvatica*) leaf, for example, contains almost four metres of finely branched veins.

Transport of carbohydrates and hormones in the phloem

The xylem tissues of the leaf are located on the upper sides of the veins, linking with the centrally located xylem of the stem. The other major conducting tissue, the phloem, is located on the undersides of the veins (Figs 1.8 and 1.9), linking with the peripherally located phloem of the stem. The phloem tissues consist largely of vessels called *sieve cells* joined end to end to form *sieve tubes*, involved in the transport of sugars, and their associated *companion cells*.

The sugars glucose and fructose produced during photosynthesis are usually converted to sucrose for transport around the plant. Once

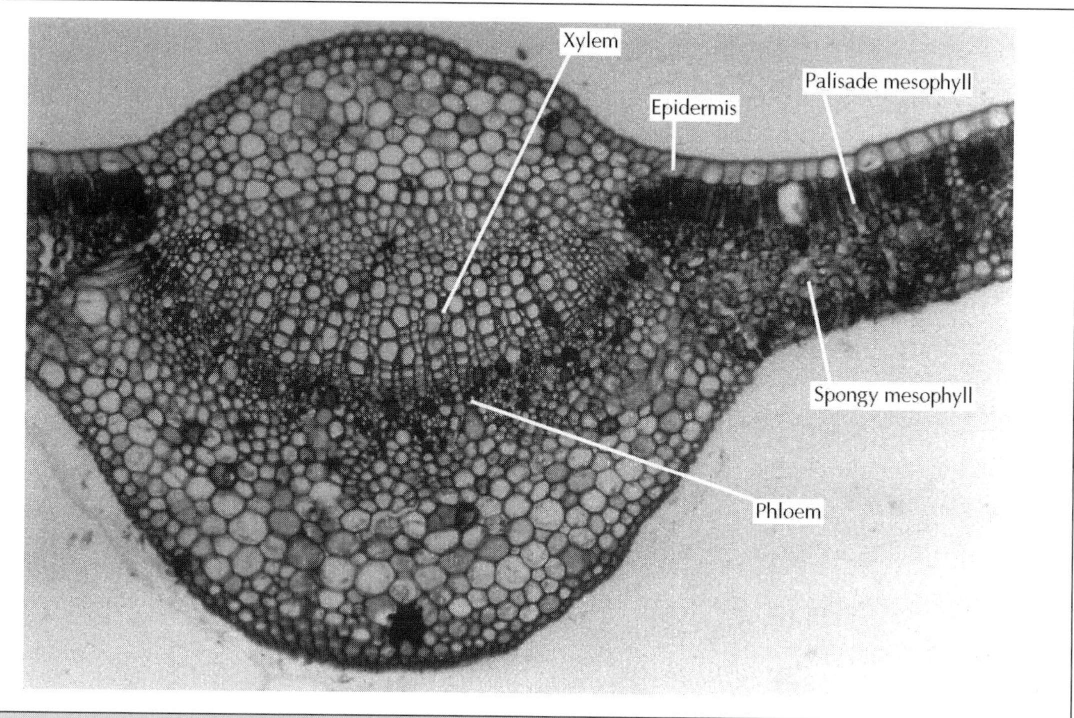

Fig. 1.9 Section of the mid-rib and part of the lamina of a leaf of camphor (*Cinnamonum camphora*) viewed with a light microscope. Clearly visible are the *epidermises*; *palisade* and *spongy mesophylls*; *xylem*; *phloem*; and *cambium*. The *bundle sheath* has not stained and is therefore indistinct. Specimen prepared and photographed by B.G. Bowes, University of Glasgow.

formed, the sucrose passes out of the palisade cells into the *apoplast*, which is the non-living space within a plant comprising the cell walls and the intercellular spaces between them. From here it is taken up by the sieve cells of the phloem (Figs 1.8 and 1.9) for transport out of the leaf. The sieve cells are elongated cells, connected to one another by their end walls, which are perforated, hence the name. These cells are alive, with intact *plasmalemma* membranes (Fig. 1.9), but their contents have liquified and they have lost their nuclei. The contents of each of the sieve cells are connected to one another at the ends by thin, membrane-bounded threads of liquid cytoplasm that pass through the holes in the end walls. Each sieve cell is also connected laterally to a companion cell, which has a nucleus and is packed with organelles and is therefore metabolically very active. It provides the energy required for the transport of sugars and other substances in the phloem.

Sucrose is pumped into the sieve cells and the companion cells by metabolic pumps located on the plasmalemmas and powered by high energy molecules of ATP. These pump protons (H^+ ions) out of the phloem cells into the cell walls and intercellular spaces and the sucrose is then carried back in with them on sucrose-proton transporter molecules. Often the area of plasmalemma available for this process is greatly increased in the companion cells by finger-like outgrowths of the wall which project into the cytoplasm and around which the plasmalemmas are folded. The sucrose loading mechanism is so effective that the concentration within the phloem cells can reach 20–30%. Since the solution of sucrose outside the phloem cells is very weak, the resulting hydrostatic pressure generated in the phloem cells as a result of osmotic movement of water into them (see caption to Fig. 1.8) can be as high as 20–30 atmospheres (2 to 3 megapascals). For this reason the walls are thickened with hoops of extra cellulose, a polymer with great tensile strength, otherwise the cells would burst.

The phloem tissues of the leaf connect, via the stem, to all the other parts of the plant, where sucrose is off-loaded to provide the energy and wall building material required for growth. Notable *sinks* for off-loaded sucrose are, of course, the root and shoot tips, where active growth is occurring, and storage organs such as roots, fruits and modified stems such as tubers and rhizomes. As sucrose is off-loaded into such tissues, the hydrostatic pressure in the sieve elements is reduced, creating hydrostatic pressure gradients in the phloem between the leaves and the rest of the plant. These gradients vary from 0.5 to 5.0 atmospheres (50 to 500 kilopascals) according to the rate of photosynthesis and the rate of off-loading into the sinks, and are sufficient to drive the contents of the sieve elements at speeds of up to 50 cm per hour.

It follows that any damage to a high pressure system such as this will result in leakage of the vessel contents. Such leakage is especially apparent, for example, when the phloem is exposed to remove the sugar sap from sugar maple trees (*Acer saccharum*) in spring. In most plants, when injury to the phloem occurs, the perforations in the end walls of the sieve tubes are rapidly clogged with filaments of phloem protein and are then plugged permanently with a polysaccharide called callose, a substance widely deployed by plant cells to repair injuries.

In addition to transporting sugars, sieve tubes are also important in transporting other organic molecules about the plant, notably hormones such as *indolyl acetic acid* (IAA) and the *gibberellins* (Table 1.2). Some minerals are also transported in the phloem as well as in the xylem. Because the loading of sucrose requires the extrusion of protons (H^+ ions), however, the phloem sap tends to be rather alkaline at pH 8.0 to 8.5. This means that iron, calcium, manganese and copper precipitate out of solution and can only, therefore, be transported in the xylem. During periods of drought, when transpiration is not occurring and xylem transport is restricted, these elements soon become depleted in the leaf tissues (see Chapter 4).

When light levels are high and water and other nutrients are plentiful, photosynthesis may proceed so rapidly that the phloem transport system is unable to cope with the removal of all the sucrose produced. If this were not dealt with, sucrose would build up in the leaf tissues, disrupting the osmotic balance of the cells (see Fig. 1.6). Excess sucrose produced under such conditions is therefore rapidly converted to insoluble starch by starch synthetic enzymes, and starch grains are laid down within the chloroplasts and elsewhere in the palisade and spongy mesophyll cells. This stored carbohydrate may later be released for transport and use during times of low photosynthesis by reconversion of the starch to sucrose. A similar process of starch synthesis and breakdown also makes possible the storage of carbohydrates in specialised storage organs such as tubers, thickened roots or seeds (see below).

Producing new xylem and phloem

So far, no mention has been made of the thin layer of small, regularly shaped and metabolically very active cells in the vascular bundles of the leaf comprising the *cambium*. This tissue occurs between the xylem and phloem, and the cambium cells are capable of dividing to produce daughter cells on their upper and lower faces, which differentiate to form new xylem and phloem cells respectively.

The bundle sheath

Finally, in most plants the vascular bundles of the leaves are completely enclosed, except at the tips, by a prominent, single layer of cells comprising the *bundle sheath* (Figs 1.8 and 1.9). This effectively provides a living seal around the bundles, ensuring that air is not sucked into the xylem vessels from the surrounding intercellular spaces, causing breakage of the water columns and thereby disrupting water transport. The bundle sheath cells perform other functions too. They are important in transferring mineral ions from the non-living xylem vessels and tracheids to the surrounding living cells of the leaf, and like the companion cells of the phloem may have inward pointing projections from the walls to increase the area of plasmalemma available for this process. Bundle sheath cells may also provide a convenient repository for unwanted waste products until these can be lost from the plant when the leaves are shed in the autumn. Moreover, in some plants, the bundle sheath cells may be involved in a special kind of photosynthesis, called C-4 metabolism, which will be discussed below and in Chapter 5.

Connecting with the stem

The leaf may be attached to the stem of the plant in a variety of ways. Most of the monocotyledons such as the grasses (family Poaceae, Fig. 1.7), bamboos (also in the family Poaceae) and spring bulbs have relatively sword-shaped (lanceolate) leaves with parallel veins, attached to the plant by a broad base which often completely ensheathes the stem. This feature is particularly noticeable in the palms (family Palmae) and bananas (*Musa* spp.) where the persistent, ensheathing leaf bases may constitute the bulk of what is loosely called the 'stem', serving both to support and protect the true stem tissues. In all cases, however, the parallel vascular bundles continue down into the stem and link there with the scattered vascular bundles that are characteristic of monocotyledons (see later, Fig. 1.13).

In most of the dicotyledons (Fig. 1.7) such as members of the rose (Rosaceae) and daisy (Asteraceae) families, and some monocotyledons, such as members of the fig family (Ficaceae), the leaf lamina narrows at its base and a stalk, the petiole, of varying length in different species, attaches the leaf to the stem. Sometimes this contains only one, crescent-shaped vascular bundle, but in most cases there are several. These again link to the vascular bundles of the stem which, in dicotyledons, are arranged in a regular pattern.

The point of attachment of the petiole to the stem is sometimes slightly swollen, as in *Oxalis* species and scarlet runner bean (*Phaseolus coccineus*), to form a structure called the *pulvinus*. Many of the cells located in the pulvinus respond to external stimuli, especially light intensity and light direction, by taking up or releasing water. This leads to changes in turgidity, causing the leaves of some plants such as *Oxalis* species to droop at night as the light levels fall, a so-called sleep movement, and to recover at dawn as the sun rises. Adaptations of this kind may originally have evolved in response to environmental extremes such as very cold night temperatures, where drooping leaves may afford some protection against frost (see Chapter 9).

Changes in the turgor of the cells of the pulvinus may also enable the leaves of some other plants like the scarlet runner bean to turn towards the sun, thereby maximising their ability to collect the energy of sunlight (see Chapters 5 and 9). The leaves of most plants, however, do not possess pulvini; leaf movement is still possible, but is much slower, being brought about by the differential rate of growth of the cells at the base of the stem in response to light coming from one direction only, as when a plant is partially shaded or grows against a wall. Such growth responses, called *phototrophy*, are not restricted to the leaf but also occur in the stem so that the whole plant may appear bent towards the direction of its principal light source (see Chapter 5).

Leaf fall

At the point where the base of the petiole meets the stem, a special layer of cells, the *abscission layer*, forms when the life of the leaf is over, to facilitate leaf fall. This occurs annually in deciduous species such as oak (*Quercus* spp.) or *Weigela* spp., but less frequently in evergreens such as holly (*Ilex* spp.). In this zone the xylem vessels may be narrower than normal or there may only be tracheids rather than vessels, to reduce the risk of air bubbles being drawn into the xylem vessels of the stem and blocking them if leaves are torn off by wind or browsing animals. When the time is reached for a senescent leaf to be lost, often in response to low temperatures or short days in temperate climates, a protective layer of corky cells forms at the base of the *abscission zone*. These are impregnated with the hydrophobic corky polymer *suberin* and are impermeable to water. The vessels of the vascular tissue are then plugged and the leaf falls. Immediately before this occurs, however, the plant breaks down the chlorophyll, starch and other useful substances in the leaf and transports them into the body of the stem. Only the waste products are left behind. Many of these are complex, highly coloured chemicals like the red, yellow and purple anthocyanins, and are only revealed in all their glory as the masking effect of the chlorophyll is removed, giving the autumn colour of so many of our garden trees and shrubs (see Chapter 8).

In some normally deciduous woody species, such as oak and beech, an abscission zone does not form in juvenile plants and the dead leaves remain attached to the stem throughout the winter, perhaps providing the buds with a degree of protection from frost and damaging winds. This phenomenon is exploited by gardeners who plant beech hedges, maintaining juvenility by annual clipping.

Fig. 1.10 The *spiral phyllotaxy* of the strawberry tree (*Arbutus unedo*). Photograph by David S. Ingram.

Leaf patterns: phyllotaxy

Finally, if you look carefully, you will see that leaves are not attached to stems in a random way, but are arranged in a regular pattern which varies from species to species. In some cases they form a spiral along the stem, as in the strawberry tree (*Arbutus unedo*, Fig. 1.10), whereas in others, such as privet (*Ligustrum* spp., Fig. 1.7), they may alternate with one another in opposite pairs. In sunflower (*Helianthus* spp.) the first leaves are in opposite pairs, but a spiral arrangement soon sets in as the plant grows. Sometimes the leaf petioles may be different lengths and the laminas different sizes according to the position of the leaf on the plant, an arrangement most easily seen in the maples (*Acer* spp.). These different patterns of leaf attachment, called *phyllotaxy*, have evolved to maximise the amount of light reaching the leaves by ensuring that each leaf shades its neighbours as little as possible. That different arrangements of leaves may be seen in different groups of plants shows that there may be a variety of structural solutions to this problem.

Variations on a theme

The basic form of the leaf described above, a flat plate with a high surface area to volume ratio, roughly oval to lanceolate in shape, covered with a cuticle and an epidermis pierced by stomata and with the cells of the various tissues arranged to maximise the internal surfaces available for gas exchange, is a magnificent compromise. It has evolved in response to the conflicting environmental and physiological constraints imposed most notably by the need to collect the maximum amount of light energy and atmospheric carbon dioxide for photosynthesis, whilst at the same time minimising heat gain and water loss. However, such a compromise is only effective under conditions of an equable climate and adequate supplies of water, light, carbon dioxide and mineral nutrients. Any deviations from these optimal conditions reduce the effectiveness of the leaf as a photosynthetic organ. Many plants have evolved, in response to a variety of environmental conditions, a number of variations on the basic leaf form (Fig. 1.7), some of which are so extreme (as with the spines of cacti, Fig. 1.11), that it requires the experienced eye and knowledge of the botanist to assure us that what we are looking at is really a leaf at all.

Variations in the shape, form and colour of leaves

One of the commonest variations has been in the evolution of divided leaves, perhaps minimising wind and rain damage by posing less resistance, as in the maples (*Acer* spp.). In other species the leaf may have holes, as in the cheese plant (*Monstera deliciosa*), caused by the death of certain cells as the leaf develops, or drip-tips as in the weeping fig (*Ficus bejamina*), to facilitate run-off of water in tropical climates. In other cases compound leaves have evolved, again perhaps minimising wind and rain damage, and these may be pinnate, as with ash (*Fraxinus excelsior*), rose (*Rosa* spp.) and date palm (*Phoenix dactylifera*), or palmate, as with horsechestnuts (*Aesculus* spp.), clover (*Trifolium* spp.) or *Trachycarpus* palm.

Such compound structures may develop by programmed, differential division and growth of the cells of the young developing leaf, as with the divided leaves of dicotyledonous plants like ash and clover. In such cases each leaflet of the compound leaf resembles a perfect simple leaf. In the palms, however, compound leaves develop in a completely different way. To begin with, the young leaf is intricately folded, with the folds packed together like a fan. As the leaf blade unfolds and expands, lines of cells along alternate folds die and break down so that the fingers of the older compound leaf separate from one another along the resulting lines of weakness.

The surfaces of leaves may vary greatly too. Dense epidermal hairs, as in lamb's lugs (*Stachys byzantina*), or scales, such as those on the lower surfaces of leaves of rhododendrons (*Rhododendron* spp.), may reduce water loss by increasing the width of the still air layer over the leaf surface. Alternatively, scattered hairs or ridges may help to break up the still air layer, facilitating carbon dioxide uptake. Waxy blooms, as in cabbages (*Brassica oleracea*) or a shiny, thick cuticle, as in cherry laurel (*Prunus laurocerasus*) may reduce the damage caused by high light levels. In some plants such as rosemary (*Rosmarinus officinalis*), glandular hairs on the leaf surface may secrete aromatic oils that render the plant unpalatable to browsing animals in the dry conditions that this plant inhabits in the wild. The glandular hairs of

Fig. 1.11 *Echinocactus* sp. The spines of cacti are modified leaves, with the stem, often ridged or flattened, being modified for photosynthesis. Photograph by Chris Prior, Royal Horticultural Society.

the leaves of sundew (*Drosera* spp.), secrete sticky mucilage and protein-degrading enzymes, enabling this inhabitant of acid, boggy soils, where nitrogenous salts are in short supply, to augment its nitrogen supply by trapping and digesting insects. In other insectivorous plants such as venus flytrap (*Dionaea muscipula*) and pitcher plant (*Nepenthes* spp.) the whole leaf, not just its surface, may be modified to create an insect trap. In nettle (*Urtica* spp.) the sharp hairs contain irritants which are released when the hairs penetrate and break off in the skin of the unwary gardener weeding the vegetable patch.

The distribution of stomata also varies from species to species. Although usually found mainly on the lower surfaces of the leaves, they are restricted to the upper surfaces of the leaves of floating aquatic plants like the water lily (*Nymphaea* spp.). In submerged aquatics, like Canadian pond weed (*Elodea canadensis*), both the cuticle and the stomata may be absent. Stomata may also be sunk in pits or grooves, as in pines (*Pinus* spp.), restricting the loss of water vapour under dessicating conditions by creating a moist microclimate with the grooves. In marram grass (*Ammophila arenaria*), which grows on sand dunes, not only are the stomata confined to grooves but the leaf is also tightly rolled.

Indeed, water stress has been one of the most powerful environmental factors in the evolution of modified leaves. In some species like lavender (*Lavendula* spp.) the leaves may be reduced in size, while in brooms (*Sarothamnus* spp.) and gorse (*Ulex* spp.) the small leaves are lost under dry conditions and photosynthesis is carried out by the stem. In cacti (family Cactaceae) the leaves may be modified as spines, although the seedling may possess relatively normal leaves. In such cases not only is water loss reduced, but the succulent stem, modified to carry out photosynthesis, is protected from grazing animals (Fig. 1.11).

Other evolutionary responses to water stress include thickened, succulent leaves, as in stonecrops (*Sedum* spp.), or the massing of leaves in a tight rosette, as in the daisy (*Bellis perennis*) and many alpines, or both, as in the houseleeks (*Sempervivum* spp.). Alternatively, a combination of woody cells within the leaf, a reduced surface area, a spiny outline and a thick reflective cuticle may create the hard, sclerophyllous leaves typical of Mediterranean trees such as the holm oak (*Quercus ilex*) or holly (*Ilex* spp.).

Lack of adequate light has been overcome by many plants by the evolution of leaves modified as climbing aids to take them to the top of the canopy in woodlands, forests or scrub; these include the tendrils of pea (*Pisum sativum*), or the twining petioles of *Clematis* spp. The availability of light is such a problem for some woodland species that they have evolved to flower and photosynthesise during the winter months, while the leaves are off the trees, and to store nutrients during the summer months in modified leaves and leaf bases, as in the bulbs of tulips (*Tulipa* spp.), daffodils (*Narcissus* spp.) and bluebells (*Hyacinthoides non-scripta*).

Lastly, the colours of leaves may vary greatly (see Chapters 3 and 8). In some cases, variegation may be caused by genetically controlled loss of chlorophyll in some but not all cells, as in variegated *Pelargonium* spp. or ivy (*Hedera helix*). In other cases, variegation of the leaf may result from infection by a virus, as in the variegated form of *Abutilon*, *A. megopotamicum* 'Variegatum'. Brightly coloured leaves without chlorophyll, as in *Euphorbia*, may cluster together at the tip of the stem to form false flowers, attracting pollinating insects to the more insignificant true flowers at the heart of this leaf cluster. And of course, the petals as well as the other components of flowers and fruits are, in evolutionary terms, simply modified leaves.

It would be possible to quote many more examples of modifications of leaf shape, form and colour that have evolved in plants in response to a great variety of environmental factors. To do so would lead to little more than a boring list. A much more interesting approach is to go out into the garden, note the endless variety of leaves and speculate, for absolute proof is rarely if ever available, on what environmental factors may have led to the evolutionary selection of each modification. At the same time, it will soon become apparent to the most casual observer that since the domestication of garden plants, selection and breeding have led to the enhancement or accentuation of the many leaf modifications that render the plant either attractive or edible (see Chapter 3).

Variations in photosynthesis: C-4 and CAM plants

Two final modifications that are not visible to the eye of the gardener, for they occur at the cellular level, are nevertheless of such importance that they deserve a special mention. These are *C-4 photosynthesis* and *crassulacean acid metabolism* (*CAM*) (see also Chapter 5). Both are modifications to the basic processes of photosynthesis, and both may have evolved in response to the stress imposed by low levels of soil water and high temperatures. Under such conditions it is necessary to close the stomata because the plant will die of water loss if they remain open. Although, as we have seen, partial closure of stomata is relatively effective in reducing water loss without significantly reducing carbon dioxide uptake, further closure to cope with severe desiccation begins to limit the availability of CO_2 for photosynthesis. This creates problems because ribulose bis-phosphate carboxylase, the key enzyme in normal photosynthesis, is very inefficient and does not function at very low CO_2 levels. C-4 plants have evolved a mechanism for concentrating carbon dioxide in the chloroplasts of the palisade mesophyll cells under such conditions using light energy and organic acids such as malic acid, containing four carbon atoms in their molecules (hence C-4) as carriers of carbon dioxide. The 4-carbon acids then pass from the palisade cells to specially modified bundle sheath cells surrounding the vascular bundles of the leaf, which contain further chloroplasts which carry out photosynthesis in the normal way. The carbon dioxide is released in its concentrated form from the C-4 acids and normal photosynthesis begins. Plants possessing such modified bundle sheaths are said to exhibit Kranz anatomy, after the person who discovered it. Many tropical and desert grasses carry out C-4 photosynthesis and exhibit Kranz anatomy. Not many such plants are grown in British gardens, but three groups that will be very familiar to gardeners are maize (*Zea mays*), bamboos (family Poaceae), and prairie grasses (also family Poaceae).

CAM plants, which carry out a similar process in a slightly different way, include many common succulents, such as stonecrops (*Crassula* spp.). In these plants stomata open for the uptake of carbon dioxide during the cool hours of darkness, when water loss will be at a minimum. Because light is not available for photosynthesis at this time, the carbon dioxide is concentrated and stored in 4-carbon acids like malic acid, and then regenerated in the same cells during the day, when the stomata are closed, and fed into normal photosynthesis. The capacity of succulents to store malic acid overnight is very limited, however, so their rate of growth is very much slower than that of C-4 plants, in which the malic acid is not stored but regenerates its CO_2 immediately.

REACH FOR THE SKY: THE STEM

The stem or shoot is the main axis of the plant, generating at its tip (the apex) the leaves which it holds aloft to collect the sun's energising light, and providing at its base a vital link to the roots and the water and mineral nutrients of the soil.

The growing point

To find the growing point of the shoot, the so-called *apical meristem* (Fig. 1.12), it is necessary to strip away the immature leaves that surround and protect it from damage. It is usually dome-shaped, a millimetre or so in diameter, and very delicate. In flowering plants (Angiosperms) it usually comprises an outer layer, the *tunica*, which is one to several cells thick, enclosing an inner mass of cells, the *corpus*. The cells of the apex are packed with cytoplasm and are constantly dividing and differentiating to produce new tissues. Such centres of cell division and growth are called *meristems*, in this case the *apical meristem*.

As the apex grows, the cells of the tunica divide in one plane only, at right angles to the surface of the plant, which means that the tunica remains as a discrete layer. It ultimately gives rise to the epidermis of the shoot and the leaves. The cells of the corpus divide in a more irregular way and by expansion and division give rise to the internal tissues of the stem and leaves. In the stem apex of conifers and their relatives (Gymnosperms) the

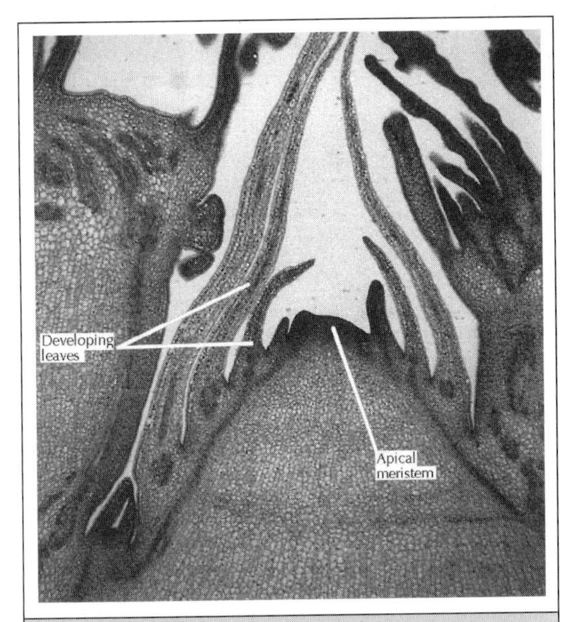

Fig. 1.12 Section of the apex of castor oil plant (*Ricinus communis*). Clearly visible are the *tunica*, *corpus* and *leaf primordia* of the developing leaves. Specimen prepared and photographed by B.G. Bowes, University of Glasgow.

apex is not clearly differentiated into tunica and corpus, although the epidermal tissues of shoot and leaf still arise from the outer cells.

Nourished by nutrients and water carried in the vascular tissues from the mature leaves and the roots, and under the control of both the environment and internally produced hormones, the meristematic cells of the apex produce both the stem of the plant and the leaves.

Forming new leaves

The leaves grow from small bumps of active cell division, called *leaf primordia*, on the sloping sides of the apical dome. These do not arise at random, their position relative to one another being determined genetically and regulated by the apex itself and the previously formed primordium. In this way a distinctive pattern develops, the *phyllotaxy* already referred to. In dicotyledons the leaf primordia may arise singly, ultimately producing a spiral phyllotaxy, in opposite pairs or in whorls of three or more. In grasses (Poaceae) and many other monocotyledons they form in two opposite rows, although angiosperm-type phyllotaxy may occur in some monocotyledon species.

Inside the body of the shoot apex, just below each leaf primordium, the cells divide longitudinally to form a strand of cells, the *procambial strand*. As the apex grows upwards the procambial strands differentiate as distinct cambiums (a cambium is a meristem capable of producing xylem or phloem), which in turn produce the xylem and phloem of the vascular bundles of the shoot. The pattern of vascular tissues of the shoot is therefore determined by the position of the developing leaves these tissues will service.

The leaves are finally formed by a series of centres of cell division, meristems, each with a fixed period of activity. First a meristem forms at the apex of the leaf primordium, forming a narrow peg about a millimetre in length. This will become the midrib of the leaf. Next, cells down either side of this peg begin to divide, growing out to form thin plates of tissue which will become the lamina. Finally, a wave of cell divisions followed by differentiation passes along the whole length of the leaf to produce the internal tissues.

The final shape of the leaf is determined by the precise position and duration of activity of the leaf meristems. For example, long, narrow leaves such as those of conifers are produced when the meristems on the sides of the leaf peg have a very short period of activity, whereas large flat leaves such as those of broad-leaved trees are produced when the active period of the marginal meristems is much longer. Compound leaves such as those of roses (*Rosa* spp.) form when the marginal meristem activity is intermittent rather than continuous. Small areas of meristematic activity, called meristemoids, provide the finishing touches, like the stomata.

The pattern of growth of leaves just described holds true for most dicotyledons and for some monocotyledons such as the grasses (Poaceae) and lilies (Liliaceae). In other monocotyledons, with tubular leaves, such as onions (*Allium* spp.) and rushes (*Carex* spp.), the meristems at the margin of the leaf are suppressed and the leaf arises from a new meristem that ensheaths the stem apex.

Most plants have buds which occur in the angle between the leaf and the stem, called the axil (Fig. 1.9). These *axillary buds* often arise very early on during leaf development, in the axils of the leaf primordia. Like the leaves, each possesses a procambial strand. Such buds do not usually grow out to form new shoots immediately, being held in a state of dormancy, perhaps by hormones secreted by the apex, until circumstances change, as when the apex is cut off by pruning (see Chapter 8). When axillary buds are not present, branch stems may form from buds that arise spontaneously from the stem tissues. These buds are said to be *adventitious*.

A tower of strength

The arrangement of vascular bundles within the main body of the stem is, as we have seen, that established in the apex. Thus in the young stem of the common dicotyledonous garden sunflower (*Helianthus annuus*), for example (Fig. 1.13), they form a peripheral ring. The cells in the centre of the stem are large, irregular and thin-walled, with large vacuoles, and form a tissue called the *pith*. Sometimes the pith cells contain starch grains and are important for the storage of carbohydrate. External to the ring of vascular bundles and between the individual bundles, linking with the pith, are smaller, irregular cells comprising the *cortex*. These may be photosynthetic, especially in young plants. Beyond this is an epidermis, the cells of which may also contain chloroplasts and be photosynthetic. Scattered stomata may be present to facilitate gas movement in and out. And finally, like the leaves, the stem is ensheathed by a waterproof layer of cuticle.

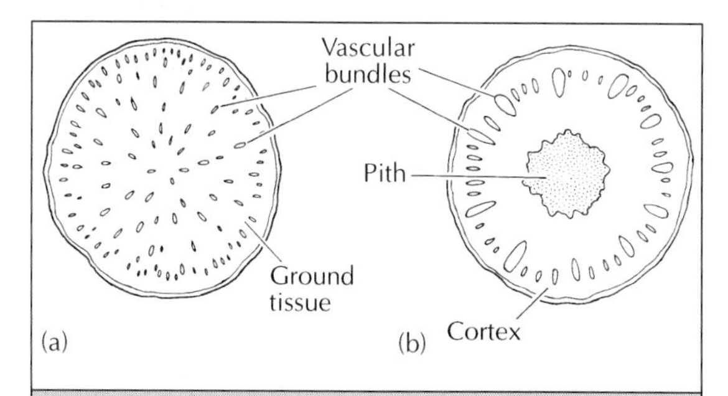

Fig. 1.13 Arrangement of vascular bundles in the stems of (a) Monocotyledon and (b) a Dicotyledon. Drawing by Chris Went.

The vascular tissues of most monocotyledons are arranged rather differently from those of dicotyledons, the bundles being scattered throughout a background tissue of large, irregular cells (Fig. 1.13). Occasionally a central pith is present, as in rushes (*Juncus* spp.).

Each vascular bundle of the stem (Fig. 1.14) comprises an outward facing layer of phloem-fibre cells, their walls thickened and strengthened with lignin. These serve to strengthen the stem and to protect the phloem beneath. The phloem itself, as in the leaf, is composed mainly of sieve tubes and companion cells, involved in the transport of sugars and organic substances such as hormones (Table 1.2). The xylem tissue, composed of lignified water-conducting vessels, or tracheids, with lignified fibres between them, forms the inward-facing region of each bundle. It is separated from the phloem by a cambium, the *fasicular cambium*, of files of small, box-shaped cells which divide continuously to produce phloem on the outer face and xylem on the inner. Eventually, as the stem matures, the cambiums of adjacent bundles link up by development of a so-called *interfasicular cambium* in the cells between the bundles. This has great significance, as we shall see, in a process called *secondary thickening* which in the maturing stem of trees and shrubs leads to the development of a mass of woody tissue.

There are many variations on this basic arrangement of the vascular tissues. For example, in some plants phloem tissues occur both outside and inside the xylem. In many monocotyledons the xylem tissue may occupy the centre of each bundle, being completely surrounded by phloem.

In the young stem it is clear that the peripheral ring of vascular bundles provides not only an efficient transport system for water, sugars and other substances, but also strength and support for what would otherwise be a rather fragile structure. Such a hollow cylinder of strengthening tissue gives much greater resistance to mechanical damage caused by the side-to-side movement of the stem on a windy day than if the vascular tissues formed a solid core in the middle of the stem.

The simple vascular anatomy described above is more complex at stem *nodes*, the points at

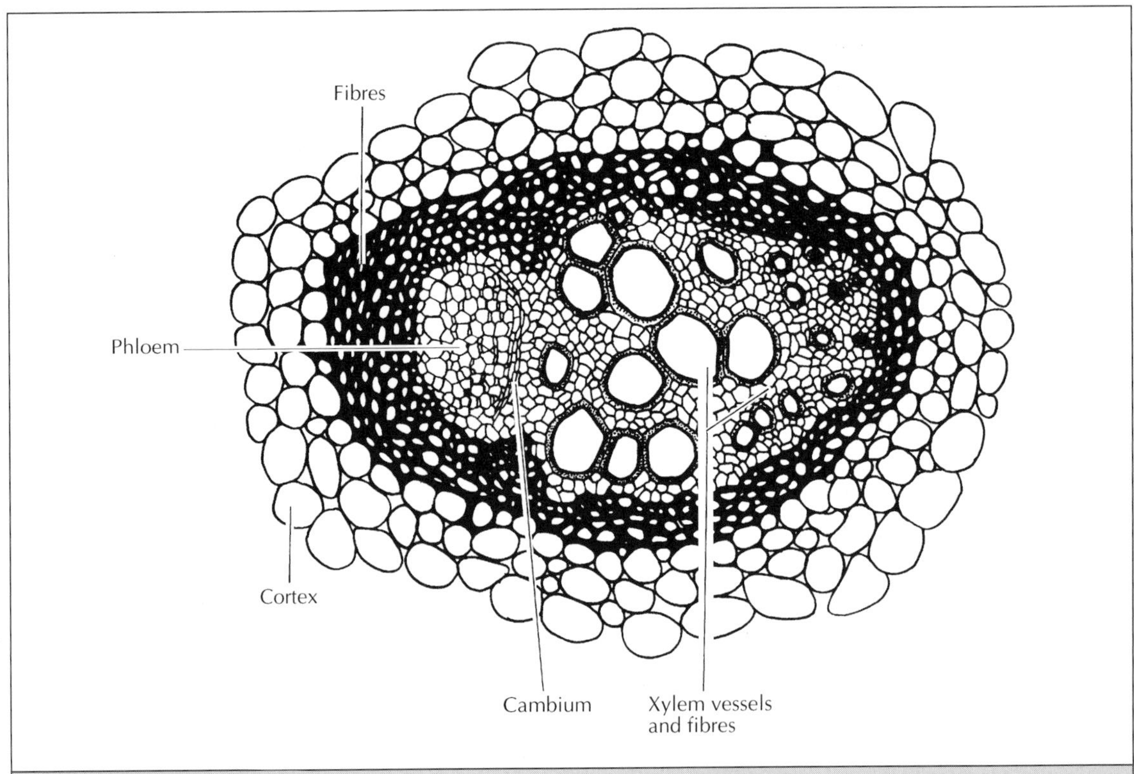

Fig. 1.14 Section of single vascular bundle from the stem of a Dicotyledon such as buttercup (*Ranunculus* spp.). Drawing by Chris Went.

which the leaves, axillary buds and branches emerge, for the vascular supply to these organs necessitates sometimes complex patterns of branching of both the phloem and the xylem tissues – a veritable 'spaghetti junction' of the plant world. Moreover, there may be cross links between the individual bundles in the internodes, the portion of the stem separating the nodes. Finally, complex repositioning of vascular bundles also occurs in the *hypocotyl*, the portion of the stem which at its base links with the root, for in the root proper the xylem comprises a central core, with the phloem tissues dispersed around it, as we shall see below.

Secondary thickening

As the stem of a dicotyledonous plant ages in woody plants, it undergoes a process called secondary thickening (Fig. 1.15), in which woody tissues are created to provide additional support, and an outer layer of bark is laid down to protect the phloem beneath.

This process begins when the cambiums of adjacent vascular bundles link up by the development of the interfasicular cambium referred to above (Fig. 1.15b). Thus a continuous cylinder of cambial cells is created, running the whole length of the stem, like the letters through a stick of seaside rock. By the end of the first season of secondary growth, this cambium has formed a complete cylinder of xylem tissue on its internal face and a somewhat thinner layer of phloem tissue externally (Fig. 1.15c). During the next season a further cylinder of xylem is produced, so that two annual rings of xylem are visible. A second layer of phloem is also formed, but in this case the first layer of phloem is disrupted to accommodate the second. The different fates of the secondary xylem and phloem are explained by

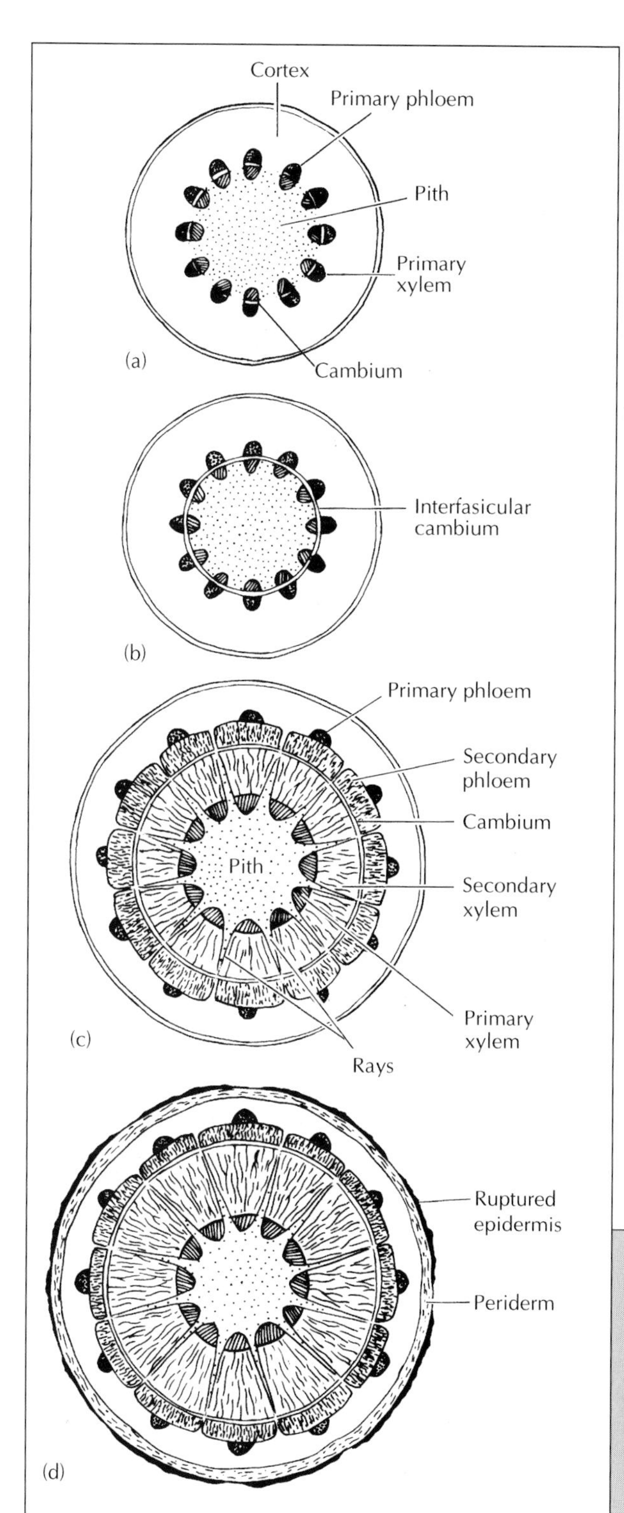

Fig. 1.15 Diagram illustrating the early stages of secondary thickening in the stem of a Dicotyledon: (a) the unthickened stem; (b) development of the *interfasicular cambium*; (c) development of the *secondary phloem* and *secondary xylem* (the *primary phloem* and *primary xylem* are pushed to the periphery and centre of the stem, respectively) and the development of the *rays*; (d) development of the corky tissue of the *periderm* and rupture of the *epidermis*.

the geometry of the thickening stem. The cambium produces xylem cells on its inner face, so that each year's ring of tissue is completely enclosed by a new ring of xylem tissue the following year. As the girth of this increases, the cambial cells divide laterally to accommodate the expansion, rather as one might put extra gussets into an old-fashioned pair of corsets to accommodate an increase in girth. The first year's layer of phloem, however, does not usually expand laterally and so it breaks up with the increasing girth of the stem, being replaced by a completely new layer of phloem tissue on the outer face of the cambium in the second year. The epidermis is also disrupted, and replaced by a layer of corky cells, the *periderm*, produced by another new cylinder of cambium – the *cork cambium* – that arises in the cells of the outer layers of the stem (Fig. 1.15d). This corky layer is disrupted each year as the stem expands and is again replaced by a new layer of corky cells the following year.

In trees and shrubs this process is repeated annually until a large woody stem with prominent annual rings is produced. The central, older layers of xylem eventually become inactive, the vessels being plugged with resinous substances and the mass of tissue being protected from attacks by rot-causing fungi by the deposition of aromatic fungitoxic substances. This *heartwood* has a darker, richer colour when seen in section than the outer, active zone of the stem, called the *sapwood*. Communication between the layers of xylem is maintained by slightly wavy radially arranged sheets of living cells running the length of the stem, called *rays*. The rays formed in the early stages of secondary thickening link the pith with the outermost layer of phloem cells and are called *medullary* rays. Those formed later simply link phloem and xylem and are called *vascular* rays.

The evidence of annual rings of xylem may be clearly seen at the cut end of a tree after felling or of a branch after pruning. The annual rings, cut tangentially, are what create the grain in wood cut into planks and planed for cabinet making. The sheets of ray tissues, also cut tangentially, create the figure, which cuts across the grain, in such prepared timber. The wood of each species has its own characteristic pattern of grain and figure. This is in part fixed genetically and is partly determined by climate and rate of growth. Fast growing trees like conifers (e.g. *Pinus* spp.) produce soft woods, whereas slower growing species such as oaks (*Quercus* spp.) or beeches (*Fagus* spp.) produce hardwoods. A thick plank of oak is ideal for seeing the various features just described. It will have a prominent dark grain, with silvery figure running across it. The variation in the darkness or density of the layers of grain results from the laying down of large water conducting vessels during the spring, when new growth and new leaves are being produced, followed later in the season by the laying down of narrower vessels and fibres for strength.

The connection between the grain and the annual rings can be seen clearly by looking at the cut end of the plank. The difference in the diameter of the vessels in the spring and summer wood of each ring will also be apparent here. It is sometimes possible to push a thin piece of wire for a considerable distance into the wider vessels of an oak plank, demonstrating the pipe-like nature of these conducting cells.

Bark

Finally, a word about bark (Fig. 1.16). This usually consists of disrupted phloem tissue, interspersed with layers of corky tissue. It thus provides a waterproof, protective covering for the woody stem, resistant to both mechanical and climatic damage. Gases such as oxygen and carbon dioxide can pass in and out, however, for at intervals there are large pores in the bark, called *lenticels*, loosely filled with large, thin-walled cells. Lenticels are especially prominent in the bark of elder bushes (*Sambucus nigra*). Also, they can be seen as long channels in wine corks, which are made from the unusually thick, corky bark of the cork oak (*Quercus suber*).

The particular pattern, form and structure of bark is distinctive for each species and depends upon the arrangement of cork cambiums which generate the corky cells and the amount of phloem and cork produced each year (Fig. 1.16). For example, in deciduous oaks (*Quercus* spp., Fig. 1.16a) and similar rough-barked trees such as elms (*Ulmus* spp.) and pines (*Pinus* spp.) the cork cambium is not a continuous cylinder, but instead comprises a series of plates re-formed each year as

Fig. 1.16 Bark: (a) oak (*Quercus* spp.), a rough-barked tree in which the cork cambium comprises a series of plates, re-formed each year, and pulled apart as the trunk expands; (b) beech (*Fagus* spp.), a smooth barked tree in which the cork cambium forms a continuous cylinder that expands to accommodate the increasing girth of the tree. Photographs by Chris Prior, Royal Horticultural Society.

more phloem is produced, thereby creating separate plates of bark tissue made up of alternate layers of corky and phloem tissue. As the girth of the tree increases the outermost plates are pulled apart, giving the tree its fissured surface. These plates are especially easy to see in pine bark. In other species, such as London plane and cherries (*Platanus* × *hispanica* and *Prunus* spp.) they may take the form of large sheets which split and peel away as the stem expands. In smooth-barked trees such as beeches (*Fagus* spp.) and limes (*Tilia* spp.) the cork cambium remains as a continuous cylinder, increasing its girth year by year, and forming only a very thin layer of phloem each year (Fig. 1.16b).

The majority of monocotyledonous plants are herbaceous and do not undergo secondary thickening. In some species, such as bamboos, secondary thickening does occur as a result of the activity of a layer of cells called a primary or secondary thickening meristem, which produces masses of woody cells. In other tree-like monocotyledons such as palms, woody tissue may be produced diffusely in the background tissue throughout the stem. The persistent bases of old leaves provide a protective covering. And in others, such as banana (*Musa* spp.) the sheathing leaf bases themselves provide support for the stem, as we have seen.

Stem modifications

As with the leaf, the selective forces of evolution have resulted in various modifications of the stem to cope with particular environmental conditions. In cacti (family Cactaceae), for example, which are native to the desiccating environments of deserts and semi deserts, the stem has become succulent and green, and ridged or flattened to increase its surface area (Fig. 1.11). It thus not only stores water, but also takes over the photosynthetic function of the leaves, which are either absent or reduced to spines. Brooms (*Cytisus* spp.) also have ridged and winged photosynthetic stems, strengthened with significant amounts of fibrous tissue. In spring, when water is abundant in the Mediterranean climates where the plants grow wild, leaves are produced and carry out photosynthesis in the normal way. As the hot dry summer advances, the vulnerable leaves are lost and the stem acts as the main photosynthetic organ. In another species, the butcher's brooms (*Ruscus* spp.), further evolution has led to a flattening of the photosynthetic stem to form a *cla-*

dode, a structure bearing an uncanny superficial resemblance to a true leaf.

Other stems may be shortened to produce rosette plants such as in common lawn weeds like daisy (*Bellis perennis*) and dandelion (*Taraxacum officinale*). The effectiveness of this modification (which perhaps evolved in response to grazing) is only too apparent when the lawnmower leaves them virtually unscathed. At the other extreme, the stems of some plants may be very long indeed, carrying the leaves high above those of competitors in the race for the sun. In these cases the stem may be supported by layers of woody tissue, as in trees, or by clinging to other plants, as with the twining plants like hop (*Humulus lupulus*) or honeysuckles (*Lonicera* spp.) or by clinging with thorns as in climbing roses (*Rosa* spp.) or brambles (*Rubus* spp.). Stem thorns may also provide a measure of protection, as in hawthorn (*Crataegus monogyna*).

Finally, stems have become modified during evolution to form underground or underwater storage organs, providing a means not only of survival during winter or very dry summer conditions, but also of propagation. The rhizomes of irises (*Iris* spp.) and water lilies (*Nymphaea* spp.), for example, are stems modified in this way. Those of water lily also contain air spaces to aid the diffusion of oxygen from the surface. Potato tubers are also modified stems (Fig. 1.17), as are the invasive long stolons of mints (*Mentha* spp.) and couch grass (*Elymus repens*). In all these cases it is possible to see modified rudimentary leaves on the stem surface, with buds in their axils. Look carefully at the 'eyes' of a potato, for example. In corms, such as those of *Crocus* spp. and *Gladiolus* spp., the stem is reduced in length and modified for storage. The old leaf bases persist as a brown fibrous covering. In bulbs such as those of *Tulipa* and *Narcissus*, where the storage function is served by the modified leaves or leaf bases, the stem is shortened and thickened to create, in essence, an underground rosette.

There are innumerable different forms of modified stem in the garden and if you need an excuse to delay getting on with the digging, it is worth trying to spot them. Diagnostic features are the presence of leaves or leaf traces, usually with a bud in each axil, and the presence of an apical bud.

Fig. 1.17 Potatoes are stems modified for storage, the 'eyes' being reduced leaves with buds in their axils. Photograph by Chris Prior, Royal Horticultural Society.

MINING FOR MINERALS AND WATER: THE ROOT

The structure and growth of the root

The most remarkable thing about roots is not their structure, but the fact that they grow and branch continually, forever exploring and exploiting new areas of soil for the minerals and water that are as essential for the health of the plant as the energy, carbon dioxide and oxygen harvested by the leaves. They are, in effect, the hunter-gatherers of the plant world. This growth occurs at the *root apex*, which is protected by a *root cap* and lubricated by mucilage (see Chapter 4 and Fig. 4.6).

The efficiency of the root as an absorbing organ depends on its absorbtive surface area relative to its volume, created by the root hairs, (extensions of the epidermis that extend out into the soil) and the complex system of branches. It has been estimated that a four-month-old rye

(*Secale cereale*) plant, for example, has a total area of over 600 square metres in contact with the soil. To put this figure into perspective, it is equivalent to the area of a lawn about 25 metres by 25 metres. How this vast structure functions and keeps the plant supplied with water and nutrients is a story that will be told in Chapter 4. Here we will consider the root as a structure for anchoring and supporting the plant, and as an organ that may have been modified by evolution and selection and breeding to carry out a variety of other functions.

The basic structure of the root body is shown in section in Fig. 1.18. The xylem tissues, consisting of vessels and fibres, form a rod or central core with the phloem tissues deployed around them (the *stele*) and provide great mechanical strength to resist the pulling power caused by wind buffeting the plant or grazing animals tugging and pulling at it as they remove its soft tissues.

The plant is even more firmly anchored by the extensive branch system of the root. Root branches arise behind the apex, being initiated deep in the central tissues, where the vascular connections are established, and then breaking through the outer layer of the *cortex* to emerge as new roots, which may branch again and again, ending in delicate, short-lived *feeder roots*. It is these that are most actively involved in the uptake of nutrients and water, and they are replaced continuously as the root system explores and re-explores the soil in which the plant is fixed. As they age their function changes to that of providing support and anchorage.

The basic pattern of branching established by the root system varies from species to species. In some garden plants, such as lupins (*Lupinus* spp.), there is a central *tap root* with side branches formed along its length. In others, such as Michaelmas daisies (*Aster* spp.), the root system

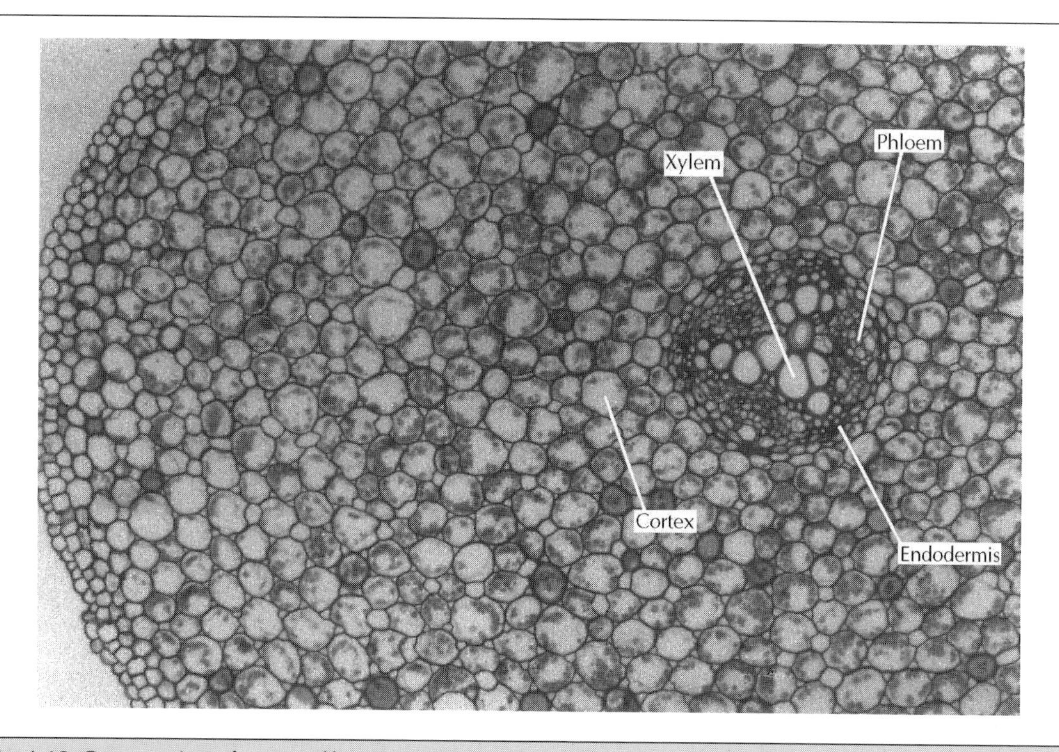

Fig. 1.18 Cross section of a root of buttercup (*Ranunculus* spp.) viewed with a light microscope. Clearly visible are the central core of *vascular tissue* (xylem and phloem), the *endodermis* and the wide *cortex*. Specimen prepared and photographed by B.G. Bowes, University of Glasgow.

has no central axis, consisting instead of a much-branched fibrous mass. Sometimes, as in maize (*Zea mays*), additional branch roots may arise from the stem base. These, called *adventitious roots*, provide additional support as well as enabling the plant to explore the surface layers of soil. Adventitious roots also arise at the base of stem or leaf cuttings, induced to form by the re-orientation of internal hormone levels induced by excision or by the application of artificial hormone rooting powder (see Chapter 7).

As roots age they may undergo secondary thickening as the stem does. In dicotyledons this is initiated by the activation of a corrugated cylinder of dividing cells, the *cambium*, which develops between the mass of xylem tissue inside and the outer columns of phloem tissue. This cambium produces rings of new xylem vessels and fibres on its inner face, and rings of phloem on its outer face. Cells impregnated with the corky substance suberin may protect the outside of the thickened root, forming a layer of bark-like tissue. As in secondarily thickened stems, lenticels allow the movement of gases in and out. In most trees and in many grasses the corky layers are formed deep in the central root tissues, thus cutting off the supply of nutrients to the outer cortex, which dies and is in consequence sloughed off. The younger roots of trees and grasses may, as a result, be thicker than older roots.

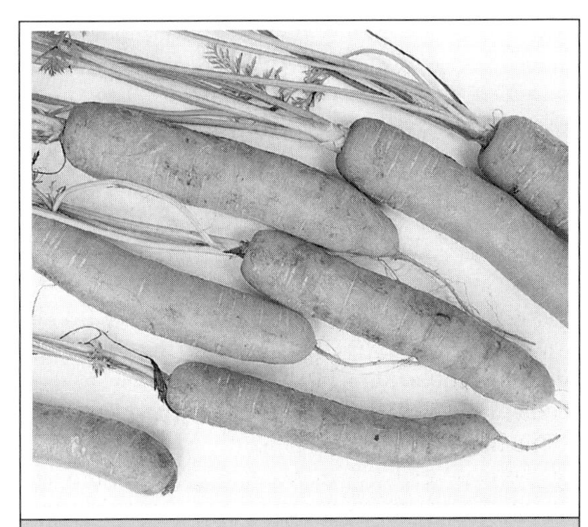

Fig. 1.19 Carrots are roots modified for storage by proliferation of the phloem tissue. Photograph by Chris Prior, Royal Horticultural Society.

Storage roots and other modifications

Roots, like leaves and stems, may have become modified during evolutionary selection in a variety of ways to perform specific functions (Fig. 1.19); the most obvious to the gardener is the over-winter storage of carbohydrates. Thus in turnip (*Brassica rapa*), for example, masses of large, thin-walled cells called *parenchyma*, packed with starch grains, are formed in the secondary xylem of the tap root as the plant ages and winter approaches. In swede (*Brassica napus*), that much-maligned winter vegetable, both the root and the *hypocotyl*, the transition tissues that connect stem and root, swell in this way. In carrot, by contrast, carbohydrates are stored in masses of secondary phloem tissue in the root, the secondary xylem being limited to the central, yellowish core. Fibrous root systems may also swell by proliferation of parenchyma cells to form tuberous storage roots, as in *Dahlia* spp. and Jerusalem artichoke (*Helianthus tuberosus*). In most cases the carbohydrate is stored as starch grains within the cytoplasm of the cells. In carrot, sugars are also present giving them a sweet taste, and in Jerusalem artichoke the carbohydrate takes the form of a starch-like compound called inulin.

Roots may have become modified in a variety of other ways too. For example, the roots of some plants that grow in trees, such as certain tropical orchids, have an epidermis that is several cell layers thick. The outermost layers of cells are dead and form a structure, the *velamen*, that absorbs water from the saturated air of the rain forests in which these species grow. Some of the roots of crocus (*Crocus* spp.) and other species that produce corms are able to contract, drawing the newly formed corm, which grows on top of the parent corm, down into the soil. Adventitious roots may be important in providing secure anchorage for climbing plants such as ivy (*Hedera* spp.) and *Hydrangea anomala* subsp. *petiolaris*, and in maize, as we have already seen, may form buttresses, helping to support the mature stem. It is interesting to look for modified roots in the garden, if you can bear to dig up some of your plants in the interest of science, and to compare them with modified stems.

Nitrogen fixation

A final group of modifications of roots, although not easily visible to the naked eye, are of vital importance to almost all plants. These are the *symbiotic* associations of roots with other organisms, mainly bacteria and fungi, to improve the supply of nutrients to the plant. The best known are the associations with nitrogen-fixing bacteria.

Nitrogen supplies in the soil are supplemented by the conversion of gaseous nitrogen from the atmosphere into nitrogenous salts by such nitrogen-fixing bacteria. Many of these live free in the soil, but others form symbiotic associations with the roots of plants in the legume family (Papilionaceae) such as beans (*Phaseolus* spp.) and clover (*Trifolium* spp.) In these associations the bacteria invade the root, causing nodules to form, and it is in these nodules that fixation occurs. Nodules can easily be seen if the roots of bean or clover plants are gently dug up and washed free of soil (Plate 3). The evidence of additional nitrogen being available around plants with nitrogen-fixing nodules can be seen in the darker green of grass in a lawn infested with clover. The bacteria are supplied with sugars by the plant. Some fungi are also able to fix nitrogen: for example, the fungus *Frankia* forms nitrogen-fixing nodules on the roots of alder trees (*Alnus* spp.)

Mycorrhizas

Less well known, but even more important for plants, are the associations between roots and beneficial fungi to form *mycorrhizas*. Mycorrhizal associations are almost ubiquitous in the plant kingdom, although a few families, notably the cabbage family (Brassicaceae) do not form them. Some say that the first plants to invade the land in Devonian times depended on mycorrhizas rather than roots to supply them with water and nutrients. There are various types of mycorrhizas, but the two most common are the vesicular-arbuscular (VA) mycorrhizas formed by most herbaceous plants, and the sheathing mycorrhizas formed by many trees. Fungi grow as thin, microscopic threads called hyphae. In VA mycorrhizal associations these hyphae grow between the living cells of the root, forming at intervals large storage vesicles containing fatty lipids and specialised, finely subdivided branches called arbuscules which penetrate living cells and form an interface for the exchange of nutrients. The hyphae then extend out into the soil, forming an extensive absorbtive network (Fig. 1.20).

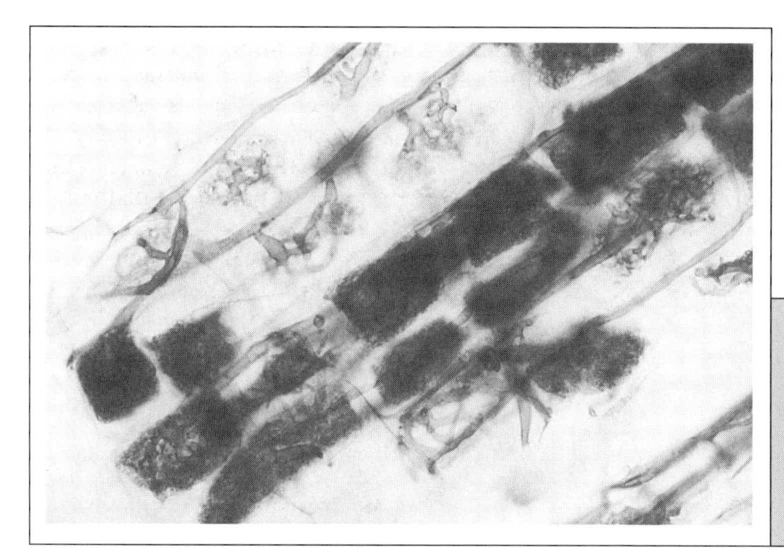

Fig. 1.20 *Vesicular-arbuscular mycorrhiza (endomycorrhiza): intercellular* hyphae *and* arbuscules *are visible in a squashed and stained root of cowpea (*Vigna *sp.). Light microscope photograph supplied by P.A. Mason, Centre for Ecology and Hydrology, Edinburgh.*

In sheathing mycorrhizas the hyphae of the fungal partner form a thick mass around the outside of the feeder roots, causing them to thicken and branch abnormally (Plate 4). The pattern of branching is characteristic of each specific association. For example, in the case of associations involving pines (*Pinus* spp.) the branches are dichotomous (y-shaped), whereas in associations involving beech (*Fagus* spp.) they are racemes (branches arising at right angles from a central axis).

The fungal sheath sends out branches on its inner face which form a network, called the Hartig net, between the cells of the cortex (Plate 5). It is here that nutrient exchange between plant and fungus occurs. The outer face of the sheath also sends out branching hyphae which grow out and form an absorbtive network in the soil. Sheathing mycorrhizal fungi often reproduce by forming toadstools in which spores, the minute equivalent of seeds, are formed. This is why characteristic toadstools are often associated with particular trees in woodlands and in the garden.

Research has shown that mycorrhizal fungi are especially involved in scavenging phosphates from the soil, for these minerals are largely insoluble, and to acquire sufficient for the plant's needs, especially in impoverished soils, a vast absorbing surface area is required. The fungi may also be involved in the uptake of other nutrients too, especially nitrates. In return, so to speak, the mycorrhizal fungi extract sugars from the plant with which to sustain their own growth. The association is thus one that is truly beneficial to both partners, enabling them to grow in habitats that would not be available to either of them growing alone.

Other specialised forms of mycorrhizas are formed by heathers and other members of the family Ericaceae and orchids (family Orchidaceae). In the latter the fungus provides the orchid with carbohydrates, released by breaking down organic matter in the soil.

SECURING THE FUTURE: REPRODUCTION

Although some plants may live for many years, or even centuries in the case of trees, all must reproduce to establish further generations or to spread to new habitats. Most plants are able to reproduce sexually by producing flowers and, following pollination, setting seed (see Chapter 6). Some hybrids, however, have lost this ability and reproduce vegetatively. And some species are able to reproduce both sexually and vegetatively, thus increasing their reproductive flexibility.

Vegetative reproduction

Examples of vegetative reproduction include the formation of clumps of individuals from a proliferating crown, as with many herbaceous perennials such as lupin (*Lupinus* spp.). Similarly, some bulb-forming plants such bluebell (*Hyacinthoides non-scripta*) and onion (*Allium* spp.) may produce new bulbs from buds located between the bulb scales of the parent. Alternatively, branches may arch over and root, giving rise to new plants, as with bramble (*Rubus* spp.), or adventitious roots may form from partially buried or broken branches, as in willow (*Salix* spp.). In some species specially modified stems called runners (*stolons*) grow along the ground, producing plantlets at the nodes, as in the case of strawberry (*Fragaria* × *ananassa*). In others, such as *Kalanchoe daigremontiana*, plantlets may form at the edges of the leaves, subsequently falling to the ground and taking root. Finally, in some hybrid grasses, such as viviparous fescues (*Festuca* spp.), plantlets may replace the sexual structures of the flower. These and the many other examples of natural vegetative propagation depend upon the ability of plants to produce roots and shoots adventitiously from vegetative tissues, usually stems, but sometimes also leaves, roots or flower parts. Some plants cannot do this, however, in particular many mature trees, for reasons we do not fully understand.

Sexual reproduction

Although there is a great diversity of mechanisms for vegetative reproduction in the plant kingdom, the feature they have in common is that they give rise to progeny that are clones, genetically identical to the parent plant. Vegetative reproduction is thus harnessed by the gardener, as in taking cuttings or layering, to produce large numbers of uniform individuals (see Chapter 7). Sexual reproduction, in contrast, involves the mixing of the genetic material from two different individuals, the parents, and thus generates diversity. Genetic diversity among progeny is important, not least because it increases the possibility of survival of at least some individuals of a species where the environment is subject to change. It also creates the rare possibility of new combinations of genetic information among progeny that may make a species more successful in an existing environment, perhaps, for example, because it can compete better with its neighbours or reproduce more effectively. It is, in other words, fitter. The reproductive advantage so acquired is the basis of evolutionary change, the reproductively fittest individuals in each generation being at a competitive advantage over the less fit individuals of the same species. Of course, evolution takes place very slowly in nature, requiring countless generations to produce significant changes. Plant breeders may speed up the evolutionary process by selecting desirable individuals with, for example, enhanced visual appeal, yield or resistance to disease, from amongst the progeny of a controlled cross between two known parents; growing them on and carrying out further crossing and selection until a new variety is produced. Most of our garden plants were produced from wild plants in this way and this whole subject of designing plants is discussed in detail in Chapter 3.

Cones and flowers

The process of sexual reproduction in seed plants involves the formation of male and female sexual structures either in cones (Fig. 1.21), as in the Gymnosperms (conifers, gingko, cycads and their relatives, Table 1.3), or in flowers, as in Angiosperms (the flowering plants, Fig. 1.22; Table 1.3). The male structures are anthers in which the microspores (pollen) are produced. Pollen grains have thick, fatty walls and are often richly ornamented. They are carried by wind or insects (or occasionally by other species such as bats or birds) to the female organs, where pollination and fertilisation occur. In wind-pollinated plants such as grasses and conifers the flowers usually lack both colour and scent, but are carried in such a way on the plant as to maximise the likelihood of pollen being picked up by the wind and carried to a receptive female. Where insects or animals are involved in pollination, the production of sweet nectar or scent in special glands at the base of the petals, or flower

Fig. 1.21 Cones of *Abies* sp. Photograph by the Royal Botanic Garden Edinburgh.

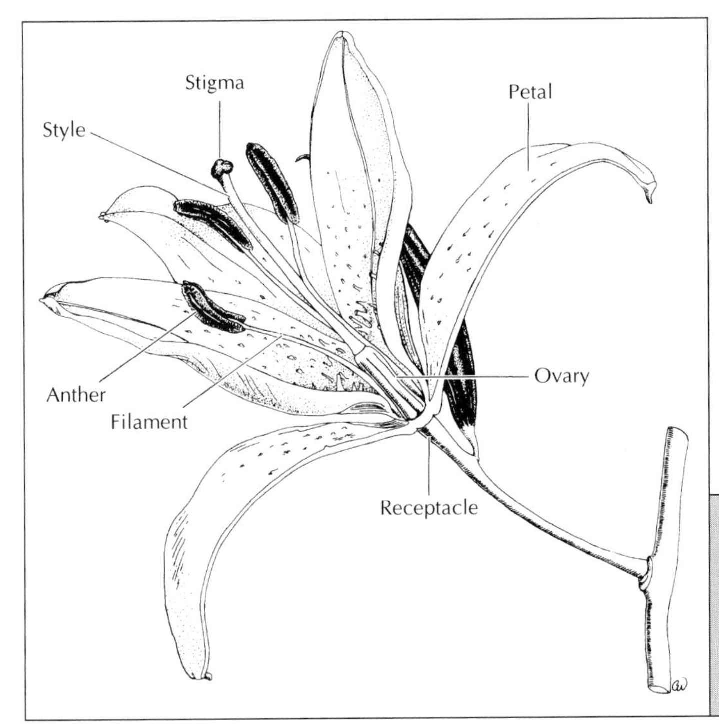

Fig. 1.22 Section through a flower of an oriental lily (*Lilium* sp.), a Monocotyledon. Note that in Monocotyledons there is only a single ring of *petals*, with no outer ring of green *sepals* as there would be in a Dicotyledon flower. Drawing by Chris Went.

colour, may have evolved to attract the appropriate pollinating species. The white flowers of different plant species, which may all appear similar to our eyes, reflect UV light in different ways. They may therefore appear as different colours to the compound eyes of insects that can detect wavelengths of light that are invisible to us. Scent is especially important as an attractant in the case of night-flying pollinators such as moths or bats. Finally, the shape of the flower (symmetrical or asymmetrical, tube-like or divided into petals, and so on) is also fine-tuned to the pollinator with which the species co-evolved. Insights into the genetic basis of such differences are given in Chapter 3.

The female structures (the *ovules*) of Gymnosperms are unprotected by other tissues and the pollen reaches them direct, immediately followed by fertilisation. In the Angiosperms the ovules are enclosed by an *ovary* with a *stigma*, supported by a *style*, for reception of the pollen (Fig. 1.22). The pollen grains therefore germinate on the stigma surface to produce long pollen tubes which grow through the surface and down through the style, eventually reaching the ovaries, where fertilisation occurs.

Male and female organs in the Gymnosperms are usually produced in separate, structurally different cones. It is easy to distinguish the soft, pollen-forming male ones and the much larger, woody, female cones on pine trees (*Pinus* spp.) and cedars (*Cedrus* spp.). The male cones fall off the trees as soon as the pollen has been released, whereas the female cones continue to grow and mature over a long period. In the Angiosperms the male and female organs may be formed in different flowers, as in hazel (*Corylus* spp.), or in the same flower, as in buttercup (*Ranunculus* spp.). However, most plants have inbuilt incompatibility mechanisms to prevent self pollination, which would lead to inbreeding (see Chapter 3), or pollination by an unrelated species, which would produce a nonsense hybrid. Normally this relies on the recognition by the cells of the stigma of specific proteins on the surface of the pollen grains. Following recognition of alien pollen, fertilisation is aborted.

Plants carry their flowers in different ways,

perhaps in part to facilitate pollination by wind or the insect and other pollinators with which they have co-evolved. Often they are borne singly, as in poppies (*Papaver* spp.); sometimes they are borne all the way up the stem, as in foxglove (*Digitalis* spp.), to form a *spike*, or with each flower on the spike having a short stalk, forming a *raceme*, as in *Delphinium* spp.; or they may be borne on a branched structure to form a *panicle*, as in roses (*Rosa* spp.). In onions (*Allium* spp.) all the flower-bearing branches arise from a single apex, forming a dense head of flowers called *an umbel*, and in *Hydrangea* spp. these branches are subdivided to form a flat head of flowers called a *corymb. In hazel (Corylus* spp.) the male flowers form a *catkin*. In the daisy family (Asteraceae) all the flowers arise from a single platform, the *receptacle*, in such a dense mass as to form a *composite* structure resembling a single flower (Fig. 1.23). Very often, as in daisy (*Bellis perennis*), the inner and outer flowers may be very different, the latter being expanded to give the appearance of petals and the former being very small and tube-like, forming the yellow centre.

Fig. 1.23 The compound flower of *Echinea pallida*, a member of the family Asteraceae. Note the outer ray florets and the inner mass of tube-like florets. Photograph by Chris Prior, Royal Horticultural Society.

Alternation of generations

The genetic information needed to co-ordinate the development of a new plant is encoded in the genes within the DNA of the cells. This DNA is packaged in elongate structures, the chromosomes, contained within the cell nuclei (see Chapter 3). The number of chromosomes per cell is fixed for each species, but varies between species. Chromosomes normally occur within the nucleus as almost identical pairs. Indeed, they are identical for genes defining all the major characteristics of the plant, but may vary for minor characteristics like rate of growth or height. Such variations are generated by random *mutations* in specific genes (see Chapter 3). Each cell of a plant therefore contains two almost, but not quite, identical sets of genetic information and is said to be *diploid*. When a cell divides in the body of a plant to produce two daughter cells, the chromosomes also divide to produce identical pairs of daughter chromosomes. This division is termed *mitosis*. The process of sexual reproduction, in contrast, first involves a specialised form of cell division called *meiosis*, which results in a halving of the genetic material of the parent to produce specialised *haploid* cells, the microspores and the ovules already described. Fertilisation of an ovule by a microspore restores the genetic complement and the resulting *zygote* is therefore *diploid*. This process is dealt with in detail in Chapter 3.

Following fertilisation a diploid seed develops from the zygote, and this is capable of germination to produce a new diploid plant which contains genetic materials from each of the parents. This alternation of haploid and diploid generations in seed plants has evolved over millions of years from a situation in more primitive plants, such as mosses and ferns, in which there is an alternation between separate, free-living sexual (*gametophyte*) and spore-bearing (*sporophyte*) generations. A knowledge of reproduction in these more primitive species makes it easier to understand the detail of the structures and processes involved in the sexual reproduction of seed plants.

For example, the fern plants (Filicinophyta, Table 1.3 and Fig. 1.24) familiar in gardens and natural habitats are diploid sporophytes, representing one generation of a two-phase life history.

Fig. 1.24 Fern sporophytes, including tree ferns, growing at the Royal Botanic Garden Edinburgh. Photograph by David S. Ingram.

Each sporophyte plant produces, on the underside of its leaves, large numbers of minute, dust-like haploid brown spores, resulting from meiotic cell divisions, which may fall to the ground or be blown vast distances by the wind.

If these haploid spores land on moist soil they germinate and each one grows into a minute, heart-shaped plantlet only a few cells thick, attached to the soil by threads called rhizoids. This haploid plant is the gametophyte generation, sometimes called the *prothallus*, and on it are formed male reproductive structures, the antheridia, and female structures, the oogonia. Haploid motile male sperm are released from the antheridia and swim in a film of water to fertilise the egg cells produced by the oogonia. This restores the diploid condition and a new sporophyte (the fern) begins to develop on the gametophyte, eventually obliterating it. If the egg cells are fertilised by sperm from a different gametophyte, genetic mixing occurs, thus generating variation among the progeny.

Because of its minute size and dependence on water for growth and fertilisation, the gametophyte generation of the fern is very vulnerable and is thus the 'Achilles' heel' of the life cycle. In seed plants, however, the risk of damage to the gametophyte generation has been reduced by its total incorporation into the sporophyte generation until after fertilisation has occurred. Thus, although there is an alternation of sporophyte and gametophyte generations, this is not apparent to the casual observer. The new sporophyte resulting from fertilisation can receive food from the parent sporophyte and can be dispersed in a complex unit called the seed. This contains the minute new sporophyte (the *embryo*), together with stored food materials, and the whole structure is protected by a seed coat (see Chapter 6).

The main difference between the two major groups of seed plants (Table 1.3), the Gymnosperms (cycads, conifers and their relatives) and the Angiosperms (flowering plants) is, as we have already seen, in the amount of protection given to the developing female structures. In Gymnosperms the ovules are completely exposed on the surface of the plant, the term literally meaning 'naked seed'. In Angiosperms, meaning 'encased seed', the ovules are protected by hollow envelopes of tissue called *carpels*. These evolved from leaf-like structures folded over and fused along the edges.

Seeds and fruits

The embryos of Angiosperms may have single *cotyledons* or seed leaves that enclose the embryo, as in the monocotyledons (e.g. the family Poaceae, grasses and their relatives) or two or more seed leaves with the embryos between them, as in the dicotyledons (broad-leaved flowering plants). The embryos and cotyledons are in turn surrounded by the *integuments*, which become the seed coat (*testa*) (see Chapter 6). Finally, one or more seeds are contained within the carpels, the walls of which thicken to become the fruit.

Peas and beans

Most of these structures can be seen very easily with the naked eye in the pea (*Pisum sativum*) or broad bean (*Vicia faba*). Take a mature pod, the fruit (Fig. 1.25) in your hand and split it open along its length. In these cases the fruits were derived from single carpels and their ancestral leaf-like form is very obvious. The 'beak' end is the remains of the stigma with its short style. The remains of the flower, now shrivelled, may still be attached to the stalk end. Attached along one margin of the pod are the individual seeds, the peas or beans. Each pea or bean will be surrounded by a tough coat, the testa, derived originally from the integuments. This can be split with the fingers to reveal the contents. Before doing so, however, look closely at the point where the pea or bean was attached to the pod, ideally with a lens. If you have good eyesight and a good lens you should just be able to see the remains of the *micropyle,* the minute slot-like hole through which fertilisation by the pollen tube growing from the pollen grains on the stigma actually occurred. Now remove the testa to reveal the two fleshy cotyledons, the seed leaves, which enclose the embryo. The embryo itself, again close to the point of attachment to the pod, will be seen to have an embryonic shoot, the *plumule*, and an embryonic root, the *radicle*. Following germination these will grow to produce a new plant.

Other fruits

There are many other forms of seeds and fruits, some more and some less complex than the bean or pea. All have evolved to be as they are to provide protection for the embryo and to facilitate the dispersal of the seed. In the weed dandelion (*Taraxacum officinale*), for example, each carpel contains only one seed, and hardens to give protection, but also develops a parachute of hairs at one end that enable it to be carried on the wind. The paired carpels of *Acer* spp., each with a single seed, also become hard and develop membranous 'wings', again to aid wind dispersal. In wild progenitors of beans (*Vicia* and *Phaseolus* spp.) and in their wild relatives such as brooms (*Cytisus* spp.) and gorses (*Ulex* spp.), the carpel walls dry and harden in such a way as to develop internal tensions. On a warm day, as the fruit ripens, these tensions cause the pods to twist and split, throwing the seeds for considerable distances. If you have a broom bush in the garden you may

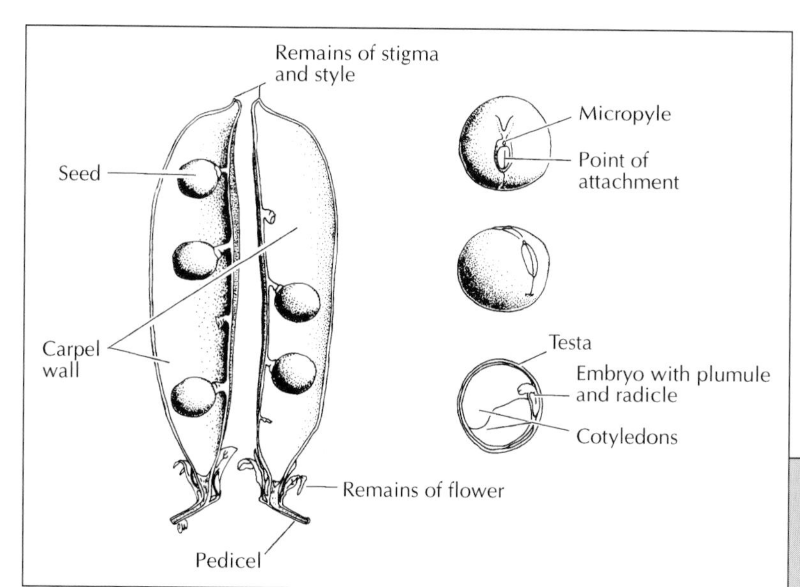

Fig. 1.25 Drawing of a pod of the garden pea (*Pisum sativum*), opened to show the internal structure of the fruit and seeds.

have heard it popping as you dozed away a summer afternoon.

In other species the carpel develops into a fleshy, often coloured and sweet structure, encouraging dispersal by animals and birds. These eat and digest the fruits. The seeds, however, which must be resistant to the digestive enzymes, pass straight through the gut and may be deposited in the faeces many miles from the place of their formation. In plum and cherry (*Prunus* spp.) (Fig. 1.26), for example, a simple seed is surrounded by a three-layered *pericarp* or fruit wall, giving the stone, the flesh and skin. In tomato (*Lycopersicon esculentum*) there may be two, three or more fleshy carpels comprising the fruit, each containing many seeds attached to a central thickened region called the *placenta*. Tomato seeds pass through the human gut unscathed and in the days before the development of modern sewage disposal works, tomato plants were, it is said, major weeds of primitive sewage 'farms'.

Fig. 1.26 Plums are 'true fruits' with a three-layered fruit wall, the *pericarp*, consisting of skin, flesh and stone. The *seed*, which has two *cotyledons* and is surrounded by a brown, papery *testa*, is contained within the stone. Photograph by Chris Prior, R.H.S.

Finally in this brief list of examples, there are fruits in which the carpels harden to provide protection to the seed, but the structure on which the flower developed, the *receptacle*, swells to become fleshy and attractive to animals and birds. In strawberry (*Fragaria* × *ananassa*) (Fig. 1.27), for example, the receptacle grows to form the familiar red pyramid structure, a *false fruit*, with the tiny woody true fruits, (*achenes*) each containing a single seed, dotted all over the surface. In apple (*Malus* spp.) and rose (*Rosa* spp.) (family Rosaceae) the fleshy receptacle grows up and completely surrounds the fruit. If you cut an apple in half you can see that the core is formed from the seeds, each with a brown testa, contained within about five horny carpels. It is this core that is the *true fruit*, while the surrounding receptacle, with the remains of the flower emerging through the end opposite the stalk, is the false fruit. It functions in the same way as a fleshy true fruit, however, being attractive to animals and birds and thereby aiding seed dispersal.

Over the centuries gardeners and plant breeders have selected plants with particular seed or fruit characteristics that bring either visual benefits, like the Chinese lantern (*Physalis alkekengi*), or are a good source of food, such as tomatoes and eating apples. This selection has exaggerated the particular parts of the seed and fruit, making it relatively easy to make out the component parts. Why not look more closely at some of the seeding or fruiting plants in the garden and see if you can sort out how they developed. Start with some of those I have mentioned and then move on to others. I guarantee you will become hooked, quite literally if you have the weed goosegrass (*Galium aperine*) growing in the garden, for here the fruit wall has minute hooks on it, making it stick like glue to clothing or animal fur – yet another aid to seed dispersal.

THE DIVERSITY OF PLANTS

The diversity of green plants is immense, ranging from microscopic *algae* (a term used by botanists to refer to a range of simple green organisms belonging to various taxonomic groups) that live out their lives in water to the complex flowering plants and trees that dominate the terrestrial landscape (Table 1.3). All are important to the

Fig. 1.27 Strawberries are 'false fruits' in which the red flesh is an expanded, fleshy *receptacle,* with the remains of the flower clearly visible around the base. The true fruits, which are small with a hard brown *pericarp,* are scattered over the surface of the receptacle. Photograph by Chris Prior, Royal Horticultural Society.

energy economy of the natural world. Indeed, the algae living in the oceans, lakes and rivers, as we saw at the beginning, are responsible for 40–50% of the earth's total photosynthetic productivity.

Algae and lichens

In gardens, algae may be found in wet places, and in ponds and lakes. The simplest and most ancient are the blue-greens or Cyanobacteria, which have already been referred to as the progenitors of chloroplasts. They exist as single cells or chains of cells which reproduce by simple division and may be seen in the garden as slime on stagnant soil or turf. Many of the blue-greens are able to fix the gaseous nitrogen of the atmosphere into nitrogenous salts and are therefore important in the cycling of this element (Plate 1). Since they have no clearly defined nucleus, the single linear chromosome being contained within the cytoplasm of the cell, Cyanobacteria are classified with the bacteria in the Kingdom Monera, within the Superkingdom Prokaryotae (Table 1.3).

The rest of the algae have a clearly defined nucleus, surrounded by a nuclear membrane, which contains the chromosomes. For these and other reasons they are classified in the Superkingdom Eukaryotae, along with most other living things. Although most of the algae possess photosynthetic pigments, and have always been thought of as plants by botanists, many of the pigments are chemically different from the chlorophylls of the plants familiar to us in the countryside and garden. Most algae have a motile stage in their life cycle, the movement being generated by one or more flagella, reproduce by means of spores rather than seeds, and are biochemically different from true plants. For these and other reasons they are now formally classified in a separate Kingdom of their own, the Protoctista (Table 1.3).

The algae vary in shape and size. The simplest are the single-celled forms that inhabit ponds, lakes, rivers and oceans, forming the phytoplankton, and are familiar to gardeners as the green particles that colour pondwater that is rich in nutrients and exposed to strong sunlight. As the nutrients are used up the algae die and fall to the bottom of the pond, leaving the water clear again. Many of these unicellular algae move about by means of flagella that beat in the water like a whip, propelling the cells forward and keeping them at the surface, where light and oxygen are most plentiful. Others contain gas bubbles to help flotation. Some, the diatoms, have richly ornamented shells made of silica.

Other algae, especially those living in fresh water, form long filaments composed of chains of cells, looking like green hair. A familiar example to gardeners is *Spirogyra* (so called because its chloroplasts are spiral-shaped), which proliferates during the summer in nutrient-rich garden ponds.

There are, in addition, more complex algae that the gardener will only encounter on holiday at the seaside, like the green sea lettuce (*Ulva*) composed of thin plates of cells, familiar in the brackish water of estuaries or close to outfall pipes, or the red and brown seaweeds that inhabit the sea shore and the oceans. Most of the latter are not made of tissues, like true plants, but are composed of long filaments or chains of cells, packed together to resemble tissues. The different coloured pigments that give the marine algae their names are more efficient than chlorophyll in harvesting the wavelengths of light that are not filtered out by the water.

Finally, some algae may associate in a symbiotic relationship with certain fungi to form *lichens*. These can eke out an existence on exposed surfaces such as rocks, walls and tree trunks, where neither partner could survive alone.

Mosses and liverworts

Most of the algae, like the ferns, have a complex life cycle with a distinct alternation of large sporophytic and relatively small and vulnerable gametophytic generations. In the next group of green plants to be described, the *mosses* and *liverworts* (*bryophytes*) there is also an alternation of generations, but the gametophyte generation forms the dominant phase and the sporophyte is attached to it and is dependent on it.

Mosses (Bryophyta) and liverworts (Hepatophyta), like the algae, lack a cuticle, and do not form a lignified xylem, so none of them has evolved to any great size. Many, however, possess specialised water conducting cells and have a specialised sugar transport system. Most require water to enable the male gametes, which are motile, to swim to and fertilise the female gametes. They are, therefore, largely restricted to relatively moist, shady places in the garden. However, many species of moss are able to survive considerable periods of dessication, entering a period of dormancy which may last for months or even years in some cases, and then springing to life again as soon as they are wetted. This physiological adaptation makes it possible for some mosses to grow successfully on roofs and paths and in lawns, remaining dormant in the hot dry summer months and the cold winter period, but growing actively during spring and autumn when rainfall is plentiful.

There are two types of liverwort, each with a distinct gametophyte form. In one the gametophyte is a flattish, lobed plate of cells (*thallose liverworts*), while in the second the gametophyte is a small plantlet with rows of thin, flattened leaves on either side of a thin delicate stem and a third row of reduced leaves on the underside of the stem (*leafy liverworts*).

The thallose liverworts (Fig. 1.28) have a shape that is reminiscent of liver, hence the name of the

Fig. 1.28 The thallose liverwort *Marchantia*. Note the flattened, green *gametophytes* with male (shaped like umbrellas) and female (shaped like the ribs of an umbrella) structures growing up from them. Male and female structures are found on separate plants. Photograph by David S. Ingram.

whole group which is derived from a Greek word meaning liver. They grow on moist, open sites such as shaded paths, while the leafy types grow amongst other closely packed plants. Thallose liverworts such as *Marchantia* and *Pellia* are relatively common in gardens, and as pot weeds in nurseries, but leafy liverworts are usually found only in wild places.

In one of the most common of the thallose liverworts, *Marchantia* (Fig. 1.28), sexual reproduction takes place in structures that grow up on stalks from the surface of the gametophyte. There are separate male and female plants: the stalked structures formed on the male plants resemble small umbrellas, whereas those formed on the female plants are more like the ribs of an umbrella that have lost their cover. Male gametes are carried to female plants in rain-splash droplets, swim to the egg cells and fertilise them. A small, diploid sporophyte forms *in situ*, attached to the gametophyte by a structure called a foot. Most of the cells of the body of the sporophyte then undergo meiosis, forming large numbers of minute dust-like haploid spores. These are spread by wind to new sites, where they germinate to form new gametophyte plants, thus completing the life cycle. *Marchantia* and most other thallose liverworts can also reproduce asexually by forming small balls of cells called *gemmae* in cups on the upper surface of the thallus. These are thrown out of the cups by rain droplets and grow into new plants. The gemmae cups can be seen quite easily on liverworts growing as weeds in moist places in the garden.

The gametophytes of the mosses are all leafy, but altogether more robust than those of the leafy liverworts (Fig. 1.29). There are two basic forms. The tuft-forming mosses such as *Sphagnum* (bog moss), have a clearly defined upright stem, sometimes branched, with the leaves arranged around it. The feathery mosses, such as those that commonly grow in lawns, have branched leafy stems, but these tend to be somewhat flattened and grow close to the ground, sometimes forming large mats. The leaves of both forms are multicellular, but lack a cuticle. In some cases the leaves may be modified to perform specialised functions. For example, the leaves of *Sphagnum* contain enlarged water storage cells. This characteristic was once widely exploited by gardeners who used sphagnum moss to line hanging baskets; *Sphagnum* moss is also a major component of peat, giving it its moisture-retaining properties. The stems often have simple water and sugar transporting systems, but lignified vessels and fibres do not form. Although moss gametophytes are usually attached to the ground by thread-like *rhizoids*, these are not true roots and serve merely to anchor the plant. Lacking a support tissue, therefore, mosses rarely reach a size greater than a few centimetres.

When the gametophytes are mature, *game-*

Fig. 1.29 The leafy, tuft-forming *gametophytes* of the common garden moss *Bryum* with horny, flask-shaped *sporophytes* growing *in situ* on long stalks from the tips of the gametophytes, where sexual reproduction occurred earlier in the life cycle. Photograph by Chris Prior, Royal Horticultural Society.

tangia (structures whose contents divide to form gametes) are produced at the tip of the main stem or of a branch. In some species male and female structures are formed on the same plant, but in others the plants are either male or female. Following transfer of male gametes to females by rain splash, fertilisation occurs and the diploid sporophyte grows *in situ*, attached to the gametophyte by a 'foot'. The sporophyte tends to be elongated, with a swollen capsule at its tip. To begin with it is green, like the gametophyte, but later it becomes brown and somewhat horny. Such mature capsules are easily spotted during summer on the minute plants of the moss *Funaria* and *Bryum* (Fig. 1.29) that grows on old garden paths and on walls. Meiosis occurs in the capsule to produce numerous dust-like haploid spores which are spread by wind to new sites. If conditions are sufficiently moist, these germinate to produce new gametophyte plants, and so the cycle is completed.

Ferns and their relatives

The life cycle of *ferns* (Filicinophyta, Table 1.3), in which there is an alternation between a relatively large, free-living sporophyte generation and a relatively small gametophyte generation, has already been described. It is important to point out, however, that the ferns are part of a larger group of plants, often referred to collectively as the Pteridophytes (Table 1.3), which also includes a range of other forms, including the *horsetails* (Sphenophyta, Table 1.3). The sporophytes, but not the gametophytes, of these plants share with the seed plants the ability to form water conducting and support tissues composed of the tough polymer lignin. They have also evolved a cuticle, with stomata to regulate the movement in and out of water vapour, carbon dioxide and oxygen. Finally, they possess roots. Thus, although the gametophyte generation is small and restricted to moist habitats the sporophytes can become very large and are able to grow in relatively dry environments. At the present time the largest pteridophytes are the tree ferns (Fig. 1.24), but in the geological past, as in the Carboniferous period when the coal measures were being formed, pteridophyte trees of various types formed vast forests. It is probable that the vulnerability of the gametophyte generation to desiccation led to the decline of pteridophytes as the seed plants evolved. Those that remain, like the present day ferns, provide gardeners with some of the most beautiful foliage plants or, like the horsetails, the most pernicious of weeds.

Seed plants

Although the seed plants were dealt with very fully in the early part of this introductory chapter, it is important to mention them again in the context of diversity. They probably originally evolved from green algae, or close relatives of this group, the rest of the algal groups, the liverworts, the mosses, the ferns and other pteridophyte groups being evolutionary dead-ends. The immense range of shape, form, size and physiological and ecological types that has evolved is a result of this development: the ability to form seeds in a life cycle in which the vulnerable gametophyte is retained within a dominant sporophyte that posesses not only efficient support tissue, and water and carbohydrate transport systems, but also a cuticle and stomata to control gas exchange and water loss. The major groupings of the seed plants, Gymnosperms and Angiosperms and, within the Angiosperms, the monocotyledons and dicotyledons, have already been discussed. These groups embrace plants as diverse as forest trees, grasses, orchids and the pond weeds, dominating almost every terrestrial landscape from the high mountains to the rain forests, the grasslands to the ponds and rivers, and every climate zone from the tropics to the arctic tundra. The only areas not colonised are the oceans, where the algae hold sway, and the most exposed areas of the high mountains and the poles.

Many of the adaptations that plants have evolved to enable them to grow and compete successfully in such a range of environments have been utilised and manipulated by gardeners and enhanced by breeders to give us the range of garden plants we now have available to us (Fig. 1.30). This theme will be referred to again and again in the chapters that follow. Finally, it should not be forgotten that plants themselves are colonised by a variety of insect and other pests and the many viruses, bacteria and fungi that

Fig. 1.30 Into the garden; a New Town garden gate in Edinburgh. Photograph by David S. Ingram.

cause disease (Chapter 11), as well as those bacteria and fungi already referred to that are beneficial symbionts.

FURTHER READING

Attenborough, D. (1995) *The Private Life of Plants*. BBC Books, London.

Bernhardt, P. (1999) *The Rose's Kiss: A Natural History of Flowers*. Island Press/Shearwater Books, Washington DC.

Bowes, B.G. (1997) *A Colour Atlas of Plant Structure*. Manson Publishing Ltd, London.

Capon, B. (1992) *Botany for Gardeners*. B.T. Batsford Ltd, London.

King, J. (1997) *Reaching for the Sun: How Plants Work*. Cambridge University Press, Cambridge.

Mabberley, D.J. (1997) *The Plant Book*, 2nd edn. Cambridge University Press, Cambridge.

Margulis, L. & Schwartz, K.V. (2000) *Five Kingdoms*, 3rd edn. W.H. Freeman and Company, New York.

Pollock, M. & Griffiths, M. (1998) *Shorter Dictionary of Gardening*. Macmillan, London.

Raven, P.H., Evert, R.F. & Eichhorn, S.E. (1999) *Biology of Plants*, 6th edn. Worth Publishers, New York.

Walters, S.M. (1993) *Wild and Garden Plants* (New Naturalist Series). Harper Collins Publishers, London.

2

Naming your Plant

- Identifying plants using floras, monographs, articles, illustrations and herbarium specimens
- The meaning and structure of plant names
- The use of Latin in plant nomenclature
- Assigning names to newly discovered plants
- The ancestry and evolution of present day plants and the use of new technology, including DNA analysis, to determine relationships among species and cultivars
- Conventions and procedures in cultivated plant taxonomy
- Concepts of distinctiveness, uniformity and stability
- Why plants change their names
- The quest for stability of nomenclature and the use of information systems to achieve stability in the future

The naming of plants begins with the simple spontaneity that we associate with the naming of other everyday objects – using 'common names' – but as soon as any precision is required this quickly leads to a need for some understanding of the biological principles of reproduction, heredity and evolution. Furthermore, to understand the complexity of variation observed amongst plants, a framework of apparently arcane (on first sight) rules has been established: the language of botanical nomenclature. But with the basic knowledge of the principles that underpin this language, a nurseryman's list of plants becomes as informative to the gardener as a musical score is to a musician.

How is a plant specimen analysed, to yield its name? The name is not an intrinsic property of the plant itself: the person naming (i.e. identifying) the plants in Plate 6 expects a name to have been coined for the plant before – by a *taxonomist* (a person who studies the classification and naming of plants) who has already studied similar plants and defined their limits of variation, as well as the scope and usage of their names. Once a name has been deduced for the plant from the available evidence, our identifier can find out more about the plant: where it grows wild, how to cultivate it, its possible uses and information on related plants. Discovering the name of a plant is no idle activity: it is the key to its literature, such as its origins and growing requirements.

HOW TO IDENTIFY A PLANT

There are a number of approaches to naming, but each requires as much information (as can reasonably be obtained) to be gathered about the specimen. The extent to which the gathered material is representative of the whole plant and its growth and reproductive characteristics, will determine the route or routes taken; as, of course, will the expertise and information resources of the identifier. It could involve postage and packing, or possibly a journey, and perhaps even a fee. All the options are easier if the specimens are as nearly complete as possible. Details of leaves, leaf arrangement, stem and bark type, presence of latex, flowers and fruits are all of importance – missing parts may result in incomplete or inaccurate identification. Associated information such as details of the origin and hardiness of the specimen are of equal importance.

How might someone attempt to identify a pelargonium (Plate 7), found upon a windowsill? Tens of thousands of pelargoniums have been named and this might suggest that a hopeless, or at least arduous task lay ahead; but the strategy which follows is regularly used by botanists to tackle such problems.

(1) By using the appropriate specialist literature, such as *The European Garden Flora* (Walters *et al.* [eds], 1984–2000), or other similar works which give as complete a coverage of plants known in cultivation as can be expected of a reference work. Regional botanical accounts (*Floras*) are useful when something is known about the plant's origin. Monographs provide a complete and structured account of entire plant groups; they are often the foundation of our understanding of the classification for the group, too. Articles in periodicals such as *The Plantsman*, published by the Royal Horticultural Society, provide up-to-date information and detailed accounts of less diverse plant groups.
(2) By making comparisons with illustrations or specimens. These can be in books or on the internet, named plants in botanic gardens and collections (for example those held by members of the National Council for the Conservation of Plants and Gardens), or even pressed specimens in *herbaria* (a herbarium is a collection of preserved, usually pressed, and named plants), such as those at the Natural History Museum, the Royal Botanic Garden, Edinburgh, the Royal Botanic Gardens, Kew, and the Royal Horticultural Society Garden, Wisley.
(3) By asking a botanical or horticultural specialist at, for example, the institutions listed above.

The literature option usually involves the use of *botanical keys* where the specimen's features are matched to each of a series of paired statements. The best keys are always dichotomous, with an either/or choice until a satisfactory solution has been reached. In most cases, keys are followed by descriptions and illustrations that should be used to check that this step-by-step approach has led to the right identification. It is easy to go wrong, especially with large groups of plants. It may even be prudent to follow up further references. The identity of a specimen successfully keyed in *The European Garden Flora* (a well-structured and comprehensive starting point) can be verified by looking up the illustrations cited. If it is a tree or shrub cultivated outdoors in the British Isles, the standard work, *Trees and Shrubs Hardy in the British Isles* (Bean, 1970–1988) might then be consulted. These four volumes and supplement contain very good descriptions and facts but, sadly, no keys and very few illustrations. For bulbous plants, there are many books with descriptions and illustrations; similarly for cacti and succulent species.

THE MEANING AND STRUCTURE OF NAMES

Prior to 1753, plants were known by 'phrase names' (for example, *Achillea foliis lanceolatis acuminatis argute serratis*) which were difficult to distinguish from simple descriptions. It was Carl von Linné (Linnaeus) who formalised a *binomial* system of nomenclature, consisting of two basic terms, the genus and species, that is still in use today.

Latin is the language of nomenclature because it was the accepted international language of science until the twentieth century. Although it may make plant names seem more obscure today to the layman, the legacy of the Latin binomial system has been a consistent and familiar language which enables us to understand the precise meaning of plant names from 1753 onwards. *Botanical Latin* (Stearn, 1992a) has helped greatly to make these names intelligible and to enable the system to be continued in a less classically-trained age.

Table 2.1 Classification of a variegated pelargonium, showing the hierarchy of categories and their names

Category	Scientific name	Vernacular name
Class	*Angiospermae*	Angiosperms; flowering plants
Subclass	*Dicotyledoneae*	Dicotyledons, Dicots
Order	*Geraniales*	–
Family	*Geraniaceae*	–
Subfamily	*Geranioideae*	–
Tribe*	–	–
Genus	*Pelargonium*	the pelargoniums (confusingly also called geraniums)
Subgenus*	–	–
Species	*crispum*	–
Subspecies*	–	–
Variety	*crispum*	–
Forma*	–	–
Cultivar	'Variegatum'	–

* Note that not all ranks are used in all cases.

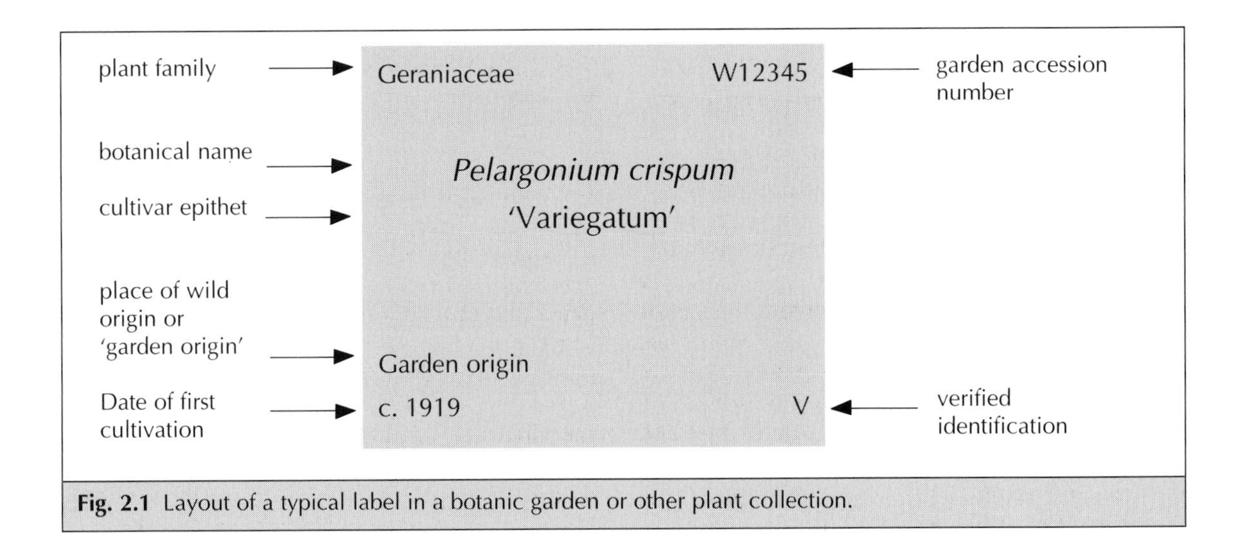

Fig. 2.1 Layout of a typical label in a botanic garden or other plant collection.

When a plant is discovered and thought to be new to science, it is given a Latin name and formally described with a Latin diagnosis according to a set of agreed international rules, the International Code of Botanical Nomenclature (ICBN). This requires that a specimen be made and chosen to serve as a reference point, the *type* of the name. This specimen will fix the use of the name unless altered through the formal procedures of the ICBN.

Just as the word 'knife' is a generic term for cutting implements, *Pelargonium* is the generic name (*genus*) for a group of plants, held together by common characteristics. Following this thread, *Pelargonium crispum* is more specific, as 'table knife' is for knives.

Genus and *species* refer to two ranks in a hierarchy of relationship, reflecting different degrees of variation (see Table 2.1). *Pelargonium crispum* belongs to the family Geraniaceae which encompasses the other genera *Geranium* (the cranesbills or hardy garden geraniums) and *Erodium* (the storksbills). At the other end of the scale, small degrees of variation (such as in flower colour or leaf variegation) might be recognised as botanical varieties or, under other circumstances, as cultivars, e.g. *Pelargonium crispum* 'Variegatum'.

It is now recommended that all botanical names of Latin form are written in italics, or underlined where italics are unavailable (Fig. 2.1). Names of species and the lower botanical ranks are not written with an initial capital letter, even when they are named after a person or place. *Cultivars* are always written with initial capital letters and placed in single quotation marks. Since 1959, in order to emphasise the different treatment of botanical and horticultural names, it has not been permissible to coin all-new cultivar names in Latin form (e.g. words with the ending *-us* or *-um*). *Pelargonium crispum* 'Variegatum', having been coined around 1919, is an acceptable epithet, however.

TAXONOMY: ORDER IN DIVERSITY

The botanical framework

The evolution of plants implies a common ancestry and, since a natural classification based on an understanding of relationships will have the most predictive value – and thus be the most useful overall – this is what botanists strive to understand and reflect in their system of naming. However, in the absence of a fossil record, much of this is based on speculation. Characters, such as spines or flower colour, have often evolved several times in different groups in order that they become adapted to new or changing environmental factors. Such '*convergent evolu-*

tion' has confused the past assessment of relationships.

Understanding the time-scale of ancestry will often indicate how closely plant groups are related but, because some groups evolve faster than others, the situation is not straightforward. Some groups have evolved very little in a long time: *Metasequoia glyptostroboides* (the dawn redwood) was first described from fossil deposits before the living tree was discovered and *Ginkgo biloba* (the maidenhair tree) is frequently referred to as a 'living fossil'. Others have shown much more rapid evolution, such as the tree species of *Echium* in the Canary Islands where different species have evolved and become adapted to the different ecological niches on the different islands.

The taxonomic botanist looks for *discontinuities* in variation to characterise the units to which names are applied. These discontinuities are brought about by *isolating mechanisms*: *reproductive*, in which plants may flower at different times; *ecological*, in which plants become adapted to separated environments (e.g. the wood and water avens, *Geum urbanum* in hedgerows and *Geum rivale* in streamside locations), or *distributional*, in which plants become stranded on geographically different mountain tops or in different valleys, with changing environmental conditions.

Hybridisation breaks down discontinuities arising from these isolating mechanisms. It gives rise to new complexities of variation for the taxonomist to unravel, a process which is often fraught with difficulties. Hybridisation occurs frequently and spontaneously in gardens where so many isolating mechanisms are removed – not simply a matter of plants from different areas being thrust together. It is also deliberately used by plant breeders in their quest for new and better garden plants.

There is an inherent conflict in trying to divide this complex and often little-understood variation of plants into the neat, hierarchical units of botanical names. Little wonder that errors of judgement are made; like plotting the outline of a plough in the stars, there are no absolutely definitive rules to dictate what is included in a group, and what is excluded.

The palette of characters available to the taxonomist has greatly expanded in recent years, and with them, a new perspective on how relationships may be deduced. Most recently, the ability to read DNA sequences has transformed our approach to the science. Resulting changes must be reflected in the classification and therefore in the naming, if the basic principles of the system are to hold true.

Two general approaches to analysis have arisen, to handle the large amount of information generated in taxonomic studies. *Phenetics* brings together a wide range of characteristics of each plant in a broad comparison of similarity. A very refined branch of statistics is required to handle the diversity of (typically) measurement data, such as leaf length. More recently, *cladistics* has dominated as an analytical approach. In this system, each character is represented by two states: primitive and derived. All characters used in the analysis are brought together in a branching diagram (like a family tree), such that the number of branches is minimised, whilst still concurring with the evidence of each character. The relationships of three plants, with one character, are easy to deduce, but the number of possible combinations will very rapidly increase as plants and characters are added. A fast computer is just as essential a tool for modern cladistic studies of DNA sequence information as it is for any phenetic analysis.

Even the most detailed taxonomic analysis is open to interpretation, and a diversity of opinion may arise in the assignment of names to the natural groups (or *taxa*, singular *taxon*) which are found amongst plants. Thus *Ulmus angustifolia* of some authors (Goodyer's elm) is recognised by others as *Ulmus minor* subsp. *angustifolia*: the same taxon, but a different name.

CULTIVATED PLANT TAXONOMY

Botanical names alone have long been used to label plants of cultivated origin. Just as hybrids occur naturally, and have been named as such (e.g. the oak, *Quercus* × *rosacea*, a hybrid of the wild species oaks, *Q. robur* and *Q. petraea*), the products of deliberate hybridisation may be named in the same way. The distinction between natural and deliberate hybridisation is blurred; Russian comfrey, *Symphytum* × *uplandicum* is a natural hybrid, but derived from an introduced plant from SW Asia (*S. asperum*) and the native *S. officinale* (and gives rise to some very garden-worthy variants). The involvement of humans is inextricably bound up in the origin of such plants. The use of a botanical epithet, however, only identifies the two parent taxa – it does not identify the *clone* (genetically identical group), or even the clone of either parent. The name applies to all independent crosses of representatives of the two parent taxa. The system is, therefore, of only limited value in identifying individual plants of merit to the gardener: an indicator of potential value, but infinitely varying between and beyond the characteristics of the parents.

The cultivar is the basic unit of cultivated plant taxonomy, which borrows the framework of the ICBN as far as possible – usually to the genus, though often to the species. A cultivar must possess the attributes of *distinctiveness*, *uniformity* and *stability* to merit recognition, and its characteristics must be retained through propagation. The extremity of form of many cultivated plants cannot usually be transferred from one sexual generation to the next, and so vegetative propagation is often required; hence most cultivars are clonal selections.

Botanical names make reference to a single specimen as a representative of the group – the 'type'. Other plants are given the same name on the basis of shared characteristics with those of the type. Cultivated plant names, however, are ascribed in a different way, usually to identify an unvarying clone – an unusual and often extreme form – and have evolved to suit the practical purposes to which plants are put in the service of humanity. A 'standard', the horticultural equivalent of a botanical type, differs from the latter in that it exhibits the precise characteristics of the cultivar, not just one example plant drawn from a pool of continuous variation. Only relatively recently has an attempt been made to impose a structured system of classification upon the diversity of plants of cultivated origin, where clonal selection is only part of the story. The utilitarian demands of cultivated plants make classification a complex affair; the system must accommodate not only individual plants, but sometimes also:

- a perpetuated single stage of a plant's growth cycle (such as the juvenile foliage of some conifer cultivars),
- a self-sustaining part of a plant (such as the lateral growths of some conifers, grown as prostrate plants: *Abies amabilis* 'Spreading Star'),
- a characteristically virus-infected plant (*Abutilon* 'Souvenir de Bonn'),
- a clonal selection of aberrant growth on a plant (such as the witches' broom, *Picea abies* 'Pygmaea'),
- an assemblage of seed-raised individuals raised from within a population, sharing the same distinctive characteristic (individuals which do not share the defining characteristic having been 'weeded out'),
- a line achieved by repeated and exclusive self-fertilisation,
- a genetically modified plant, or
- a graft chimaera (where the plant is made up of cells originating from both of two united plants; but not a simple graft, where scion and rootstock may be separate cultivars).

The *F1 hybrid* is a familiar type of cultivar: two different parents (i.e. of different species) are cross-pollinated to give rise to the (sometimes sterile) cultivar; as the cultivar does not breed true, it can only be maintained by repeated crossing of the parents. F1 hybrids are common because they tend to be vigorous plants.

Distinctiveness

Although essential to the definition of a cultivar, distinctiveness may relate to rather obscure attributes: for instance, long season of fruiting or resistance to stem collapse ('lodging' – important for grain crops), as well as the more familiar attributes of colour, shape and scent. As variation (ideally) is eliminated within a cultivar, distinctiveness may be assessed to the finest degree –

and consequently may make cultivars far more difficult to distinguish than botanical taxa. In common with the definition of a botanical taxon, however, the description of distinctiveness can only relate attributes to those of other known taxa: *flowers purple* may adequately distinguish a cultivar today, but tomorrow's new plant with *glossier purple flowers* blurs the distinction.

Uniformity

Non-uniform plants fall outside the definition of the named cultivar (even if wholly derived from it) – and uniformity of a cultivar need not necessarily be spontaneous. Seed-raised cultivars are often only maintained through selection of plants that match the cultivar's physical description, others being discarded; *sports* (non-typical individuals or parts of individuals, such as a white flower on an otherwise red-flowered carnation) from otherwise uniform clones cannot be assigned the same name.

Stability

The characteristics that make a cultivar worthy of attention must also be stable in the selected individual to merit naming. A variegation may be interesting at the time, but will not be worthy of a name unless it can be sustained by the plant. Attributes maintained by the gardener's attentions through the life of the plant do not count: a pleached lime is taxonomically not different from its free-grown counterpart, whereas the fastigiate beech, *Fagus sylvatica* 'Dawyck', maintains its shape naturally.

WHY PLANTS CHANGE THEIR NAMES

The reasons for name changes fall into three categories: taxonomic, nomenclatural and misidentification.

Taxonomic changes

These are usually the result of advances in botanical knowledge, such as the case in which a species is found to have been classified in the wrong genus or a genus placed in the wrong family. For example, a new species of conifer discovered in Vietnam in 1999, *Xanthocyparis vietnamensis*, has given rise to a new genus, and *Chamaecyparis nootkatensis* has been transferred to this genus. As this is one parent of the famous × *Cupressocyparis leylandii*, this is also reclassified and must now be called × *Cuprocyparis leylandii*.

Some changes reflect taxonomic opinion and it is difficult to draw a line between matters of fact and opinion when making taxonomic judgements. Some botanists prefer not to recognise the rank of subspecies, elevating these taxa to species and sections of genera to genera. Such people are 'splitters', a more radical tradition which has often been favoured by Russian and East European botanists. The user is left with a choice of classifications and names between which there are no botanical rules to arbitrate.

The improvement of classifications may be judged by their benefit to the widest range of users of the plants, including plant breeders and phytochemists. Name changes will be inevitable as we move in this direction.

Nomenclatural

These changes arise where plant names are contrary to the ICBN. They can only be avoided by the action of international committees which continually decide cases to avoid unfortunate name changes. In recent years, it has been shown that the annual chrysanthemum (a species native to SW Morocco) is a very different genus from the shrubby chrysanthemums of the Canary Islands, and very different from the European cornfield weed species and the Japanese ones, so well-known to horticulturists.

The ICBN required the annual chrysanthemum to be called *Ismene* according to the strictly fair rule of priority by earlier publication: an important nomenclatural principle, but one which is not readily appreciated by most users of plant names. Thus the Canary Island ones are called *Argyranthemum*, the weedy ones remain *Chrysanthemum* and the Japanese cultivated ones were called *Dendranthema*.

After an outcry from horticulturists, the international committee on names has decided that the type should be changed, so that the Japanese species retain the use of the name *Chrysanthemum*. The weedy species, including our native *Chrysanthemum segetum* become *Xanthophthalmum*, certainly upsetting European weed scientists. Committee members involved take all views on board in an effort to obtain nomenclatural stability, though clearly not everyone's needs can be satisfied all of the time.

Misidentification

Often species are introduced into cultivation under the wrong name: they have simply been misidentified at some stage; names such as *Corydalis decipiens* and *Achillea taygetea* have become well-established after such early confusion. The latter name is likely to have been inaccurately applied to *A. clypeolata*, which was itself not well known at the time; and generations of hybridisation have separated the garden plant still further from the rare, isolated plant of the Greek Mount Taygetos, which would have little place outside the alpine house. The take-up of a name in horticulture can indeed be surprisingly unquestioned, sometimes for a considerable amount of time, and naturally, some upset is caused when the true identification is revealed.

To help with the stability of names, the Royal Horticultural Society has set up an Advisory Panel on Nomenclature and Taxonomy to help solve problems. This Panel meets regularly and has played a very considerable role in stabilising and correcting plant names in the RHS Horticultural Database and its associated publications, particularly the *RHS Plant Finder*. If discrepancies are found in nomenclature and classification, it is recommended that the names in use in this work be followed.

THE QUEST FOR STABILITY AND LINKING INFORMATION SYSTEMS FOR THE FUTURE

Gardeners are the most vocal users of plant names to seek stability, because their needs are often the least demanding of nomenclature: they seek a straightforward and distinct 'label' for a plant, and expect to gain no more than an indication of its characteristic value in the garden. In this sense, *Actaea* carries a meaning distinct from *Cimicifuga* – meaning which is equally apparent from the common names 'baneberry' and 'bugbane'. Whilst most other users have greater demands for a name to convey some underlying meaning, this desire for stability strikes a chord – to a greater or lesser extent – with all. The problem has lain in deciding upon a single, functional and universally acceptable set of names.

It is not only gardeners who want to be able to cross-reference botanical and horticultural literature. Botanists often turn to the rich heritage of horticulture for conservation purposes: some species that are genetically depauperate in the wild are well represented in cultivation, such as the cedar of Lebanon. Plants may have been taken from long-vanished wild populations, making them an invaluable genetic resource today.

Trade in plants is an area of increasing legislative activity, and future government regulation is only predictable in that it will expand – leading to increasing interest in names to define those plants affected. Breeders protecting their intellectual property rights through Plant Breeders' Rights (PBR) or Plant Patent, restriction of international trade in endangered species, computerisation of stock control systems and consumer attachment to 'brand names' all push toward 'standardisation' of plant nomenclature. PBR is becoming familiar across a wide range of ornamental and food plants, protecting a breeder's investment in careful selection and manipulation of the plants; without such protection, a competitor is free to propagate and distribute the same plant, which all too easily undermines the profitability of the original breeding work.

Botanical plant name stability

Index Kewensis is a comprehensive guide to plant names (except those of specific horticultural application) and their first place of publication, and has been relied upon by many as a standard reference to plant nomenclature. It is not, however, the reference to accepted nomenclature that it is so often taken to be. *Index Kewensis* (now part of the larger International Plant Names Index initiative, http://www.ipni.org) does help to prevent the appearance of de-stabilising, earlier-published names for well-known plants which the avid taxonomist might rescue from the obscurity of a little-read scientific journal.

Index Kewensis and other such publications follow 'after the event', tracking the publication of names as they appear. A proposal for registration of botanical names has been made, which would prevent their appearance in obscure journals, there to lie as a time-bomb to de-stabilise nomenclature in the future. However, this proposal for compulsory registration of new names has been resisted by many botanists, who are wary of its potential effect upon their intellectual freedom.

Since 1991, the Names in Current Use (NCU) project has provided a platform of stability for genus names. This has helped avoid a worrying tendency for botanists to discover obscure references which invalidate or take precedence over commonly-accepted genus names. The NCU vascular plant list was based largely upon the standard published by Kew, *Vascular Plant Families and Genera* (Brummitt, 1992).

A new approach to stability has appeared with the advent of the internet. It is now possible to link information without necessarily adopting a standard naming system: thus the same information about plants might be viewed using different names by different users. Major information systems such as IPNI (www.ipni.org), Species 2000 (http://www.species2000.org), and ILDIS (http://www.ildis.org) have already been established as a first foray into this sphere. The RHS Horticultural Database, currently offering an independent perspective on naming, will link into other such systems in the future.

Cultivated plant name stability

Stability in the naming of cultivated plants is affected by many of the same factors influencing botanical nomenclature: priority, changing taxonomic perceptions, common usage and orthography (spelling). However, the influence of market forces brings new challenges to the naming of cultivars.

Voluntary registration of cultivated plant names first arose in 1955 as an attempt to avoid the unnecessary duplication of cultivated epithets, and has been successfully implemented for most of the major cultivated plant groups – though not as universally adopted by those who name plants as might have been hoped for. A list of International Cultivar Registration Authorities (ICRA) may be found in the International Code of Nomenclature for Cultivated Plants (1995), and an updated version is available online. Amongst popular groups of plants, there is considerable pressure to allow re-use of certain names with fashionable associations or other marketable qualities. The name 'Wedding Day' may help sell a cut flower, but the temptation to re-use the name time and time again can only cause confusion.

The trend towards use of coded cultivar names adds further confusion; by adoption of a meaningless combination of letters and numbers as the 'cultivar name', the seller is able to attach other names to the plant without infringing any legal or other naming requirement. This allows nursery industries to protect their intellectual property rights over new plants without 'wasting' good, marketable names before time. These 'other names' are known as trade designations or selling names, and should be distinctively styled in print: *Choisya* Sundance ('Lich') is a well-known example. Roses, with their particularly rich horticultural heritage, labour under the greatest proliferation of coded cultivar names. Because roses are often named after well-known personalities of the time, a rose's breeder cannot risk applying the name before the plant is ready to be sold in large numbers, as the namesake may lose popularity.

Since 1987, *The Plant Finder* (later known as the *RHS Plant Finder*) has offered a standard and readily-available reference to the names of plants and their annual availability in Britain, and usefully bridges the divide between cultivated and botanical nomenclature – with some emphasis on adopting names of practical value to gardeners.

Despite the need to ensure precision in naming, cultivated plants are no more amenable to our rules than their wild counterparts. As a cultivar becomes more established in cultivation, and passes through successive generations, variation may creep in. At what point can new names be applied to describe this creeping variation? For the sake of stability, the cultivar-group name has been introduced. Thus *Achillea* 'The Pearl', a seed-raised cultivar known since before 1900, has become so variable that a cultivar-group name (*Achillea* The Pearl Group) has been coined, and selections made therein (e.g. *Achillea* The Pearl Group 'Boule de Neige').

Vernacular or common, descriptive names have a rich tradition in the service of cultivated plants, and are laden with the same difficulty as those applied to wild plants – imprecision. Nevertheless, the cultivar names we employ today are largely founded upon descriptive names, and a careful transition has been made to reduce confusion. Descriptive names relate principally to the place of origin, the raiser's name, or obvious characteristics of the plant (e.g. Fortune's double yellow rose, now treated as *Rosa* × *odorata* 'Pseudindica' or *R.* 'Fortune's Double Yellow'). The Cultivated Code now seeks to prevent misunderstanding through the avoidance of obviously descriptive terms associated with common names for cultivated plants, and the terms 'form' and 'variety', which are reserved with stricter meaning by the Botanical Code.

The Cultivated Code has been introduced to help stabilise names of plants in cultivation. Whilst providing a sound, voluntary code of practice for those giving and using such names, it must give leeway to developments in the industry and the new ways in which plants are produced, marketed and sold. Horticultural taxonomists are needed to unravel this complexity for the sake of gardeners, who show an increasing interest in the wonderful diversity of plants available to them.

REFERENCES AND FURTHER READING

Australian Geranium Society (1978, 1985) *A Check List and Register of Pelargonium Cultivar Names, Part 1 A–B; Part 2 C–F.* Australian Geranium Society, Sydney.

Bean, W.J. (1970–1988) *Trees and Shrubs Hardy in the British Isles*, 8th edn. edited by Sir George Taylor and D.L. Clarke, *& Supp.* edited by D.L. Clarke, 4 volumes and supplement. John Murray, London.

Bricknell, C.D. (ed.) (1989) *The Royal Horticultural Society Encyclopedia of Plants and Flowers.* Dorling Kindersley, London.

Bricknell, C.D. (ed.) (1996) *The Royal Horticultural Society A–Z Encyclopedia of Garden Plants.* Dorling Kindersley, London.

Bridson, D. & Forman, L. (eds) (1992) *The Herbarium Handbook*, revised edn. Royal Botanic Gardens, Kew.

Brummitt, R.K. (1992) *Vascular Plant Families and Genera.* The Royal Botanic Gardens, Kew.

Cullen, J. *et al.* (1986–2000) *The European Garden Flora*, 6 volumes. Cambridge University Press, Cambridge.

Greuter, W., Barrie, F.R., Burdet, H.M., Chaloner, W.G., Demoulin, V., Hawksworth, D.L., Jørgensen, P.M., Nicolson, D.H., Silva, P.C., Trehane, P. & McNeill, J. (eds) (1994) *International Code of Botanical Nomenclature* (*Tokyo Code*). Koeltz Scientific Books, Königstein, [*Regnum Vegetable* **131**].

Griffiths, M. (1994) *Index of Garden Plants.* The MacMillan Press Ltd, London.

Huxley, A. (ed.) (1992) *The New RHS Dictionary of Gardening*, 4 volumes. The MacMillan Press Ltd, London.

Kelly, J. (ed.) (1995) *The Hillier Gardener's Guide to Trees and Shrubs.* David & Charles, Newton Abbot.

Lord, W.A. (Comp.) (1999) *The RHS Plant Finder 1999–2000*, 13th edn, Dorling Kindersley, London.

Mabberley, D.J. (1997) *The Plant-book*, 2nd edn. Cambridge University Press, Cambridge.

Miller, D. (1996) *Pelargoniums: a Gardener's Guide to the Species and their Cultivars and Hybrids.* B.T. Batsford, London.

Phillips, R. & Rix, E.M. (1989) *Shrubs.* Pan, London.

The Royal Horticultural Society (2001) *The RHS Colour Chart*, The Royal Horticultural Society, London.

Stearn, W.T. (1992a) *Botanical Latin*, 4th edn. David & Charles, Newton Abbot.

Stearn, W.T. (1992b) *Stearn's Dictionary of Plant Names for Gardeners*. Cassell Publishers Ltd, London.

Stockdale, J. (Comp.) (1999) *The RHS Plant Finder Reference Library*. CD-ROM, J. Stockdale, Lewes.

Trehane, P., Bricknell, C.D., Baum, B.R., Hetterscheid, W.L.A., Leslie, A.C., McNeill, J., Spongberg, S.A. & Vrugtman, F. (eds) (1995) *International Code of Nomenclature for Cultivated Plants – 1995*. Quarterjack Publishing, Wimborne.

Vaughan, J.G. & Geissler, C.A. (1997) *The New Oxford Book of Food Plants*. Oxford University Press, Oxford.

Walters, S.M. *et al* (eds) (1984–2000) *The European Garden Flora*. Cambridge University Press.

3

Designing Plants

- Structure and functioning of DNA, genes, chromosomes and genomes
- Inheritance of characters
- Variation: recombination, allelic variation, mutations, chimeras
- Hetero- and homozygosity
- Dominant and recessive alleles
- Breeding systems: inbred lines; open pollinated populations; clonal propagation; F1 hybrid breeding
- Wide hybridisation: somatic hybridisation; recombinant DNA technology
- The future

ADAPTATION AND DESIGN

The majority of plants that we cultivate in our gardens are wild plants that have been 'designed' (or 're-designed') to meet human purposes. The re-design process is necessary to convert a plant, adapted to survive and compete in a natural environment, into a plant adapted to meet human requirements for decoration, food, fuel or raw materials (chemicals, medicines, timber, fibre).

Many of the common adaptations that enable plants to compete effectively in the wild render them unsuitable as crops in the managed environments of field, garden or glasshouse. Plant characteristics that are typical of adaptations to a wild environment are:

- sprawling growth habit
- vigorous vegetative growth
- very short or very extended flowering periods
- low abundance of flowers at any one time
- high levels of seed loss from ripe seed heads and pods
- limited colour variation
- erratic germination
- long or erratic dormancy of seeds and
- toxic chemicals in the plant and/or harvested product.

Contrast this list with the characteristics that are of importance in garden crops:

- the size and uniformity of the plants
- hardiness
- growth habit
- evenness of seed germination
- early or late flowering and
- pest and disease resistance.

In ornamentals we look for prolific (or delayed) flowering, extended and/or repeat flowering periods, foliage and flower colour and shape; in vegetable varieties, prolific fruit and seed set, good seed retention at maturity, large root size, bolting resistance, flavour and colour and earliness or lateness of maturity. Fruit varieties may be required to set a lot of fruit but with a low seed content and the size, colour and shape of fruits are important. In crops grown for food or raw materials, the harvested product (as it occurs in the wild plant) may also have qualities that make it unsuitable, or less than ideal, for its intended use.

Until recently it was thought that the designing, or 're-designing', of plants for cultivation had typically taken place within the natural range of the wild species. If the re-designed plant was successful, and generally desirable, then it was exported to new locations, where it might have undergone some further re-design to adapt it to local habitats and climatic conditions. (The extent of this process would be largely determined by the degree of similarity between the plant's original and its new habitat.) However, it is now clear that the source of a crop's wild ancestors, its area of domestication and the area of continuing cultivation and evolution may coincide, but equally may not. Thus, we now understand that designing plants is a dynamic and fluid process. Rather than a 'finished' crop type being developed in a relatively small and well-defined geographic location and then exported, it seems that prototype crops travelled from their centres of origin, with humans, and were domesticated along the way.

About 10 000–12 000 years ago the earliest farmers began to make major changes to the landscape by creating fields through the burning and clearing of forests and heath and breaking up the cleared ground. They also began designing plants through the process of selection and domestication. The earliest records of what is undoubtedly domesticated plant material come from sites in the Near East (wheats, flax and pulses) and the New World (beans, *Phaseolus vulgaris*, and *Curcubita* spp.) dated to 6000–8000 BC. These crops were developed by cultivating wild plants and choosing, year-on-year, seed from those that performed well – according to the criteria of the people who grew and used them. These seeds were then planted in the following season. Unbeknown to the farmers, they were selecting for particular genes, forms of genes and gene combinations that together provided the characteristics they required in their crops.

Plant re-design was probably not restricted to food and raw materials crops. Certainly the Greeks and Romans understood the concept of gardens for pleasure rather than utility. The earlier cultures of Egypt and Babylon also seem to have cultivated gardens, sometimes for vegetables but also for pleasure, in which case a garden was probably a status symbol. Thus we can imagine that early gardeners were also designing plants by selecting for improved flowers, foliage and form.

Initially, the spread of new plant types was relatively slow. But the advent of extensive sea exploration and trade facilitated the discovery and movement of plants that had an economic or social value. Those who probably made the greatest contribution to the distribution of plants around the world are the plant hunters. Beginning in the sixteenth century they have collected enormous numbers of plants from remote regions of the globe to feed the curiosity of professional and amateur plantsmen and women, and the economic interests of plant breeders, nurserymen and seed producers. Thus, the trade in novel plants for food, raw materials, flowers and foliage has distributed exotic species and new plant types all over the world.

Plant design entered a new phase in the early 1900s with the birth of modern genetic science. It was a gardener, Gregor Mendel, studying the inheritance of traits in the common garden pea (*Pisum sativum*), who first established the fundamentals of genetic science in the 1850s. Mendel was Abbot of the monastery at Brunn in Austria (now Brno in the Czech Republic) and an amateur scientist. Through his experiments in the monastery garden he uncovered the principles of heredity, although he knew nothing of genes, DNA, or any of the other key biological components involved in transferring hereditary information from one generation to the next. Precisely what Mendel understood from his discoveries is still debated, but his major contribution was to recognise that there are 'units' that carry inherited information from parents to offspring. These units, and the information they carry, do not blend in the offspring but are passed from generation to generation unchanged (we will see later that some changes to the information do occur, but not due to blending). Mendel realised that it is the interaction between these units of inheritance in the offspring that account for the similarities and differences between members of different generations.

The history of genetics is a fascinating biological detective story. But in order to understand how genes work, how they account for plant characteristics and how genetics has been, and still is, used to design new plant types, it is easier to begin from a practical rather than an historic perspective.

GENES

Plants depend on enzymes (see Chapter 1) to carry out the processes that are essential for their life cycle. Assembling the appropriate enzymes in the correct amounts and locations (within the cell and in relation to one another) enables a cell to manage all the chemical reactions it requires at a particular developmental stage, or under particular internal or external conditions, in order to operate efficiently. The plant requires a system to store and control the use of all the information that is needed to assemble this complex biological system. This is the function of the genes.

DNA

In common with the majority of living organisms, the genes of plants are made from the molecule DNA (*deoxyribonucleic acid*). There are some viruses that use a closely related molecule RNA (*ribonucleic acid*) to encode genetic information. RNA also has a role in the reading of DNA based genes. As DNA and RNA function in the same way, in terms of encoding genetic information, the fact that they are chemically slightly different is unimportant here.

DNA is made up from four chemical building blocks called bases. Each base is made up of two parts, a sugar molecule (this is the same in all four bases) linked to a nucleotide. There are four nucleotides that are used in the DNA molecule, giving four different bases; the initial letter of the nucleotides – Adenine, Cytosine, Guanine and Thymidine are commonly used to identify the bases. Any base can be joined to any other by linking them together through the sugar component of the molecule (Fig. 3.1). In this way bases

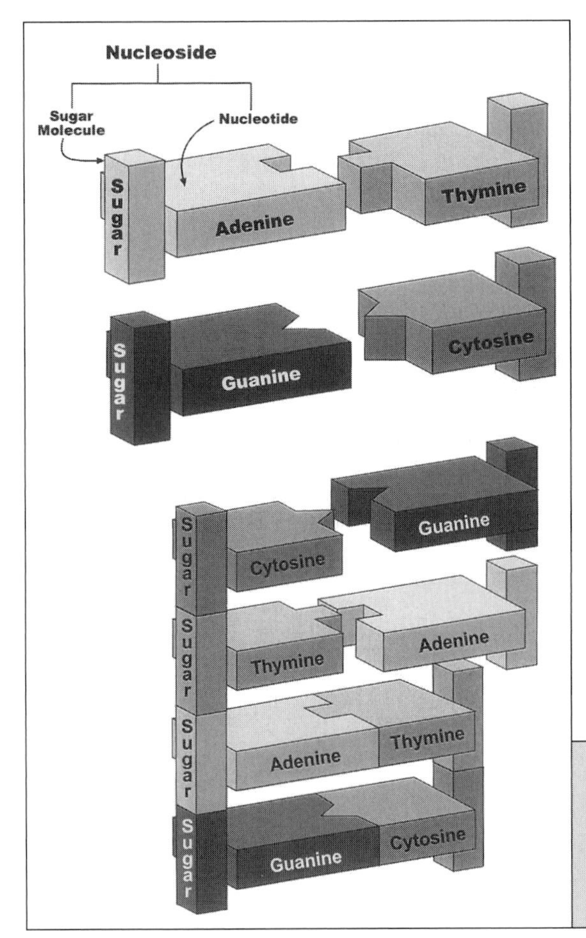

Fig. 3.1 Schematic representation of DNA structure. Nucleotides can be joined in any order through the sugar molecule part of the nucleoside. The physical size and chemical reactivity of the nucleotides ensures that only Adenine can pair with Thymine and Cytosine with Guanine.

can be assembled into strings of tremendous length. For example, if unravelled, the entire DNA content of a wheat cell would stretch to nearly 1.7 metres. The length of the DNA molecule reflects the large quantity of information it has to store, in the case of wheat we estimate this 1.7-metre string carries about 100 000 genes.

The way in which DNA encodes genetic information is both elegant and simple. The four nucleotides function like an alphabet of four letters from which 'words' and 'sentences' can be constructed.

The genetic code encodes two sorts of 'word'. The first are simple three-letter combinations (called *codons*) which represent amino acids. Amino acids are the molecular building blocks for making proteins. A protein consists of a string of amino acids (anything from around one hundred to several thousand). There are twenty naturally occurring amino acids and each is represented by at least one (sometimes more) distinct codon(s) (Fig. 3.2). By reading along a DNA molecule a cell can assemble a sequence of amino acids according to the sequence of bases in the DNA. Different sequences, and amounts, of amino acids produce different proteins with different chemical and biological activities. There are also codons that act as punctuation in the genetic code; one codon (which also encodes the amino acid methionine) identifies the start of a gene and there are several codons that can mark the end of a gene (Fig. 3.2).

The second sort of 'word' in the genetic code is longer and more complex. These sequences of bases convey instructions about how, when and where the coding regions are to be read. All of a

Second letter in codon

First letter in codon	T	C	A	G	Third letter in codon
T	TTT - Phenylalanine	TCT - Serine	TAT - Tyrosine	TGT - Cysteine	T
	TTC - Phenylalanine	TCC - Serine	TAC - Tyrosine	TGC - Cysteine	C
	TTA - Leucine	TCA - Serine	TAA - Stop	TGA - Stop	A
	TTG - Leucine	TCG - Serine	TAG - Stop	TGG - Tryptophan	G
C	CTT - Leucine	CCT - Proline	CAT - Histidine	CGT - Arginine	T
	CTC - Leucine	CCC - Proline	CAC - Histidine	CGC - Arginine	C
	CTA - Leucine	CCA - Proline	CAA - Glutamine	CGA - Arginine	A
	CTG - Leucine	CCG - Proline	CAG - Glutamine	CGG - Arginine	G
A	ATT - Isoleucine	ACT - Threonine	AAT - Asparagine	AGT - Serine	T
	ATC - Isoleucine	ACC - Threonine	AAC - Asparagine	AGC - Serine	C
	ATA - Isoleucine	ACA - Threonine	AAA - Lysine	AGA - Arginine	A
	ATG - Methionine	ACG - Threonine	AAG - Lysine	AGG - Arginine	G
G	GTT - Valine	GCT - Alanine	GAT - Aspartic acid	GGT - Glycine	T
	GTC - Valine	GCC - Alanine	GAC - Aspartic acid	GGC - Glycine	C
	GTA - Valine	GCA - Alanine	GAA - Glutamic acid	GGA - Glycine	A
	GTG - Valine	GCG - Alanine	GAG - Glutamic acid	GGG - Glycine	G

Fig. 3.2 Codon usage in DNA encoding amino acids. Codons, made up of three nucleotides, encode the 20 naturally occurring amino acids. There is some duplication of coding, for example TTT and TTC both encode the amino acid phenylalanine. Specific codons also encode start and stop signals for the enzymes that 'read' the genetic code. In the RNA molecule Uracil replaces Thymine.

plant's cells contain the same genetic information. So for a plant to be able to produce a variety of different cells, tissues and organs and to respond to its internal and external environment it must be able to regulate the activity of individual genes or groups of genes. Regulation is achieved by specific base sequences that control gene activity; they effectively tell plant cells when in development, under what conditions and in which tissues a gene is to be switched on or off. They also regulate the level of gene activity by controlling how frequently a gene is read by the cell. This (along with other factors) determines the amount of gene product in the cell at any time. Because of their function in regulating gene activity these sequences are called 'regulatory' or 'control' sequences, to distinguish them from the protein 'coding' sequences.

The DNA is 'read' by a cluster of several enzymes that physically move along the DNA strand and, under the direction of the regulatory sequences, make RNA copies of the coding regions of genes. This copying process is called *transcription*. As RNA and DNA function in the same way, the cell is in effect making multiple, identical photocopies of the information content of the gene. These RNA copies are then distributed to another type of enzyme cluster, which read along the RNA molecule, recognise each group of three nucleotides and match the appropriate amino acid to each codon. These enzymes are *translating* the genetic code into proteins (Fig. 3.3).

There are a few genes that do not make proteins but instead produce some of the key chemical components in the transcription/translation apparatus.

Not all proteins are enzymes. There are at least three other important classes of proteins: structural proteins, storage proteins and transcription factors. Structural proteins contribute to the physical structure of the cell and include proteins found in the cell membranes and the cell skeleton. Storage proteins are typically found in the seed and other storage organs, where they form a reserve of amino acids, and other constituent elements, for use in the growth of the germinating seedling or sprouting plant.

Transcription factors

Transcription factors have a very important role in gene regulation. As their name suggests they are involved in the process of transcription, namely the copying of DNA to RNA prior to protein synthesis. How transcription factors work and their importance in biology is revealed by two examples from research on the control of flower structure.

The basic structure of any flower is four concentric rings of floral organs: an outer ring of sepals, a ring of petals, a ring of stamens and, in the centre of the flower, the carpels. There are many common flower types where this standard pattern is disrupted. Double flowers, where stamens or sepals are replaced by extra petals, are common. The flowers of single and double roses, or single and double 'Shirley' poppies, clearly

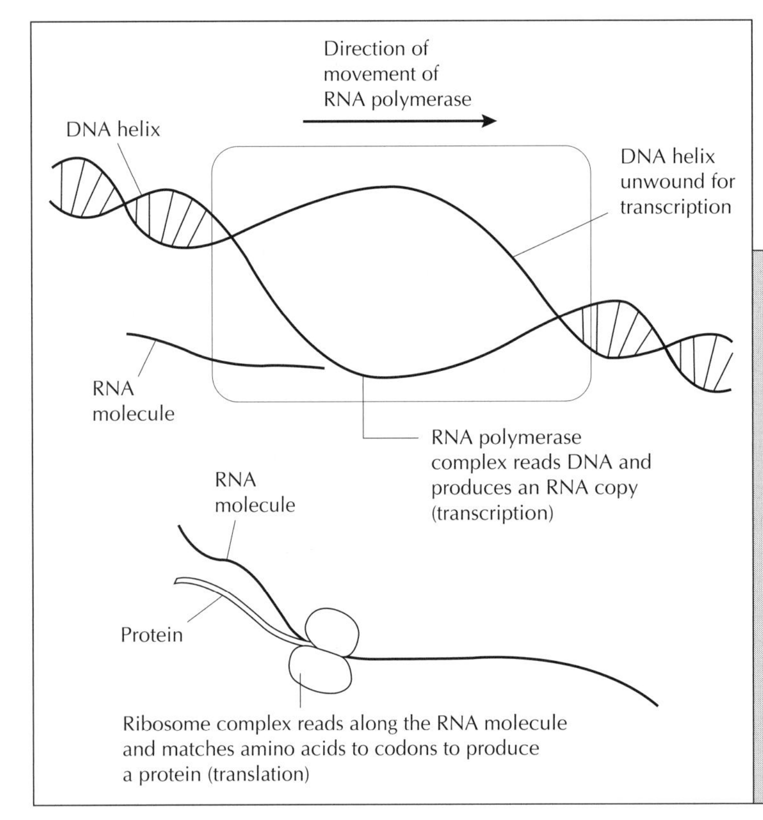

Fig. 3.3 DNA transcription and translation to produce proteins. In order to 'read' a gene the DNA double helix in the region of the gene is unwound and the two strands separated from one another. This allows access to the gene for the RNA polymerase enzyme complex that reads the DNA and makes an RNA copy of it. This process is called transcription. The gene may be read many times to produce multiple RNA copies. In the RNA molecule the nucleotide Uracil replaces Thymine. The RNA is then read by an enzyme/protein complex (a ribosome), which translates the code by matching amino acids to codons to produce a protein.

show reduced numbers of stamens and increased numbers of petals in the double flowers. Other flowers have different floral organ substitutions, for example, in the viridiflora rose (*Rosa* × *odorata* 'Viridiflora') all the floral organs are replaced by sepals. These changes are caused by gene mutations (discussed later) and studying mutations that cause novel flower structures has enabled scientists to understand that three key genes determine the organisation of the organs within the flower. These genes have been identified in several plants by different investigators. The model of how they operate is called the ABC model, so for convenience we will use A, B and C to label the genes.

A, B and C are genes that produce proteins that control the activity of other genes; they are transcription factors. The products of A, B and C interact to produce their effect, which is to establish an invisible map that defines the four different zones in the group of dividing cells at the shoot tip (floral meristem) that will produce sepals, petals, anthers and carpels. The genes are active very early in flower bud development before the primordia, which develop into the flower organs, become visible on the surface of the floral meristem. Gene A is active (at a low level) throughout the meristem, but is most active in the meristem's outer zone. Gene C is active in the middle of the meristem. Let us imagine that the product of gene A colours the zone where it is active red, and that the product of gene C is blue. Gene B is active in a zone that overlaps the inner edge of zone A and the outer edge of zone C. Let us imagine that its product is yellow.

The activity of these genes, and the presence of their products, divides the meristem into four zones of (imaginary) colours. Where only the product of gene A is present (red) the developing primordia will form sepals; where only the product of gene C is present (blue) the developing primordia will become carpels; the products of A and B together (red + yellow = orange) induce petal formation; and the products of C and B together (blue + yellow = green) induce stamen formation (Fig. 3.4a).

A change in the activity of one or more of these genes will result in a change in the pattern of floral organs in the flower. For example, a mutation that renders gene B inactive will result in flowers that have no petals or stamens but extra rings of sepals and carpels (Fig. 3.4b). Loss of B and C function will result in flowers with only sepals because the product of gene A is present in all the cells of the meristem, even though the level is so low as to normally be of no significance in determining flower structure (Fig. 3.4c).

Genes that produce transcription factors also determine flower shape. Flowers exist in two basic shapes. Those like tulips (*Tulipa*) and roses (*Rosa*), have radial symmetry; when looking into the flower an imaginary line can be drawn in any direction across the centre of the flower to obtain two mirror images. The other basic flower type has bilateral symmetry; examples are snapdragon (*Antirrhinum majus*) and pea (*Pisum sativum*). In these flowers there is only one line that will cut the centre of the flower and produce two mirror images. The difference between these two basic flower types is not just aesthetic. Bilateral symmetry is thought to be the more advanced evolutionary form as it provides the opportunity for co-evolution of plants with animal pollinators and the development of sophisticated pollination mechanisms.

In species that normally have flowers with bilateral symmetry such as snapdragon (*Antirrhinum majus*), toadflax (*Linaria vulgaris*), African violet (*Saintpaulia* spp.) mutations are sometimes found that result in plants producing flowers with radial, rather than bilateral, symmetry (Plate 8). Remarkably, these dramatic changes in flower shape can be caused by damage to one gene. Recent studies have revealed that this key gene (called *cycloidea*) is switched on very early in the development of the flower bud, before the development of the primordia that will form the floral organs. The gene is active in only one part of the bud, the region that will become the uppermost (dorsal) part of the flower. This localised activity of the gene establishes a polarity and the region where the gene is active will become the 'top' of the flower (Fig. 3.5). In the cells where the gene is active, the protein it produces (a transcription factor) interacts with the regulatory regions of other genes involved in determining flower structure and organ formation to switch them on and off.

To understand how this system might work let us imagine that, once polarity is established, the

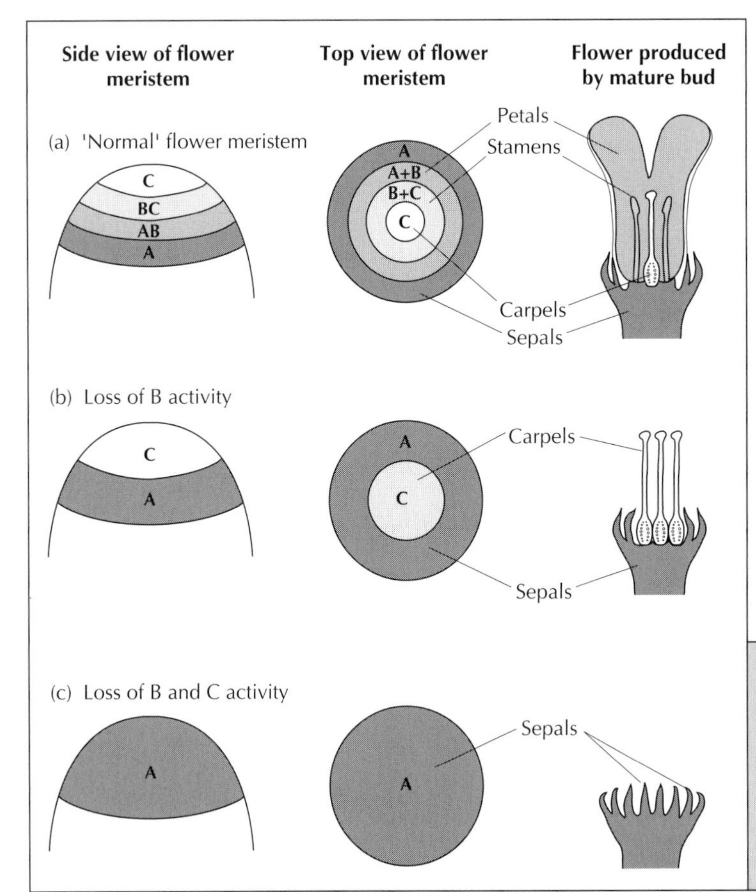

Fig. 3.4 Controlling the position, number and type of floral organs: the ABC model of flower development. Gene activity early in flower development creates a series of invisible zones in the shoot tip that determine which regions will produce sepals, petals, stamens and carpels.

cells in the top region are stimulated to produce a chemical signal that diffuses into the other cells in the bud. Imagine this signal as black. We now add a second signal, with the imaginary colour white; which is produced by all the cells that are *not* in the 'top' region. However, the amount of white signal is reduced if the cell producing it receives the black signal and the stronger the black signal (the nearer the 'top') the more the production of the white signal is reduced. We now have four regions in the bud: the top – where the black signal dominates; the bottom region – where the white signal dominates; and the two sides – where the signal is grey, a combination of black and white signals.

If the signals affect the switching on and off of genes, then we have a mechanism for establishing patterns and for determining where particular organs develop. For example, the genes involved in producing lateral petals are only switched on in the 'grey' regions (Fig. 3.5). A fascinating and readable account of the control of biological development and pattern formation is to be found in *The Art of Genes – How organisms make themselves* (Coen, 1999) from where is borrowed the idea of 'invisible colours'.

Chromosomes

The DNA, and genes, in a plant cell are largely contained in the nucleus, but a small proportion of the cell's total DNA is found in the *chloroplasts*, and the *mitochondria*.

The nuclear DNA of any cell does not consist of one continuous DNA fibre, but of several shorter pieces, the chromosomes. These are not broken fragments but are defined pieces whose ends are marked by specific DNA base sequences.

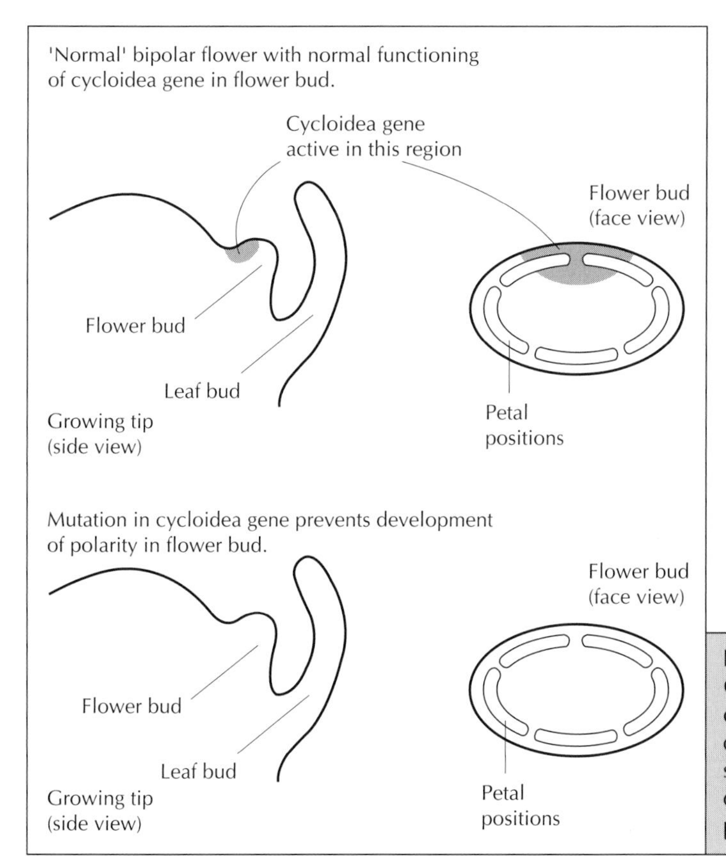

Fig. 3.5 The control of flower symmetry. Gene activity early in flower development creates invisible zones in the shoot tip that determine the polarity of the meristem and so enable the flower organs to develop differently from one another based on their positions.

During most of the cell's life the nuclear DNA exists as fibres, visible only under powerful electron microscopes at magnifications greater than 30 000 times. However, as cells prepare for cell division the DNA fibre undergoes a complex 'packing' procedure. All along its length, loops of the DNA fibre are wound around small particles of protein, two loops per particle. The partly wound molecule is then wound again to create a spiral, which in turn, is wound into another spiral. The result is that, immediately prior to cell division, the DNA becomes condensed into compact packets, the chromosomes, which are usually visible at magnifications of 200–400 × under a standard laboratory microscope. Following division the DNA is unpacked.

DNA replication

We have seen that bases are linked in long strings (through the sugar molecule that is common to every base) and that any base can be joined to any other. Bases can also pair with other bases, but their physical size and chemistry allows pairing only of Adenine with Thymidine and Cytosine with Guanine (Fig. 3.1). The DNA molecule exists as the famous double helix, which is arranged like a spiral ladder, with the paired bases forming the rungs and the long, sugar-linked, strands the styles. It is possible to create two new copies of a double helix by breaking the pairing between the bases, to produce two separate strands. Pairing new bases with the bases exposed on the single strands produces two new double strands. This is called *semi-conservative replication*, because it uses an existing string of bases to create a new double strand. This method of copying information is extremely effective, and is backed-up by sophisticated error-checking and repair systems.

GENOMES

The complete genetic information content of an organism is called its *genome*. This includes all the genes, but may also include DNA that does not encode genetic information or control the reading of the code.

There are some important DNA sequences that define the 'middle' (*centromere*) and 'ends' (*telomeres*) of the chromosomes. These have a very significant role in the biology of the plant because they are important in cell division.

However, there are other regions of the genome where the DNA does not have an obvious function – it does not encode information, and is not associated with control, centromeric or telomeric regions. This DNA is typically an assortment of sequences, some of which can be identified as fragments of sequence from plant viruses, plant genes and transposable elements (discussed later). It has the appearance of a genetic scrap or junk yard and for this reason is often described as 'junk' DNA. The amounts of this type of DNA vary enormously among different species, but it typically occurs as long stretches of 'junk' amongst clusters of active genes.

The scientists' shorthand description of this DNA as 'junk' is potentially misleading as it suggests that it is useless. However, biological systems rarely retain material and structures that are entirely without purpose because natural selection operates against inefficient biological systems. Although its function is unknown, scientists have proposed that junk DNA may have a role as 'packing material' that contributes to chromosome structure and size. Another explanation may be that these large stretches of non-coding, and therefore non-critical, material reduce the proportion of the genome that contains active genes (critical information). In the event of a virus or transposable element (discussed later) inserting itself into the genome (the insertion points for these sorts of events seem to be randomly distributed around the genome) the probability that the insertion will occur in an active gene is reduced. Thus junk DNA may reduce the likelihood of certain types of damage occurring to active genes.

The size of the plant genome, in terms of total DNA content, varies dramatically between plants with no particular relationship between genome size and plant size. For example, the insignificant weed *Arabidopsis thaliana* (thale or mouse ear cress) contains half the DNA of the horse chestnut tree (*Aesculus hippocastanum*) while the broad bean (*Vicia faba*) contains 200 times the amount of DNA found in *Arabidopsis*. Nor does genome size reflect a difference in the number of genes; plants contain between 25 000 and 100 000 genes (approximately), which in itself is only a four-fold difference.

One other factor has a major effect on the total DNA content of the genome; the *ploidy* level of the plant. The cells of most flowering plant species contain two complete copies of the genome, one copy from the male parent and the other from the female parent; these cells are called *diploids*. As we will see later, the gametes (egg and pollen) contain only one copy of the genome and are called *haploids*. During their evolution some plants have undergone a process of *polyploidisation*, so that they contain multiple copies of the genome. The most common forms are *triploids* (tri = three copies of the genome), *tetraploids* (tetra = four copies of the genome) and *hexaploids* (hexa = six copies of the genome). The mechanisms by which polyploids arise are complex but basically are the result of errors in cell division and gamete formation. Treating plants with chemicals that interfere with cell division and so increase the likelihood of polyploid formation can be used to induce polyploidisation deliberately. (Colchicine is a naturally occurring toxic compound, found in the autumn crocus (*Colchicum autumnale*), that is used for this purpose.) Polyploidy creates its own problems but plants that are stable polyploids can be both physically larger and genetically more robust than their diploid relatives and for this reason breeders have used polyploidisation to try to produce new varieties, for example in *Antirrhinum*, lettuce (*Lactuca sativa*) and tomato (*Lycopersicon esculentum*). Polyploidy is often associated with larger flowers, but may be also associated with later flowering. Many cultivated plants are natural polyploids, for example potato (*Solanum tuberosum*), cyclamen (*Cyclamen persicum*), daffodil (*Narcissus* cvs. 'Cheerfulness' and

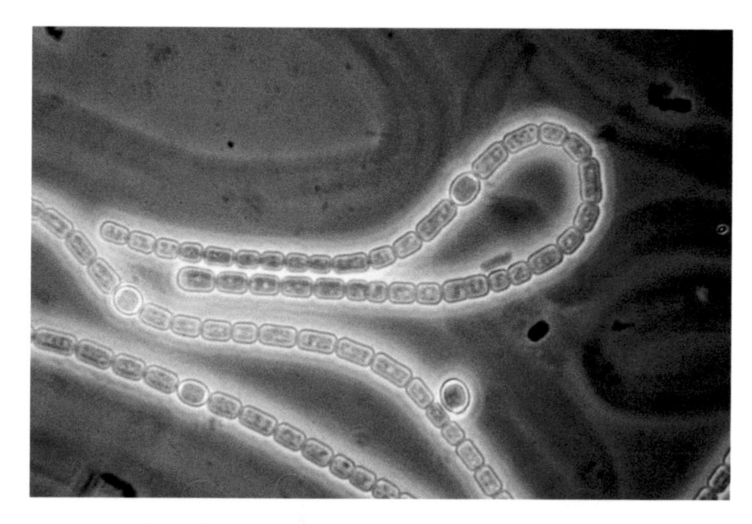

Plate 1 (*left*) A chain of cells of the Cyanobacterium (blue-green) *Anabaena*. Light microscope photograph by Patrick Echlin, Multi-imaging Centre, University of Cambridge.

Plate 2 (*below*) *Chloroplasts* in the cells of a moss leaf. Light microscope photograph by Patrick Echlin, Multi-imaging Centre, University of Cambridge.

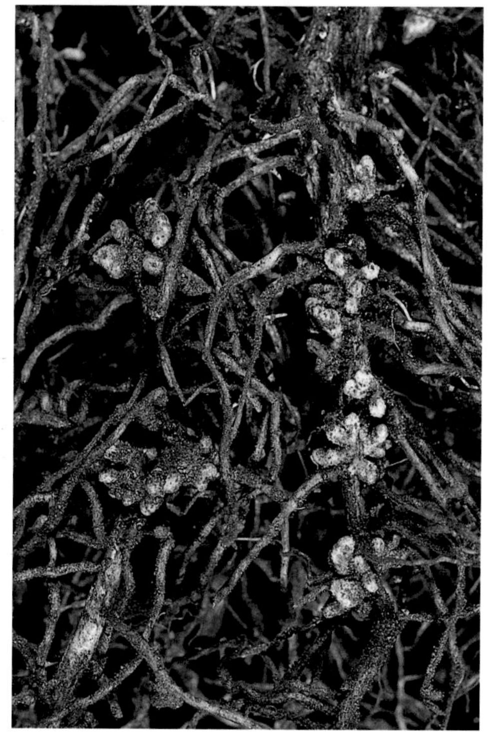

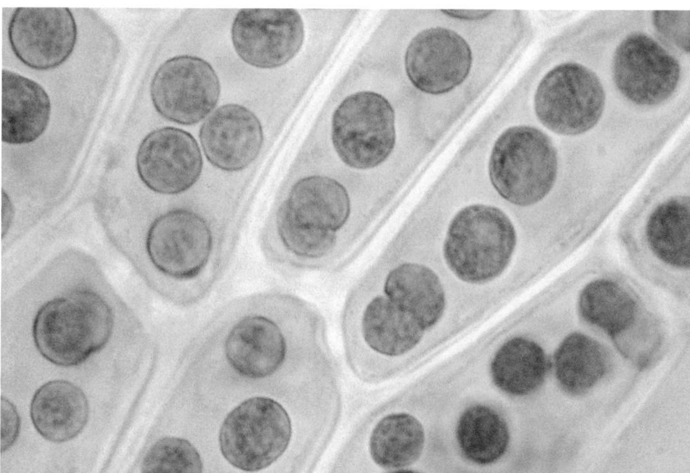

Plate 3 (*above*) Nodules containing *nitrogen-fixing bacteria* on the roots of broad bean (*Vicia faba*). Photograph by Debbie White, Royal Botanic Garden Edinburgh.

Plate 4 (*above*) *Ectomycorrhiza*: swollen and dichotomously branched infected roots of pine (*Pinus* sp.). Photograph supplied by P.A. Mason, Centre for Ecology and Hydrology, Edinburgh.

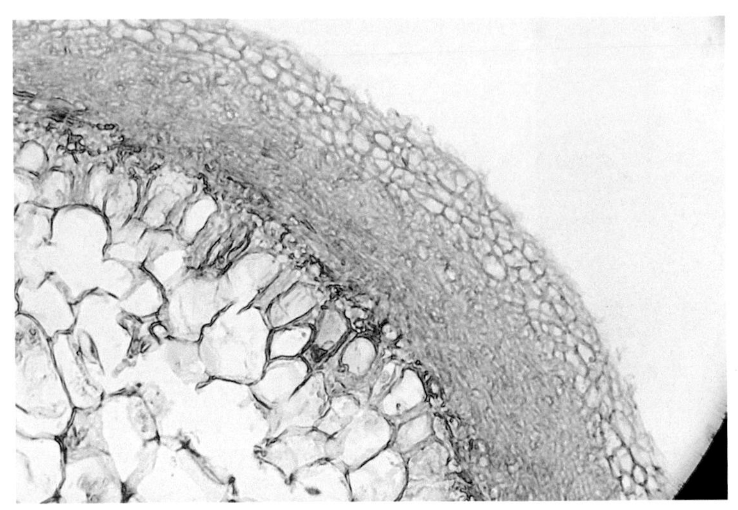

Plate 5 (*left*) *Ectomycorrhiza*: section of an infected root of beech (*Fagus* sp.). A two layered fungal *sheath* can be seen around the outside of the root with the hyphae of the *Hartig net* growing between the cells of the cortex. Specimen prepared and photographed by H. J. Hudson, University of Cambridge.

Plate 6 (*right*) Identifying plants.

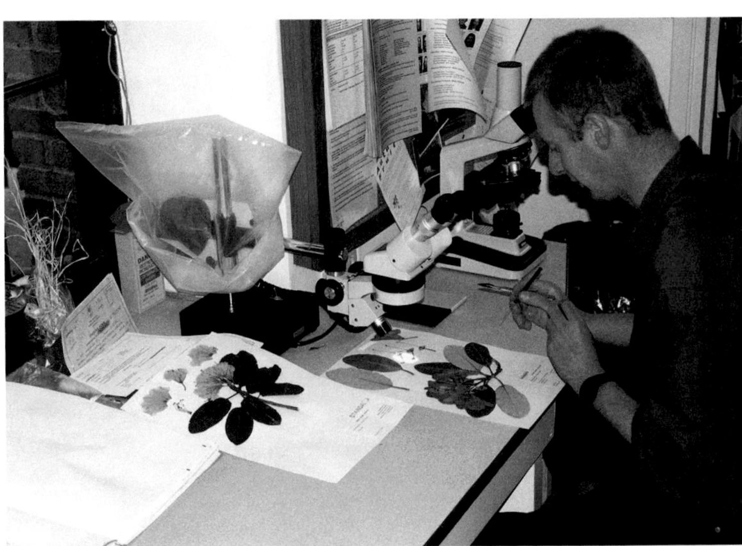

Plate 7 (*left*)
A plant of *Pelargonium crispum* 'Variegatum'.

Plate 8 Common species exhibiting flowers with radial and bilateral symmetry.
(a) *Antirrhinum majus* (wildtype, *left*, with bilateral symmetry and mutant with radial symmetry).

(b) *Linaria vulgaris* (wildtype (*top*) with bilateral symmetry and mutant with radial symmetry).

(c) *Saintpaulia* hybrid (varieties 'Kazuko' (*top*) with bilateral symmetry and 'Nada' with radial symmetry).

Plate 9 Transposon activity demonstrated in the flowers of common plant species.
(a) *Rosa mundi* (*left*) (photograph provided by Peter Beales Roses)
(b) *Pelargonium* spp. (variety Raspberry Ripple) (*right*).

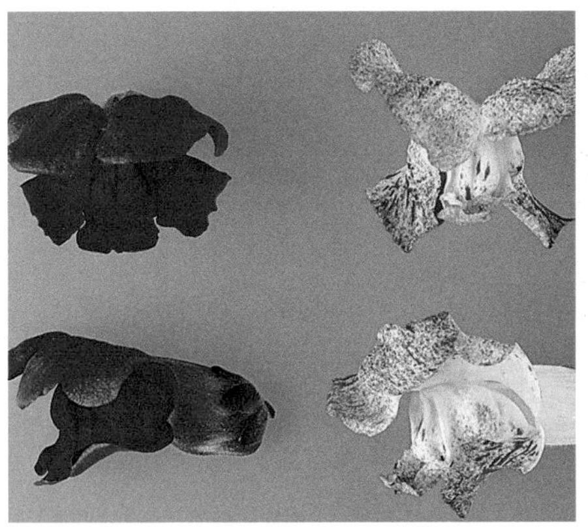

(c) *Antirrhinum majus* (wildtype – solid red and variety with active transposon – red spots on a white background).

Plate 10 (a) (*below left*) *Euonymus japonica* 'Aureus' showing a chimeric structure in which the outer cell layers of the plant are green (due to chlorophyll) but the cells in the inner core of the plant are unable to make chlorophyll and are therefore colourless. This results in a colourless (golden) leaf centre bordered by green.
(b) (*below right*) A variegated variety of *Ficus benjamina* showing a chimeric structure which is the reverse of that described above.
(c) (*bottom left*) A branch tip from the Gymnosperm *Chamaecyparis thyoides* 'Variegata'. The meristem structure and layering of the plant's anatomy is not so regular as in Angiosperm species. Consequently, the green/white patterning, due to the inability of some cells to make chlorophyll, is random.

(d) (*bottom right*) 'Pinwheel' African violet (*Saintpaulia*) showing a chimeric structure in the flower. The cells in the outer layers of the plant are unable to make petal pigments so that the petals have a pigment-less (white) outer margin and pigmented centre.

'Golden Dawn'), *Primula* × *kewensis* and *Freesia* hybrids.

Gamete formation

Most plant cells contain two complete copies of the genome. In the developing flower bud, at the time of egg and pollen formation, the cells that are destined to become the *gametes*, or *germ* cells, undergo a particular type of cell division. This *reduction division*, or *meiosis*, ensures that at cell division only one copy of the genome is passed on to each of the daughter cells. Consequently, these daughter cells have only one, rather than two, copies of the genome. At fertilisation the haploid pollen and egg cell fuse to produce a fertilised egg or *zygote*, which now contains two copies of the genome, one from each parent. The diploid fertilised egg undergoes a form of cell division called *mitosis*. This common form of cell division, by which the plant produces the majority of its cells during its growth, replicates the entire complement of genetic information in the parent cell and passes on identical copies of both copies of the genome to each of the daughter cells at division. Eventually development of the zygote will produce a new plant consisting of diploid cells in which the genomes contributed by both parents are present.

Recombination

Meiosis has a second function. It ensures that the genetic information from the two parental genomes is mixed, before gamete formation, through a process called *recombination*. During the early stages of meiosis equivalent chromosomes from each of the parent genomes pair up, so that throughout their entire length the two genomes lie side by side. The cell then cuts and rejoins the DNA in such a way that equivalent stretches of the genomes are switched. This is recombination and results in two 'new' genomes, that are still complete sets of genetic information, but are now made up from a mixture of the genomes that the plant received from its parents. Immediately following recombination, reduction division occurs so that the new genomes are distributed into separate germ cells. The points in the genome at which the cuts and splices occur, and the number of events, varies from cell to cell, so the germ cells produced by any one plant will contain a wide range of recombined genomes. Recombination ensures that only in very rare cases, those where no recombination occurred, will germ cells contain a genome that is identical to that of one or other of their 'grandparents'.

Allelic variation

Why does recombination occur and why is it important to mix fragments of parental genomes?

We have already seen that genes can exist in different forms, the simplest example being where an active gene is damaged and rendered inactive on one chromosome of a pair, but not the other. The active and inactive forms of the gene are *alleles* and the difference in the trait controlled by the two alleles is called *allelic variation*. However, damage to a gene may not render it completely inactive but increase or decrease its activity or the activity of its product. In a situation where several gene products interact to control a trait, then the products of different alleles may interact in subtly different ways and so produce equally subtle effects on the trait.

Recombination produces new combinations of alleles in each generation and, through these new combinations, potentially more competitive individuals (in a wild environment) or more attractive/productive individuals (in a managed one). Both plant hunters and plant breeders are, in different ways, searching for new sources of allelic variation.

MUTATION

Mutations are familiar to gardeners as 'sports'. Despite the sinister connotations associated with the word, mutation is a natural process and describes any change to the genetic code. There are several types and causes of mutation. Natural and man-made radiation and particular chemicals can cause mutations. Although the mechanism for replicating DNA is extremely accurate there are occasional copying errors and these can generate mutations. Large chromosome fragments, or entire chromosomes, can be lost during cell division if the cellular machinery malfunctions; finally, mutations can be generated by the activity of viruses and transposable elements.

Mutations can occur anywhere in the genome. They range from a single base change in one codon of a gene, the equivalent of adding, deleting or swapping one letter in one word of a sentence, to the loss or duplication of large pieces of genetic information representing a significant percentage of the genome. The larger the change the more likely it is that it will so damage the plant's metabolism that it will be lethal. However, large duplications and deletions of genomes are not uncommon in polyploids where they are reasonably well tolerated, as genes lost by deletion may well be duplicated in other copies of the genome.

Breeders subject seeds to radiation and chemical treatments to increase the frequency of mutations in their breeding material. Increasing mutation rates in the parent plants increases the frequency of new genetic types in the offspring. The process of mutation is random, so the breeder cannot determine which genes will be mutated and hence which characteristics will be changed. However, if conducted on a large enough scale using thousands of plants, it is possible to find interesting changes in plant traits that can then be used in subsequent breeding programmes. This approach has been used with some success, for example in breeding programmes for fruit colour changes in apples, compact growth in apples and pears, and flower colour changes in several ornamentals (*Alstromeria* spp., *Chrysanthemum* spp., *Dahlia* spp. and *Streptocarpus* spp).

Transposons

Transposons are naturally occurring fragments of DNA that are able to move around the genome by cutting themselves out of the DNA and splicing themselves in again elsewhere. They typically consist of a small piece of DNA encoding a few genes with a special sequence of bases at each end. Typically the genes on the transposon encode the enzymes that actually carry out the cutting and splicing. These enzymes recognise the special sequences that mark the ends of the transposon and use them as the site for cutting and splicing. Transposon insertion is random and so a transposon may, by chance, insert itself into, and disrupt the activity of, a gene. If the transposon subsequently moves, it may cut itself out of the gene so precisely that the gene is reassembled intact and will function correctly. However, the process is not always precise, in which case the gene remains permanently damaged. The activity of some types of transposons causes higher than normal frequencies of breaks in chromosomes and thus can lead to losses of large fragments of the genome. Some plant viruses are also able to insert themselves into the genome and in so doing may disrupt gene activity. All of these activities are described as mutation.

The activity of transposons can be seen in several common garden plants. Indeed, they have enjoyed some popularity in recent years with the fashion among breeders to produce flower varieties with randomly striped and spotted petals. The beautiful random red stripes and spots on the petals of one of our oldest rose varieties, *Rosa mundi* (*Rosa gallica* 'Versicolor'), are the result of transposon activity. In fact careful examination of any collection of old rose varieties is likely to reveal some specimens with similar random patterning in the petals. The ancestor of *Rosa mundi* was a red rose that was carrying a transposon. At some point this transposon happened to insert itself into one of the genes involved in red pigment production, damaging the gene and preventing the production of pigment. This event must have happened in a germ line or meristem cell as it has been possible to propagate the resulting 'sport', a 'white' rose without pigment in the petals. However, this transposon has continued to move, and each time it does so it may restore the structure and activity of the pigment gene. As transposons are active within individual cells, only those cells where the transposon moves, and any daughter cells they produce will be capable of producing red pigment. If this happens during flower bud development in one of the cells that is involved in producing a petal, then when the flower opens a red spot will mark the presence of that cell and any daughter cells it produced after the transposon moved. If the transposon moved early in flower bud development then the affected cell will have produced many daughter cells, resulting in a large red stripe or spot on the opened petal. If the transposon was active late in bud development the cell will have divided only a few times before petal formation was complete and consequently only a small red spot will be evident. If the transposon is highly active, moving frequently, then there is a high probability that it will move in several cells in each petal during the period of bud development, and we would expect to see several spots and stripes on each mature petal. A less active transposon will move less frequently, the probability of it moving in more than one or two cells per petal during bud development will be small and so we would expect to see only a few spots and stripes per petal (Plate 9).

The snapdragon (*Antirrhinum majus*) and pelargonium (*Pelargonium* spp.) are common garden flowers in which transposon activity can be seen as striping and spotting of the petals. Although their activity is most obvious and attractive through their effects on flower colour, transposons can affect the activity of any gene (Plate 9).

It was not until the 1960s and 1970s that scientists really began to understand how transposable elements worked, with the groundbreaking work of Barbara MacLintock on the transposable elements found in maize. Her interest was aroused by the presence in maize of unstable genetic traits. Her far-sighted and accurate explanation of the phenomenon was initially treated with some incredulity, but laid the foundation for subsequent studies that have culminated in our now being able to use transposons to search for other genes. This technique is called transposon-tagging and relies on the fact that transposons sometimes move in germ line cells (see below) and the mutations they cause are inherited. The disruption of a gene by a transposon may be seen as a change in a plant trait. A relatively simple genetic test enables scientists to demonstrate whether the mutation is caused by insertion of a transposon. If it is, they can isolate the transposon, since the DNA sequences of the transposons found in cultivated plants are known, and thus use the transposon to obtain the gene that it inserted itself into. Once isolated from the plant the transposon can be cut out of the gene and the original gene reconstructed for use in further studies on the genetic control of the character it affects.

Transposons occur in many of the plants studied so far, although in some cases they seem to be inactive, their presence being detected not by their biological activity but by their characteristic DNA sequences. They are fascinating genetic phenomena that are of interest to gardeners because of their inherent instability, generating an endless supply of unique patterns in flower and foliage colours.

Somatic and germ line mutations

So far we have assumed that genetic changes are inherited. However, only a few cells in the plant will ever have the opportunity to pass on genetic material (and any mutations that have occurred during their lifetime) to the next sexual generation of plants. These are the germ line cells, the relatively few cells in the plant whose route through growth and development will eventually result in their producing eggs or pollen. The majority of cells in the plant are somatic cells (*soma* = body). They will not produce germ cells and their genetic material, and any mutations it carries, will not normally be inherited by the next generation. So while mutations occur in both somatic and germ cells, somatic mutations are not transmitted from one sexual generation to the next.

Chimeras

Somatic mutations will, however, be transmitted to any daughter cells produced by the cell carrying the mutation. Plant chimeras provide some good examples of somatic mutations and provide an interesting insight into how the internal anatomy of plants is organised.

Named after the chimera of Greek mythology, which was part lion, part goat and part snake, plant chimeras are made up of genetically distinct groups or layers of cells. Chimeras can be one of two basic types. 'Graft chimeras' are familiar to many gardeners, perhaps the best-known being +*Laburnocytisus adamii*, the result of grafting purple broom (*Cytisus purpureus*) and laburnum (*Laburnum* spp.). The first recorded chimera was the Bizzaria orange, produced in 1644, when a seedling orange scion was grafted onto a citron stock. The Bizzaria orange tree produced leaves, flowers and fruits of both citron and orange, as well as some fruit that was a blend of citron and orange. Similarly, +*Laburnocytisus* is distinct from either parent but frequently produces branches that have reverted to one or other parental type.

These chimeras can develop and be maintained because flowering plants have an internal anatomy that is layered (see Chapter 1). The cells in the growing tips (meristems) of monocotyledonous plants (grasses, tulips, irises) are organised as an inner core and an outer layer. In dicotyledonous (broad-leaved) plants there is a third layer which is sandwiched between the outer layer and inner core (Fig. 3.6). The small group of actively dividing cells that form the meristem are pushed forward as they produce daughter cells. The layered structure laid down by the meristem is maintained by the daughter cells as they divide, grow and differentiate into the various tissues and organs of the mature plant. The outer layer becomes the epidermis of the plant, the internal core forms all the internal tissues and the intermediate layer, when present, produces a layer of cells immediately below the epidermis (Fig. 3.6). This layering makes it possible for one partner in a graft to 'grow over' the inner layer(s) of the other partner, resulting in a plant made up of cell layers from two different species. For example, +*Laburnocytisus* has outer cell layers of *Cytisus* and an inner core of *Laburnum* cells.

A less extreme type of chimera results when a mutation occurs in a dividing cell in one layer and this mutant cell type is then propagated through that cell layer by cell division. The precise position of the cell in the meristem will determine not only which layer(s), but also how much of the layer(s), is affected. Mutations in meristem cells will be transmitted to many of the cells in that layer, while a mutation that occurs later in development will affect fewer cells.

Chimeras are extremely common, but not often apparent, as the mutation must be in a gene

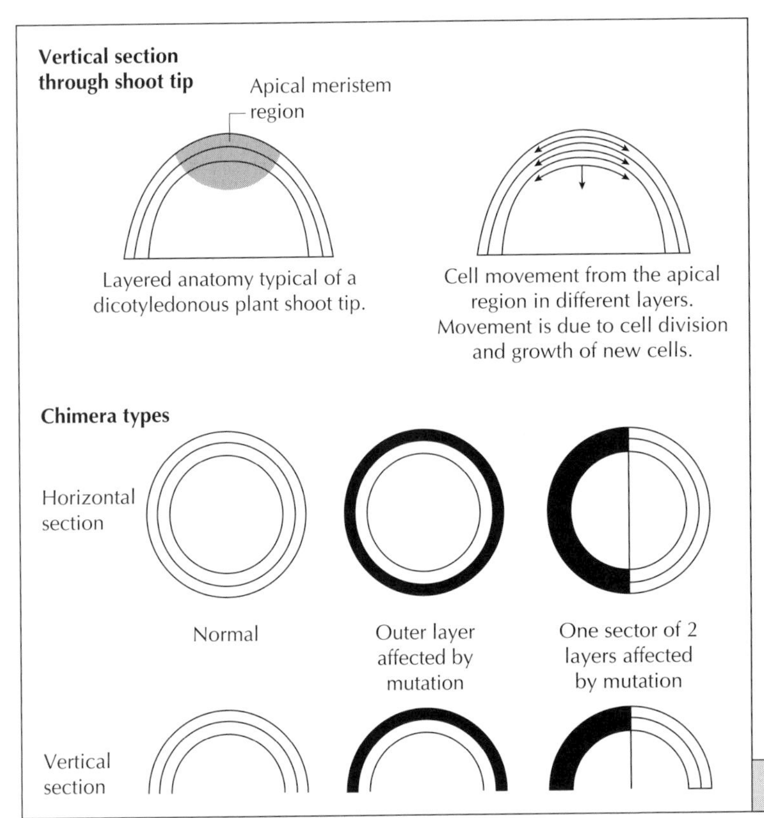

Fig. 3.6 Chimera structure in plants.

affecting an obvious trait such as flower or foliage colour. Several house and garden plants are chimeras of this type. Studying such chimeras has enabled scientists to understand a great deal about plant growth and development. For example, the cells produced by the outer layer of the meristem cover the entire outer surface of a leaf or petal and contribute all of the cells of the leaf and petal edges. The cells produced by the meristem core make up most of the rest of the leaf and petal.

If the cells in the outer layer of the meristem carry a mutation that prevents them from making chlorophyll, then the plant will have leaves with a pale margin and a green centre. The leaf margin is made up only from the chlorophyll-less cells of the outer layer. The leaf centre is also covered with a colourless 'skin' of cells from the outer layer, but the effect of the mutation is masked by the green cells underneath. Variegated varieties of *Hosta undulata*, *Pelargonium* × *hortorum* and *Ficus benjamina* are this type of chimera (Plate 10b). If the situation is reversed, the cells in the meristem core are unable to make chlorophyll but the cells of the outer layers are normal, then the leaves will have a pale centre with a green margin. *Euonymus japonicus* 'Aureus' and *Elaeagnus pungens* 'Maculata' are this type of chimera. Euonymus is particularly interesting as it exists in two common forms that are the chimeric reverse of one another: the leaves of one are white with a green margin, the other green with a white margin (Plate 10a). Typically the chlorophyll-less regions of chimeras are not white, but cream, golden or yellow. This is because leaf and stem cells usually produce other coloured pigments, which are generally masked by the green chloropyhll but are visible in tissues that have no chlorophyll. Although the layered structure of the plant meristem is not so well defined in conifers they also produce chimeras e.g. *Chamaecyparis lawsoniana* 'Albovariegata' and *Chamaecyparis thyoides* 'Variegata' (Plate 10c).

The pinwheel African violet (*Saintpaulia* spp.) is a chimera where the cells of the outer meristem layers are unable to make petal pigment, so these plants produce petals in which the margins are colourless (white) and only the centre is pigmented (Plate 10d). Chimeras also account for the flower colour patterns of the 'Sim' group of glasshouse carnations (*Dianthus caryophyllus*).

The pattern and extent of variegation in leaves and flowers of chimeras is determined by how many of the meristem layers, and how much of each layer, is affected by a mutation. Interesting effects arise if a few cells in one sector of a meristem layer are affected by a mutation. In this case a streak or sector can arise that runs down one side of a plant, including all leaves, the stem and side shoots derived from cells in that sector (Fig. 3.8). The demarcation between layers may not be strictly maintained during cell division, so there may be a limited, and variable, amount of 'migration' of cells from one layer into another which can add further variation to the patterns produced in chimeras.

The mutations which generate the chimera affect only the cells in one or two layers, and so the chimera will not be inherited. This is because the germ cells (egg and pollen) usually originate from only one of the cell layers and, therefore, either carry the mutation or do not. Chimeras demonstrate an important distinction between two types of mutation:

(1) germ line mutation in which the change may be inherited because it has occurred in a cell that will eventually give rise to an egg or pollen cell and,
(2) somatic mutation in which the change occurs in a somatic cell and therefore is very unlikely to be inherited via a seed generation.

However, it may be possible to maintain and propagate somatic mutations if the mutation arises in a piece of plant tissue that can be propagated vegetatively by taking cuttings, or by grafting. However, the tissues used for propagation must include all the cell layers from the donor plant, and the new shoots that develop must also include all the cell layers from the donor, otherwise the original chimeric effect will not be present in the propagated plants. For example, a variegated (yellow leaf edge) mother-in-law's tongue (*Sanseveria trifasciata* 'Laurentii') can be successfully propagated by division. However, if cuttings are taken from leaf pieces, the variegation will not be present in the new shoots that are produced as these originate from the green cells in the core of the leaf and do not include the chlorophyll-less cells from the leaf surface and edges.

New mutations and chimeras can be uncovered in vegetative propagation systems that depend on inducing somatic cells (say in the leaf) to undergo cell division and to form new meristems and hence new plants. If a cell carrying a somatic mutation is the founding cell, or one of the founding cells, of a new meristem, then the mutation will be inherited by all (or a proportion) of the cells in the plant(s) produced from that meristem.

HETERO- AND HOMOZYGOSITY

We have already seen that most plant cells contain two copies of the genome. In most cases (cross species hybrids are one exception) these genomes will be equivalent in having the same genes, controlling the same characteristics, in the same locations along the DNA molecule. However, the alleles present at each gene location (*genetic locus*) may be different in the two genomes. The more closely related two plants are the more likely it is that the majority of the alleles in their two genomes will be the same. The more distant the relationship the greater the number of differences that would be expected.

Plants in which the two genomes carry identical alleles at all gene loci are described as *homozygous* (homo = the same). Genetically homozygous plants will be *true-breeding* as the recombination events that occur between genomes, at gamete formation, will simply shuffle two identical genomes containing identical sets of alleles. All the eggs and pollen will carry identical genomes and self-fertilisation will generate offspring that in every case are genetically identical to one another and their parents. When these plants produce gametes, these too will be identical, and the next generation of plants will be very similar/identical to their parents and their grandparents.

This also holds true at the level of plant populations. By allowing only plants within a particular population to fertilise one another, and then selecting for plants that are of a uniform type at each generation, a plant breeder can establish a population of plants which are homozygous, that is they are genetically nearly identical to one another. Any hybridisation between plants in this population will generate offspring that are very similar to their parents because no new allele combinations are being produced when gametes are produced or when fertilisation takes place.

However, if a plant is *heterozygous*, that is the two parental genomes it contains are very different, then at gamete formation a large number of new allele combinations will be generated by recombination between the genomes. Self-fertilisation of this plant may not result in offspring that are very different from one another or the parents, but in subsequent generations the genetic heterozygosity will become apparent as variation in traits between related plants. Similarly, hybridisation between plants in a heterozygous population will result in many new combinations of alleles, potentially generate new variation for traits, and will maintain the heterozygosity of the population in the next generation.

Dominant and recessive alleles

One of the key observations that Gregor Mendel made in his experiments with peas was that when he hybridised two different inbred (homozygous) varieties with one another, the offspring from the hybridisation (the F1 or first filial generation) were all alike. They had a mixture of traits, some from one parent and some from the other. If he self-pollinated these plants then their offspring (the F2 or second filial generation) did not always breed true to the parent type, but sometimes exhibited characteristics from their 'grandparents' that had not been obvious in their parent.

This observation led Mendel to understand that alleles are not 'equal'. Typically, pair-wise comparison of alleles for a particular gene will reveal that some alleles are *dominant* over other, so-called *recessive*, alleles. There are several physiological explanations for dominance and recessivity, but at the genetic level the effect is relatively straightforward. The presence of a dominant allele will mask the presence of a recessive allele. Hence, only in an individual where, in both genomes, the locus for a particular gene has the same recessive allele will the effect of that allele be seen in the plant. Such a plant is referred to as 'homozygous recessive' for that gene. In a situation where the plant is heterozygous at the gene locus, that is, has a dominant allele in the locus on one genome and a recessive allele in the equivalent locus in the other genome, then the dominant allele will mask the effect of the recessive allele. Heterozygous plants are often identical, in terms of the character expressed by the gene, to the third kind of plant: these are homozygous for the dominant allele, that is have the dominant form of the allele at the gene locus in both genomes.

This illustrates an important concept in genetics. The *genotype*, the genetic make up of the plant, is a major contributor to the *phenotype*, in other words, how the plant appears. But different genotypes can give rise to similar phenotypes. For example, if the allele for red flower colour is dominant over the allele for white flower colour then both homozygous dominant plants (both genomes have the allele for red flower colour at the gene locus for flower colour) and heterozygous plants (one genome has the red flower colour allele and the other the white flower allele) will have red flowers. The effect of the white flower colour allele will only be seen in homozygous recessive plants (both genomes have the allele for white flower colour at the gene locus for flower colour).

However, the dominant/recessive relationship is not always this straightforward and allele relationships can be complex. Alleles can show semi-dominance or co-dominance, meaning that the dominant and recessive forms are not completely dominant or recessive with respect to one another. Consequently, the phenotype of characteristics controlled by alleles with incomplete dominance is not either/or, but a blend of the dominant and recessive character. Just as the dominant/recessive relationship between different allele pairs differs, so semi-dominance can vary and this will cause the phenotype for that character to shift according to the dominance relationship of the different alleles.

So far we have dealt with the relationship between genes and traits as a simple 'one gene controls one trait' relationship. Many characters are of this type. We have seen, for example, that a mutation in a single gene in *Rosa mundi* (caused by transposon inactivation) is sufficient to switch off pigment production in the petals. However, it is possible for a plant trait to be controlled by more than one gene. For example, where the colour of a flower's petals results from a mixture of pigments then a change in the activity of any of the genes involved in producing the pigments could result in a change in flower colour.

Other genes interact in more subtle ways. For example, the protein products from several genes may be required to assemble particular enzymes, enzyme complexes or structural elements of cells, which themselves account for a particular plant characteristic. Many different aspects of cell and plant development can affect some traits, such as yield, and consequently a large number of different genes contribute to this characteristic of the plant. Intuitively we would expect that the more genes that are involved in determining a particular trait the less likely it is that a change in any one gene will have a major effect on that character. However, when there are many genes controlling a character, it is usual to find that a few genes (so-called major genes) have relatively major effects, while the remainder have only minor effects on the trait.

Genes that have major effects are typically little affected by the environment, so the phenotype is closely related to the genotype. However, traits controlled by genes with minor effects, or by several genes, may well be more susceptible to the effects of the environment. Thus, variation in plant phenotype can have both a genetic and an environmental component. The activity of some genes is affected by specific environmental conditions such as day-length (See Chapter 9). Environment also has more general effects. For example, a plant with the appropriate genes to produce a tall and vigorous individual, under good growing conditions, might be short and weedy under drought or poor soil conditions.

BREEDING SYSTEMS

Most plant species are naturally heterozygous. However, homozygous species do occur naturally indicating that both of these genetic/reproductive strategies are competitive in the wild. Whether a plant is naturally hetero- or homozygous is largely determined by the mating system of the species, as it is this that largely controls the movement of new alleles in to (and out of) the breeding population.

Heterozygous species are typically outbreeders. They also tend to be perennials. They have mating systems that promote outcrossing and inhibit self-fertilisation and inbreeding. Enforced inbreeding in these species often results in a dramatic loss of vigour, size and fertility over one or a few generations, called *inbreeding depression*. Inbreeding depression results from the accumulation, in inbred plants, of recessive deleterious genes whose presence is masked in heterozygous plants. A plant population that contains (distributed among individual plants in the population) several alleles for any particular gene locus may have an adaptive advantage in a wild environment. For example, different alleles for seed dormancy could ensure that seeds from that plant population germinate over an extended period and this may increase the probability of some germination occurring during a period that is followed by favourable growth conditions. Heterozygosity within individual plants ensures that their progeny represent a variety of new allele combinations, resulting in a wide range of genetic and phenotypic variation for plant characters. This gives the species and individuals an opportunity to respond to changing conditions. The downside of this strategy is that the increased frequency of new gene combinations increases the probability of deleterious combinations occurring and thus the frequency of less 'fit' individuals in the offspring of parents. To some extent this is offset by perennial habit; the parent plants produce seed over several years so compensating for the higher proportion of unsuccessful progeny produced each year.

Homozygous species are inbreeders. They are adapted largely to self-pollinate and do not suffer from inbreeding depression even when inbred over many generations. These species contain a pool of alleles that are highly adapted to local conditions, making them very competitive. Mechanisms to ensure self-pollination restrict the flow of new alleles into the population from outside and reduce the frequency of new allele combinations. This reduces genetic variation in the offspring making the population more vulnerable to changing conditions. However, for as long as conditions do not change, the low level of variation ensures that the population produces a very high proportion of offspring that are adapted to the local environment.

In the wild, outbreeding is the basic state and inbreeding has evolved from it, probably on several separate occasions. The breeding of crops and ornamentals from wild plants has led to a general reduction in heterozygosity and increased homozygosity in plants and populations of cultivated varieties. This is because breeders, and before them farmers and plantsmen, have selected for uniformity in their plants by rejecting 'off types'.

Homozygosity has the advantage that characteristics are 'fixed'. The plants thus breed true generation after generation. This greatly increases the ease of selecting for improved plant types and of producing seed that will breed true in the next generation. However, this is not always possible (because of inbreeding depression), desirable or indeed necessary. There are four basic plant-breeding schemes and these largely reflect the constraints imposed on breeding by the mating systems of the plants being bred.

Inbreeding annual plants (which do not suffer from inbreeding depression) are bred as inbred, pure (homozygous) lines. Outbreeding annuals (which suffer inbreeding depression if forcibly inbred) are bred as open-pollinated populations or as F1 hybrids. Perennials (usually outbreeders) are bred as open-pollinated populations or as clones (if they can be readily propagated vegetatively).

Breeding inbred lines

Hybridisation of two different homozygous lines will result in a uniform *F1* population (first filial generation) but self-pollination of these plants will give a very variable population of plants in the *F2* (second filial) generation. Subsequent rounds of self-pollination of plants will reduce the variation seen between the offspring of any one parent and by the fifth or sixth generation each selfed line will effectively be homozygous. During the early stages of the breeding cycle the breeder normally selects only for characters which are highly heritable and/or easy to assess. Characters with low heritability or that are difficult to assess are generally not selected until the plants are more-or-less homozygous, that is, when these characters have been 'fixed'.

The 'pedigree' system is the classical method of conducting an inbred line breeding programme. Suitable parents are hybridised and the seed produced is collected and sown to produce an F1 population of plants. These are allowed to self-pollinate and the resulting seed sown to give the F2 generation of plants. Each of these plants will produce enough seed to sow and produce a family of closely related plants. Breeders usually begin to select for best types at this stage and seed from selected plants (in different lines) is used to produce the next generation of families. Growth and selection of families continues for several more generations, until homozygosity is reached in the F5–F8 generation. At this point traits will be 'fixed' and a few of the best plants will then be selected and retained for larger scale trials and evaluation.

Breeding open-pollinated populations

Breeding by this method is typically used with plants that have mating systems that encourage out-crossing and inhibit self-pollination. Such plants usually suffer from inbreeding depression if forced to inbreed, making it is impossible to establish homozygosity and so 'fix' the required alleles and the characteristics they determine. In open-pollinated plants, breeders must select at the population level for the alleles that they require and against those they do not, while maintaining a relatively high level of heterozygosity. In this way they can enrich a closed population with desired alleles and traits. Individual plants from this closed population will not produce seeds that breed true-to-type but, at the population level, the variety will have the improved characters selected by the breeder and can be maintained indefinitely by random-mating within the population.

This approach establishes a plant variety by shifting allele frequencies in a population through selection. Alternatively, the variety can be produced as a hybrid between two parents, or as a mixture of hybrids between several different parents. By careful selection of the parents it is possible to ensure that the population of hybrid plants contains the mix of alleles and the qualities required by the breeder.

Clonal propagation

It is not necessary for plants that are vegetatively propagated (see Chapter 7) to be homozygous since there is no gamete formation and hence no re-assortment of alleles that might generate new genetic variation within the variety. Thus plants selected for particular characteristics, and propagated vegetatively, will produce progeny plants that are true-to-type, even if they are highly heterozygous.

Breeding of clonally propagated plants largely consists of making new hybrids, selecting for improved types in subsequent generations and then 'fixing' the best performing types by propagating them vegetatively.

F1 hybrid breeding

It is possible to produce a uniform population of plants from seed even though the plants themselves are heterozygous. This is so-called F1 hybrid breeding and is based on Mendel's original observation that hybridisation of two inbred lines results in a uniform F1 generation. The disadvantage of F1 breeding is that the F1 plants do not breed true, so that each F1 seed generation has to be recreated by hybridising the two inbred parent lines. This means that the grower cannot retain seed year-to-year and that the seed is more expensive to produce than seed from inbred or open-pollinated breeding systems. Despite this double expense to the grower, F1 breeding is employed because, in many species, heterozygous plants are more vigorous and robust than homozygous types. (F1 hybrid ornamentals may produce significantly larger flowers while F1 hybrid vegetables may be heavier cropping.) This phenomenon is called *hybrid vigour* or *heterosis*. Its genetic basis is complex and still debated. In practice, heterosis means that if a comparison is made of the performance of the offspring from pair-wise hybridisations among a group of parental (typically, but not always, inbred) lines, then certain parent combinations will be found to result in offspring with higher levels of vigour (heterosis) than others. The specific level of heterosis between particular pairs of parents is called 'specific combining ability' (SCA). By creating hybrids between parents with high SCA, breeders are able to produce F1 seed that results in plants with better performance than either of the parent lines and the best inbred lines of the crop.

A major problem in F1 hybrid breeding is the need to ensure that the seed harvested from the parent plants is indeed hybrid seed and not the product of self-pollinations. Several techniques are employed to prevent self-pollination. In tomatoes the flowers of the maternal parent are hand-emasculated to prevent selfing. In onions a *cytoplasmic male sterility* (CMS) system (discussed below) is used to ensure that the female parent does not produce pollen, while a self-incompatibility (SI) system is used in some brassica vegetable breeding. At various times 'chemical emasculation' using selective gametocides has been tested, but the effective use of a chemical spray to disrupt pollen development in maternal parents has proved difficult in practice. Spraying squashes (*Cucurbita pepo*) and cucumber (*Cucumis sativus*) with the ethylene-producing chemical, 2-chloroethylphosphoric acid, has been reported to induce sprayed plants to produce only female flowers. Whatever the technology used, the intention is to prevent seed set as a result of selfing and to force hybrid seed production.

As its name suggests, cytoplasmic male sterility (CMS) prevents plants from making viable pollen. Consequently, CMS plants will only set seed if the stigma receives pollen from another (male fertile) plant. The CMS character is genetically determined by the DNA of the mitochondria, which are present only in the cytoplasm, hence the name CMS. CMS is relatively common and has been identified in sweet pepper (*Capsicum* spp.), sunflower (*Helianthus annuus*), carrot (*Daucus carota*), petunia and tomato. Genetic male sterility may be determined by nuclear genes, and this form of male sterility is found in fennel (*Foeniculum vulgare*), lettuce (*Lactuca sativa*) and broad beans (*Vicia faba*).

Self-incompatibility (SI) is a genetically determined system to prevent or deter self-fertilisation. There are some subtle differences in how different SI systems work but essential to all of them is the principle that only certain combinations of SI alleles in the male and female gametes (or parent plants) permit successful fertilisation. A crude analogy is of a lock and key system, where only a particular combination of lock and key permits fertilisation. Pollen is unable to develop properly on an 'incompatible' stigma and so will fail to fertilise an egg cell. The genetics of the system usually means that 'self' pollen and stigmas are incompatible, precluding self-pollination.

In either case, the cultivation of a mixture (usually separate rows or blocks) of the two parents will ensure that seed set on the female (CMS) parent, or in the case of an SI system both parents, will be hybrid and not 'selfed'.

Wide hybridisation

Hybridisation is central to designing plants as it is the means both to generate new allele combinations and to introduce new alleles into the existing pool represented in the breeding material. Among plants there are several natural barriers to hybridisation. Pollen exchange between otherwise compatible plants may be prevented by physical barriers, such as geographic separation, non-coincidence of flowering times and adaption to different pollinating species, where size and shape of flowers may preclude natural cross-pollination. Plants also have specific biological mechanisms for recognising the compatibility/incompatibility of gametes and thus regulating cross-fertilisation. These incompatibility reactions can occur on the stigma at pollination, in the style immediately prior to fertilisation (as described above) and in the developing seed after fertilisation.

In seeking to produce new allele and gene combinations that might lead to improvements in plants, scientists have developed several methods to overcome these barriers and so generate novel hybrids across incompatibility barriers. For example, treating the stigma with dilute salt solution and then carrying out a controlled pollination with genetically incompatible pollen can defeat the brassica SI system. This system can also be overcome by treating flowers with elevated levels of carbon dioxide after pollination, or by pollinating the stigmas of unopened buds, resulting in hybrids that are prevented in the wild by natural barriers to hybridisation. In some plants, e.g. *Alstroemeria* and *Lilium* spp., it is possible to graft a 'compatible' style onto an 'incompatible' ovary and thus overcome incompatability barriers that operate at the pollen-style interface.

Although pollination and fertilisation of distantly related plants may be successful, the embryo or the endosperm may abort during seed development, resulting in a non-viable seed. Excising the immature hybrid embryos and growing them on nutrient agar (see Chapter 7 for techniques) can enable continued embryo development and permit the recovery of a novel hybrid plant. Numerous hybrids have been produced by embryo culture for potential use in breeding programmes. For example, the lily hybrids *Lilium speciosum* × *L. auratum* and *L. lankongense* × *L. davidii* do not produce viable seed but can be recovered by culturing immature embryos. Hybridisation of *Brassica campestris* × *B. oleracea*, to produce synthetic versions of the natural hybrid *B. napus* (oilseed rape and swedes), is only successful if the embryos are rescued in culture. *Lycopersicon peruvianum* can be hybridised with *L. esculentum* (tomato) but the hybrid embryos must be rescued to obtain plants.

Somaclonal variation

In vitro, or tissue, culture is not restricted to embryo rescue techniques and new plants can be generated from fragments of many different plant tissues (see Chapter 7). In some instances it is possible to stimulate the somatic cells of the explant to produce embryos, rather than meristems.

Populations of plants produced from tissue culture often show a range of new genetic variation. This 'somatic variation' occurs even when the *explants* (small pieces of tissue taken for propagation) are from genetically homozygous plants. The causes of somatic variation are still debated. It is generally agreed that mutations in the cells of the explant play a role in producing the new variation. It has also been suggested that the tissue culture process itself in some way induces mutations (see Chapter 7 for details of the effects of repeated sub-cultures). Some of the somatic variation that has been observed is unstable and with time the variant plants revert to parental type, or to another variant type. Stable somatic variation has been explored as a route for designing plants. The technique of meristem culture results in many fewer mutations and 'sports' than when other tissue fragments are the starting point, as it relies on the proliferation of existing organised meristems rather than on the formation of new meristems within an undifferentiated mass of callus cells (see Chapter 7).

Somatic hybridisation

An extreme means of overcoming barriers to sexual hybridisation is to use *protoplast fusion*, also called *somatic hybridisation*, techniques. The walls of plant cells can be removed by treating plant tissue with a mixture of enzymes that specifically break down the cellulose and other major components of the cell wall. The naked cells that are released, called protoplasts, can be induced to re-grow a cell wall if they are cultivated in an appropriate mixture of nutrients and plant hormones. These cells, again with the appropriate culture technique, can be induced to grow and divide, eventually producing new plantlets.

If protoplasts isolated from two different plant species or genera are mixed, it is possible, by applying a chemical or electrical shock, to cause the protoplasts first to stick together, and then to fuse and their contents to merge. The fusion process is typically conducted on millions of protoplasts at one time and is largely random. Consequently, of the fusions that take place, a proportion will be between protoplasts of the same species, others will involve large numbers of protoplasts, but in some instances the fusion events will be between single protoplasts from different parents. In this case the hybrid protoplast will contain two copies of the genome from each of the parent plants.

It is, in theory, possible to induce these hybrid protoplasts to regenerate a cell wall, divide and grow to produce plants that are hybrids between the parental species. In practice there are a number of obstacles to be overcome and some hybrid combinations are not viable.

Somatic hybridisation has been used to generate new hybrids for use as potential sources of novel genetic variability in plant breeding programmes. For example, hybrids have been produced between *Dianthus chinensis* × *D. barbatus* and sexually incompatible tomato, potato and tobacco species.

RECOMBINANT DNA TECHNOLOGY

Finally, we come to what some regard as the ultimate in plant design – recombinant DNA (rDNA) technology. The development of these techniques over the last 20 years has provided scientists with the tools to identify, isolate and characterise single genes from plants, animals and microbes. By combining classical genetics and rDNA techniques scientists are now able to explore how the genetic code determines allelic variation for particular biological traits. Three new opportunities of potential interest to plant breeders have been made available. Firstly, the modification and movement of single genes, so called *genetic modification* or *genetic engineering*. Secondly, the identification of the entire genetic code of an organism, called *genome sequencing*. Thirdly, the opportunity to 'link' a particular phenotype with a specific DNA sequence, called *marker-assisted breeding*.

Scientists are, with increasing ease, able to detect the presence (or absence) of particular DNA sequences in the genomes of plants. This has enabled the development of marker-assisted breeding techniques. If a plant breeder can identify a particular DNA sequence that is always present in the plant when a particular phenotype is present then the breeder does not have to look for the phenotype but only for the presence/absence of the 'marker' DNA sequence. This may initially seem counter-intuitive, as it ought to be easier to look for a trait than to isolate DNA and test for the presence of a particular DNA sequence. Generally, this is true. However, tests for some traits are difficult to perform reproducibly. Disease-resistance tests may be difficult because the incidence of disease depends on weather conditions and the level of disease inoculum carried over from the previous season. Thus the challenge to the plant's disease resistance may vary from year-to-year making it difficult to interpret the actual level of genetic resistance. We have also seen that some traits may be determined by several genes/alleles that individually have relatively small effects. Recognising which alleles have significant, beneficial effects on a particular trait and then identifying markers for each allele means that a breeder need only select for plants that include all the markers to be able to produce a plant that performs well with regard to that particular characteristic. The marker sequence may be, but does not have to be, in the gene itself, but it must be physically close to the gene on the DNA. Otherwise the process of recombination, at gamete formation, will separate the gene and the marker at a frequency that will make the marker an unreliable indicator for the presence of the gene.

The production of DNA markers for particular genes and alleles is an expensive process and the markets occupied by many horticultural plants are of relatively low value. It is therefore unlikely that this technology will receive a great deal of use for garden plants unless its cost can be dramatically reduced.

In December 2000 an international scientific team reported the completion of the first complete genome sequence for a plant, thale cress (*Arabidopsis thaliana*). The plant is familiar to, but overlooked by, many gardeners. It is a common, insignificant and ephemeral garden weed, found around the world in many different habitats. As far as research scientists are concerned, *Arabidopsis* is a wonderful weed as it combines several important characteristics:

- it is small, so very large numbers can be grown in a relatively small space,
- it has a rapid life cycle and prolific seed set, so several generations can be produced in a year, and most importantly,
- it has one of the smallest plant genomes known.

Nonetheless, the *Arabidopsis* genome proved to contain 114 million letters of DNA code and 26 000 genes.

Arabidopsis is a crucifer, a member of the Brassicaceae family and so a very distant relative of the brassica crops, but its potential usefulness extends far beyond this single group of plants. Even before its genome sequence was complete, scientists were beginning to extract information from *Arabidopsis* that shed new light on plant genetics and biology. For example, about 50 years ago a genetic mutation was discovered in wheat. The mutation dramatically affected plant height by preventing the normal elongation of the cells in the stem, with the result that plants with the mutation are dwarf. The mutation causes insensitivity to gibberellic acid, a plant hormone that promotes cell elongation. The genetic complexity of wheat has prevented scientists from actually isolating the gene and studying its action more closely. However, a mutation with the same phenotypic effect and biological activity was found in *Arabidopsis*, where the relative simplicity of the genome made it easier for scientists to isolate the gene. The dwarfing gene isolated from *Arabidopsis* was then used to find the equivalent gene in wheat. When the mutant gene was introduced, by genetic modification, into tall varieties of rice and chrysanthemum, it caused dwarfing, demonstrating that the gene is controlling a process fundamental to plant development. Similarly, genes that control the time taken for plants to flower or a plant's response to cold treatments (Chapter 9) have been identified in, and isolated from, *Arabidopsis* and then used to explore both the basic biology of the genes and their role in important cultivated crops.

Genetic modification is an important tool that allows scientists to introduce modified genes back into plants to study the effects of modifying the gene structure on the gene's function and the plant's biology. It is also a route to producing new genetic variation in cultivated plants.

There are several methods for introducing genes into plants. Protoplasts can be mixed with DNA and then subjected to an electrical or chemical shock that causes them to absorb fragments of DNA. Some of this DNA will become integrated into the DNA of the protoplast and regeneration of the protoplast to a plant will result in a genetically modified plant that carries one or more copies of the gene(s) that was on the DNA that the protoplast absorbed.

Tiny beads of gold or tungsten can be coated with the gene of interest and then shot into pieces of plant tissue. Some of the cells in the tissue will not be hit by particles, others will be so damaged that they die, but others will survive being shot and in a few of these some of the DNA from the beads will be integrated into the DNA of the plant cell.

The most efficient, and popular, method for introducing DNA into plants is to use a natural plant pathogen *Agrobacterium tumefaciens*, the causal agent of crown galls on fruit and other trees. *Agrobacterium* is a soil-living bacterium that is able to infect many plant species by entering wounds, which may be caused by pruning or naturally by animal grazing, wind or insect damage. The bacterium is motile and is attracted to wounds by some of the chemicals that the plant releases when wounded. On entering the wound the bacteria are stimulated to transfer a specific piece of their DNA into the plant cells at the wound surface. This fragment can be integrated into the DNA of the plant cell.

Bacterial cells typically contain a single small chromosome and multiple copies of 'plasmids', which are small circles of DNA carrying a few genes. *Agrobacterium* contains a particular type of plasmid, the Ti (tumour inducing) plasmid; it is a small piece of DNA, the t (transfer) DNA, from the Ti-plasmid that is transferred into the plant cells exposed in the wound. The t-DNA carries a few genes that produce plant hormones, such as cytokinins; these induce the cells in the wound to divide and grow to produce a callus, the 'crown gall'. The t-DNA also includes genes for the production by the plant of specific molecules, called opines, that the bacteria can use as a source of carbohydrates and nitrogen as they live and grow in the developing gall.

Scientists have replaced the bacterial genes in the t-DNA with genes that they wish to transfer into plants. They put the Ti-plasmid, with its modified t-DNA, back into *Agrobacterium* where the plasmid will replicate itself, resulting in multiple copies per bacterial cell. The modified *Agrobacterium* can be grown in large numbers and then used to infect plant cells by dipping pieces of plant (usually seedling leaves, hypocotyls or segments of mature stem or leaf – depending on the plant species) into the bacteria culture. The *Agrobacterium* respond to the exposed cells on the cut surfaces of the plant tissue as if they are a 'natural' wound and transfer t-DNA into the plant cells, where it becomes integrated into the plant's DNA.

Plant tissues in culture can be stimulated to regenerate new shoots and roots, if given the appropriate hormonal signals (see Chapter 7). The cells in the *Agrobacterium*-treated explants, or in tissues that have had genes shot into them, can also be induced to divide and eventually produce new shoots. However, both means of gene introduction are random processes and not all the cells exposed to the treatment will take up, or integrate, the introduced DNA. Simply stimulating shoot regeneration from treated tissues will result in the production of a very large numbers of plants, few of which contain the gene of interest. Scientists use selectable 'marker' genes to ensure that only cells that contain the introduced gene will give rise to new plants. These markers are typically on the same small fragment of DNA as the gene of interest, so they are physically linked. The most common kinds of marker are genes for resistance to an antibiotic or tolerance to a herbicide that would normally kill plant cells. The process of inducing shoots from treated explants is carried out in the presence of antibiotic/herbicide; this prevents the growth of any cells that do not contain the marker gene for resistance. Because of the physical link between the marker and the gene of interest the majority of cells containing the marker will also contain the gene of interest.

Genetic modification has been explored as a means to change specific characteristics in a range of ornamental and vegetable species. For example, studies are in progress on improving the storage characteristics of pea, insect resistance in chrysanthemum, virus resistance in *Osteospermum*, fungal disease resistance in petunia and increased flower production and axillary bud breakage in chrysanthemum.

As our understanding of plant genetics and biology has increased, so we understand how human intervention has adapted and exploited natural phenomena to design and re-design plants for decorative, food, medicinal and other uses. Over the years, with increased understanding of genetics and biology, our interventions have become more focused. Yet, however sophisticated the technology, the challenges for breeders and plantsmen and women remain the same: to generate the allelic combinations that will result in plants which are better adapted to their intended end use in the managed environments of fields, glasshouses or gardens.

REFERENCES AND FURTHER READING

Coen, E. (1999) *The Art of Genes*, Oxford University Press, New York.

Gonick, L. & Wheelis, M. (1991) *The Cartoon Guide to Genetics*, HarperCollins, New York.

Griffiths, A.J.F., Miller, J.H., Suzuki, D.T., Lewontin, R.C. & Gelbart, W.M. (1976) *An Introduction to Genetic Analysis*, W.H. Freeman and Company, New York.

Simmonds, N.W. (2000) *Principles of Crop Improvement*, 2nd edn, Blackwell Science, Oxford.

Tilney-Bassett, R.A.E. (1986) *Plant Chimeras*, Edward Arnold, London.

4

Soils and Soil Fertility

- Recognising key features of the soil
- Factors affecting soil type: time, climate, parent materials, organisms, topography
- Properties of materials constituting soils: physical and chemical properties; biological properties
- Managing soils in the garden
- What plants require from the soil and why
- Root structure, growth and activity
- Cultivating soils and essentials for planting
- Managing soil nutrients: manures, composts and fertilisers
- Managing pH
- Managing water

RECOGNISING KEY FEATURES OF THE SOIL

Most of us are usually interested in the performance of parts of plants growing above the ground, such as flowers, leaves, or harvestable products such as beans. It is important to realise though that plants function as whole organisms, so that restrictions to growth below ground, in the soil, will produce effects above ground, and vice versa. An example of this is the production of abscisic acid (ABA) by roots in response to dry soils, leading to reduced growth of leaves and the closure of stomata (see Chapter 5). Because the responses of shoots and roots are closely coupled, horticulturists should know something of the root environment so that they can manage it effectively to produce the plants they want.

Factors influencing the type of soil

Like the environment above ground, the basic features of a soil at a particular place are determined by factors outside human influence. Nevertheless, sympathetic intervention by people can have a marked effect on soil properties that can benefit plants. The type of soil that has developed at a particular location is a result of the interaction of five factors.

Time: the duration of the various processes that result in soil formation

Soil is a dynamic medium and changes with time. Some changes can occur in a matter of minutes, such as the production of a hole by an earthworm; others such as the formation of humus from plant remains take a few years, while yet others such as the weathering of rocks and the formation of clay minerals can take several centuries. The rates at which processes occur in soils today are not the same as they were in the past, because of the interactions occurring with the other four factors below.

Climate: temperature and precipitation

Temperature and rainfall affect the rates at which rocks and minerals weather to form the basic solid materials that constitute a soil. These are essentially long-term processes, but temperature and rainfall also affect short-term processes, such as the leaching of nutrients from the soil and the rates of various biological processes, such as the decomposition of organic materials.

Parent material: the mineral matter from which the soil develops

Differences in the types of rocks and sediments which constitute the parent materials of a soil affect the end products, especially the types and sizes of mineral particles that form the soil. For example an igneous rock such as granite, formed from material deep in the Earth's core, will weather to give large amounts of quartz, micas, feldspars and gibbsite; whereas a sedimentary rock such as limestone, formed by the settling of particles in water, will weather to form calcite and dolomite. In the UK, the variety of rocks available for weathering covers almost the whole geological time course, but the soils at a particular place are rarely dominated by the underlying geology. This is because glaciers in past ice ages have moved materials around, and wind erosion from continental Europe and North Africa has deposited materials over the original rock deposits. In many places, then, drift and soil materials eroded from elsewhere are the dominant precursors of the soil mineral matter.

Organisms: the vegetation, animals and microorganisms

Soils contain a vast number of living organisms, most of which are not visible to the naked eye. A teaspoon of moist garden topsoil will contain several million bacteria, a million or so fungi and ten thousand or so protozoa, among others. These organisms live mainly on each other and on plant and animal residues which they convert into many chemical forms. The decayed and decaying organic matter in soils results from the combined activity of the plants, animals and soil fauna and flora, which, together with the burrowing and mixing activities of live animals, mean that the soil is a continually changing, living medium. The interaction of the living organisms with the dead organic materials produces a vast array of chemicals, some of which promote weathering of soil mineral particles, while others join particles together, or form a reserve of nutrients which biological activity can make available to plants.

Topography: the shape of the ground surface

The topographic location also influences the type of soil that forms, mainly because of the drainage conditions. Generally, flat areas allow more rain to permeate the soil than sloping areas, where lateral movement is possible. At the base of slopes water may accumulate, either at depth or at the surface to form waterlogged soils (known as gleys). If surface water persists for prolonged periods, peat bogs are formed.

The end result of all these interacting factors is to produce soils of almost infinite variety. Maps of major soil associations are available for most countries and indicate that at a large scale there are numerous different soil types. Even in gardens of fairly modest size there are likely to be areas noticeably different from one another that need to be managed individually if the best results are to be obtained.

A feature of soils is that they usually show some horizontal banding, most obviously of colour, but also of other properties. Usually the upper 20 cm or so has the darkest colour because of the presence of organic matter. The lower parts of the soil frequently accumulate smaller particles and other materials washed downwards. These *horizons* of the soil indicate the dominant soil-forming processes. Most gardeners only ever look at the topsoil (the top horizon), but it is the subsoil (the second and subsequent horizons) that often makes a difference to the growth of long-lived plants because of its contribution to root growth and supplies of nutrients and water. These aspects of the subsoil will be explored later in this chapter.

In addition to all of these soil-forming factors must be added the special influence of one particular organism – humankind. People contribute in many ways to the formation of new soils by altering the drainage, moving soil materials to create deeper soils, adding waste products such as manure to promote biological activity, and adding fertilisers and manures to increase the nutrient status. Horticulturists consciously engage in all of these activities and there is a long history worldwide of on-site changes by humans to sustain vegetable and flower production, especially close to major conurbations and where money is not a limiting factor.

Less well appreciated are the off-site effects of human activities that also have an effect on the development of soils. For example, industrial activity and the burning of fossil fuels ensure that the soils of most of Europe and North America are receiving inputs of nitrogen (N) and heavy metals. In southern England, the average input of N is about 25 kg ha^{-1} (25 kg per ha per year) per year; this is as inorganic N and is equivalent to applying 36 g (a large spoonful) of Growmore fertiliser to a square metre. Similarly, the cadmium content of the topsoil at Rothamsted, Harpenden, Hertfordshire has increased by 40% since 1850. These and other industrial inputs to soils will have significant consequences if they continue unabated.

PROPERTIES OF THE MATERIALS CONSTITUTING SOIL

In simple terms, soils are made up of mineral matter and organic matter with the spaces between these solid materials filled with either water or air. The organisms live on the surfaces of the minerals and in the spaces. Soils, then, are complex media that have physical, chemical and biological properties. The aim of horticulturists must be to manage these to the best advantage of plants. To achieve this effectively demands some understanding of the properties of each of the individual soil components and how they interact with each other.

Physical properties of soils

One of the most commonly used ways to define a soil is in terms of its *texture*. This is a summary term describing the relative quantities of mineral particles of less than 2 mm diameter. Particles larger than 2 mm in diameter are usually referred to as stones, gravel or pebbles, and have no part in this description. The distribution of particles less than 2 mm in diameter cannot easily be altered, and influences many other important soil properties such as aeration, water-holding characteristics and the ease of cultivation. This explains why this method of definition is so widely used; it is a stable, shorthand description of a range of properties.

For convenience, soils are allocated to a textural class depending on their content of sand-, silt- and clay-sized particles. Sand-sized particles are typically 0.05–2 mm diameter; silt, 0.002–0.05 mm; and clay, less than 0.002 mm diameter. With a little practice, it is possible to determine the textural class from about half a handful of moist soil (with roots, stones, animals and so on removed) worked between the fingers and thumb (see Fig. 4.1). If the soil contains substantial additions of organic matter, then the scheme shown in Fig. 4.1 will not work because silt particles and finely divided organic materials both feel silky to the touch.

Sand particles usually consist of quartz, are rounded or irregular in shape (depending on the amount of abrasion they have suffered) and are not sticky when wet. Sands cannot be moulded, possess good drainage and aeration, but may be prone to drought. Silt particles are intermediate in size between those of sand and clay, and consist essentially of micro-sand, so they too are dominated by quartz. These particles give the sand some cohesion and plasticity, and may cause the soil surface to become compacted and to form a crust when dry. The clay-sized particles are dominated by clay minerals rather than by quartz. Such particles have a very large surface area in relation to their unit mass – about 10 000 times more than that of sand. The large surface area allows the adsorption (binding) of water and nutrients, so that wet clays are sticky and cohesive.

Most soils are a type of 'loam'. An ideal loam contains a mixture of sand, silt and clay particles in about equal proportions. Note that the term 'loam' indicates nothing about the quantity of organic matter present, although in everyday language gardeners often erroneously refer to soils with moderate amounts of organic matter as 'loams'. Another issue that can lead to misunderstanding is that the word 'clay' is used here to describe the size of particles; however, it is also used (see the next section) to describe a wide range of different minerals, with different chemical properties. So, for example, clayey soils found in the Weald of Kent have distinct properties because the dominant types of clay mineral in them are different. The terms 'light', 'medium' and 'heavy' are also commonly used to describe the texture of soils, and relate to sandy, loamy and clayey respectively. These descriptions reflect the water content of the soils when drainage has ceased, but in other respects, such as the mass of soil particles in a given volume, give a misleading impression of soil properties.

The texture of a soil is not easily modified and apart from the incorporation of large amounts of sand to improve the physical properties of some soils for horticultural purposes, is not changed by cultural management. However, the physical properties of a soil can be substantially altered by management practices such as cultivation, drainage and manuring, through the effects on soil structure (especially in the topsoil). The term *structure* describes the way in which the sand, silt and clay particles are grouped or arranged together into units known as *aggregates*. Within a soil, different structures are often present in different horizons, and these influence the movement of water, gases and heat. The addition of organic matter can exert a major influence on structure, due to its ability to act as a bridging agent between particles.

The macrostructure of soils can be seen with the naked eye. Spheroidal structures (granules and crumbs) are typical of many surface soils, particularly those under grass, and soils high in organic matter. Essentially, it is only these aggregates that are altered by most horticulturists. In the subsoil, platy, columnar, prismatic and blocky structures usually dominate, ranging in size from 1 to 20 cm or more. Between the structures are spaces called *pores*; these are normally filled with air, except in waterlogged soils. The exact process of how the structures form is unknown, but living organisms and the movement of fine materials play key roles. Plant roots tend to compress soil particles into small aggregates as they elongate through soils, and may even compress larger aggregates. Similar compression and contraction occurs as plants take up water, and soils are wetted and dried.

Plant roots and microorganisms also exude sticky organic compounds (mainly long-chain sugars known as polysaccharides), that can bind

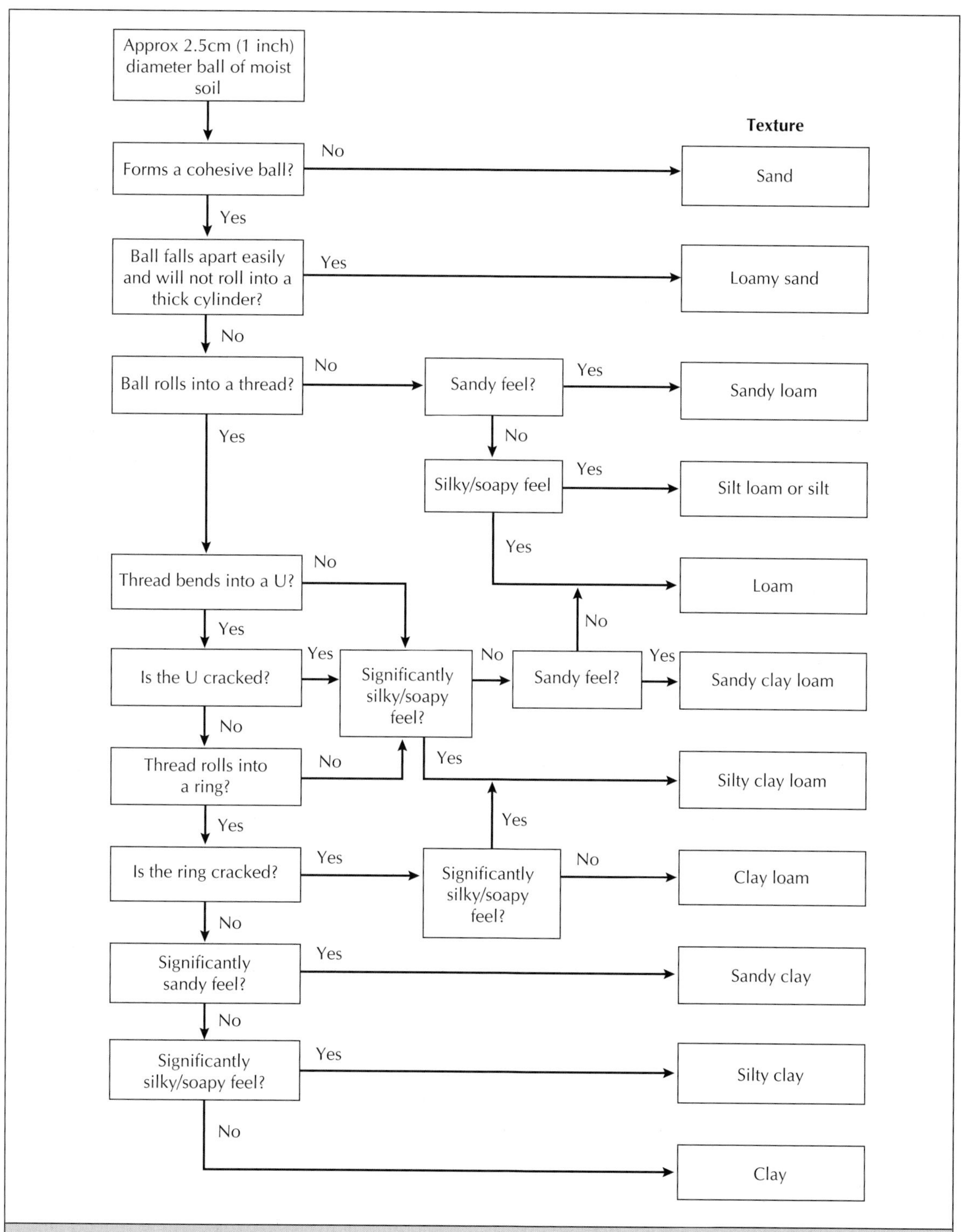

Fig. 4.1 A simple guide for assessing the texture of a soil using properties measured by rubbing a small ball of soil between fingers and thumb.

particles together. Decomposition of organic materials caused by microbes also produces other organic compounds that interact with clays to cement the particles together. Thus, clay-sized particles and fine organic materials are important components of aggregates (see Fig. 4.2), especially in topsoil, where living organisms and organic matter are most frequently found. In the subsoil, the downward movement of organic materials, clay minerals, oxides of iron and aluminium, and salts such as calcium carbonate can all act as cements under different climatic, topographic and vegetative conditions.

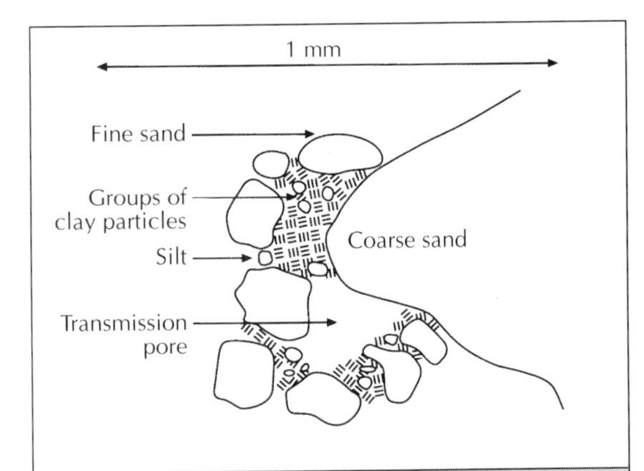

Fig. 4.2 Diagrammatic representation of soil microstructure examined with a light microscope, showing sand and silt particles joined together by groups of clay particles to form a structure containing pores. Spaces between the groups of clay particles and the surfaces of the solid particles contain organic matter, but these cannot be seen clearly by a light microscope.

The pores are an integral part of a soil and are filled with water and air. The proportions of water and air change as rain falls, as plants dry out the soil, and as evaporation occurs from the soil surface. The size, shape and continuity of the pores determine how much of the rain falling on a soil drains, is held for plant use, or is held so firmly that plants cannot use it. Table 4.1 summarises a useful classification of pores in terms of size and function. In crude terms, if a pore can be seen with the naked eye, then it will not hold water in a freely draining soil; a microscope is necessary to view pores that will retain water.

In general, sandy soils have slightly less total pore space (35–40%) than clays (50–70%), with loams in between (45–55%) and peats having about 80%. However, it is the sizes of the pores making up the pore space that differ significantly between soils of different textural classes. In sandy soils, most of the pores are typically the larger transmission pores with only a few of the very small residual pores and some storage pores. This contrasts with clayey soils, which have fewer transmission pores, but three to four times the volume of residual pores compared with sands. Sandy soils, then, drain easily, hold only modest amounts of water available to plants, and retain little water after drying by plants; whereas clays do not drain readily, but hold more water available for use by plants, and contain water even after prolonged dry periods.

The critical limits for the three categories of pores are:

- If transmission pores form less than 10% of the soil volume, then drainage may be a problem.
- If storage pores form less than 15% of the soil volume, then water availability is likely to be restricted.
- If residual pores form less than 20% of the soil volume, then mechanical difficulties are likely, because the soil will be plastic and sticky when wet, but hard when dry.

Water is held in soils against the force of gravity because of forces acting between water and the surfaces of soil particles. These forces result in water being held in small pores, at the necks of larger air-filled pores, and as thin films on particles enclosing air-filled pores (see Fig. 4.3).

Chemical properties of soils

To a first approximation, the sand- and silt-sized components of the soil consist of quartz and are, therefore, chemically inert, apart from the finer materials that are smeared across their surfaces. It is the clay-sized fraction (comprising clay minerals, amorphous alumino-silicate materials and sesquioxides of iron and aluminium, and humus)

Table 4.1 Classification of pores by size and function (from Rowell, 1994).

Pore class	Pore diameter (μm)	Pore function
Transmission (macropore)	>50	Drainage after saturation. Aeration when soil is not saturated. Both aeration and drainage require interconnection of pores. Root penetration. Root diameter is typically <200 μm.
Storage (micropore)	50–0.2	Store water available for plant use.
Residual (micropore)	<0.2	Hold water so strongly that it is not available to plants. This water controls to a large extent the mechanical strength of the soil.

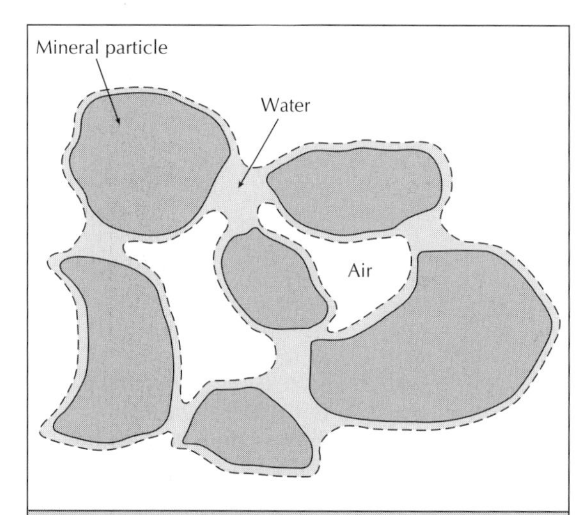

Fig. 4.3 Water, air and mineral particles in a freely-drained soil. When drained (i.e. not saturated), air occupies the larger transmission pores and water is present in the smaller storage pores and as films on the surface of soil particles.

that is chemically active. These materials are small (less than 2 μm in diameter, requiring the use of an electron microscope to view individual particles) and electrically charged enabling them to absorb ions (charged atoms and molecules), including some that are plant nutrients; to absorb water, with consequent swelling and plasticity; and to disperse as a consequence of repulsion between charged surfaces. The clay-sized fraction consists of three types of colloidal material.

Clay minerals

These are crystalline, plate-shaped structures. Each clay particle consists of a series of plate-like layers stacked one on top of the other like the pages in a book. Every layer comprises horizontal sheets of silicon, aluminium, magnesium and other positively charged ions (*cations*), surrounded and held together by oxygen and hydroxyl groups. The sheets are of two types: tetrahedral sheets composed of silicon bound to oxygen atoms; and octahedral sheets composed of either aluminium or magnesium, each bound to six oxygen atoms or hydroxyl ions.

The sheets are combined together into layers of clay minerals of three main types (Fig. 4.4):

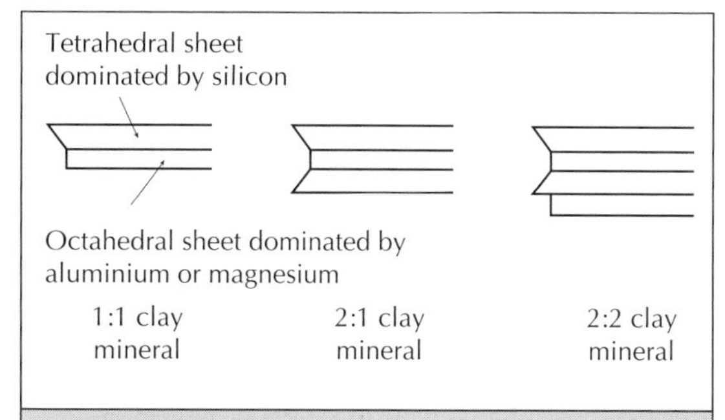

Fig. 4.4 Diagrammatic representation of the chemical structure of various types of clay minerals making up the clay fraction of soils. Three basic types of clay mineral are produced depending on the arrangement of the tetrahedral and octahedral layers.

1 tetrahedral sheet plus 1 octahedral sheet (e.g. kaolinite), 2 tetrahedral sheets either side of 1 octahedral sheet (e.g. smectite) and 2 tetrahedral sheets plus 2 octahedral sheets (e.g. chlorite). All clay minerals are small in size and this results in a large surface area that can adsorb ions, water and gases. However, there is another important property that distinguishes clay minerals, and is of vital importance to plants. During the weathering processes that lead to the formation of clay minerals the perfect chemical structures shown in Fig. 4.4 rarely occur, especially in soils. Impurities are often found in both the tetrahedral and octahedral layers and this produces clay minerals with a permanent negative charge. These charges are always balanced by positively charged ions (cations) that are adsorbed onto the surfaces of the clay minerals. In general, calcium, magnesium, potassium and hydrogen balance the negative charge, but they are interchangeable, depending on their availability in particular soils. These ions are known as exchangeable cations, the total charge as the *cation exchange capacity* (CEC), and the process as *cation exchange*. The CEC and the balance of ions are important to horticulturists because they determine such important features as the ability of a soil to replenish some nutrients to the soil solution, from where they are taken up by plants, and the pH (acidity or alkalinity) of the soil. Together, the various processes producing clay minerals result in a suite of materials with different properties and for this reason, not all clayey soils behave in the same way. For example, a clay soil dominated by smectite will swell when wet and produce big cracks on drying whereas a soil dominated by kaolinite will not.

Alumino-silicate materials and sesquioxides of iron and aluminium

Some of these materials are crystalline while others are not. Unlike the clay minerals, however, the crystalline materials do not swell or shrink and there is no permanent negative charge. At the broken edges of the crystals, though, it is possible for a charge to develop associated with the hydroxyl group. This charge is variable depending on the pH of the surrounding soil solution. At high pH (alkaline soils) the charge is negative, but at low pH (acid soils) the charge is positive and the soil has an *anion exchange capacity* (AEC) satisfied by negatively charged ions such as chloride, sulphate and nitrate. In temperate soils this AEC is rarely evident, but soils in tropical gardens, where kaolinite and sesquioxides dominate the clay-sized fraction, may have this property.

Humus

Humus is a mixture of complex compounds, not a single material, that has been synthesised by microorganisms in the soil from breakdown products or alteration of the original plant and animal materials. It is typically brown or dark brown in colour, and gives many temperate soils their characteristic brown to olive coloration. Humus colloids are not crystalline and are composed essentially of carbon, hydrogen and oxygen rather than silicon, aluminium, iron, oxygen and hydroxyl. They are small, and have a pH-dependent negative charge (never a positive charge), so that humus, like clay minerals, adsorbs cations.

Biological properties of soils

Living organisms are an essential part of the soil-forming process. Initially organisms that can photosynthesise to produce carbon compounds are needed (for example, lichens on rock surfaces), but once cells are formed, they can become a food source for other organisms. In soils that have been formed for some while, a vast array of organisms is present, living on each other and on dead organic matter returned to the soil from plants. *Microorganisms* (organisms that cannot be seen without the assistance of a microscope) convert organic residues, eventually forming a resistant residue, humus; this combines with living organisms, dead cells and mineral particles to form the solid component of a soil. The term *soil organic matter* is used to refer to all the organic material in soil, including humus. Soil organic matter comprises living material (soil animals, plant roots and microorganisms) and dead material including shoot and root residues and humus.

The amount of organic matter present in a soil depends on:

- its physical properties (clay soils generally contain greater amounts than sandy soils);
- the climate (temperate regions generally have greater amounts than Mediterranean regions); and
- the type of vegetation (woodlands and grasslands generally have more than cultivated soils).

In the UK and much of Northern Europe, only about 2–5% of a cultivated topsoil is soil organic matter, and of this, only a small proportion is living. For example, a cultivated topsoil may contain 4.5 kg of organic matter per square metre, of which only about 2.5% (112 g) will be living microorganisms (the microbial biomass). Although small, this component is vital, because almost all of the plant and animal residues must pass through it in order for nutrients to be released for plant use.

Some of the more important groups of organisms present in soils are shown in Fig. 4.5; an indication of numbers and biomass is given in Table 4.2. The microorganisms are the most numerous, and have the highest biomass. Since activity is generally related to biomass, the microorganisms, together with earthworms, dominate biological activity in the soil, with

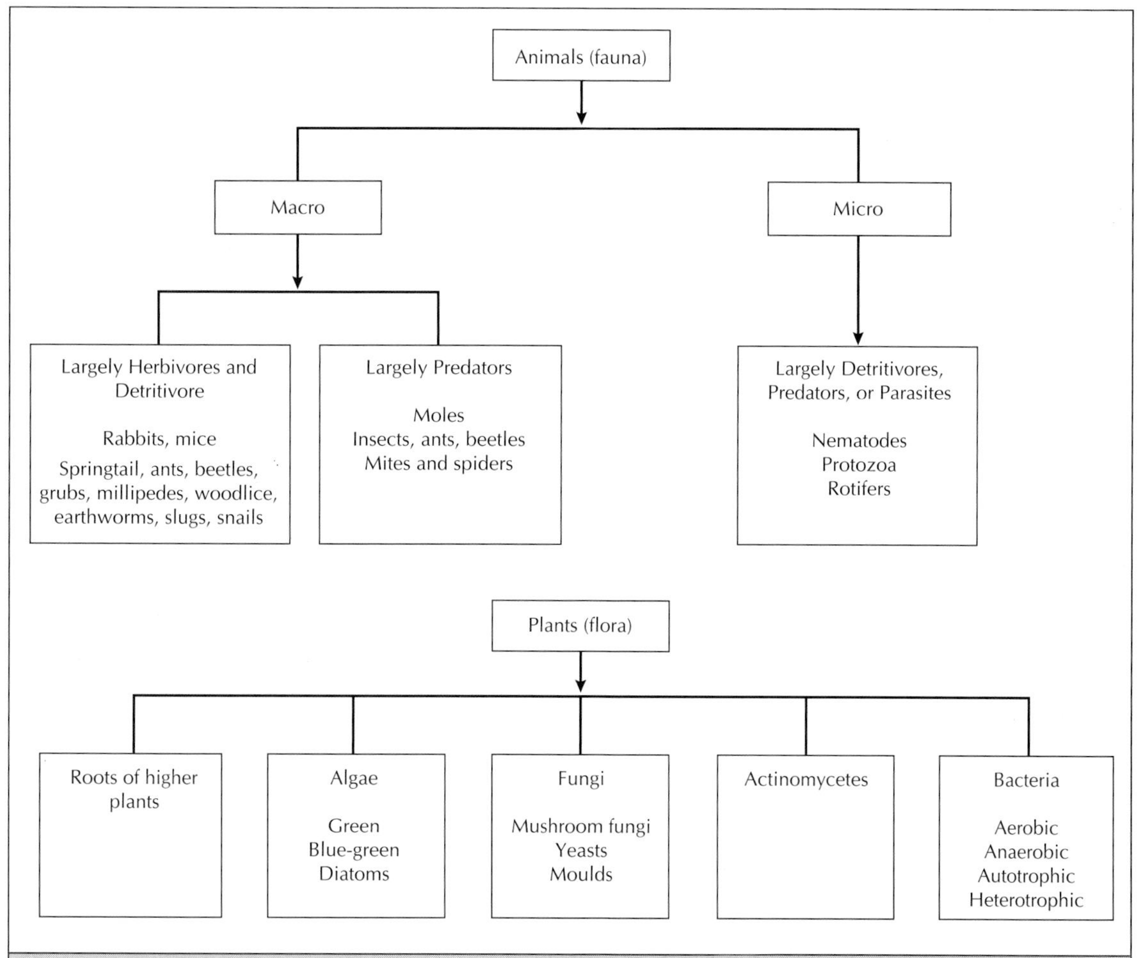

Fig. 4.5 A general classification of some of the more important groups of organisms that are commonly present in soils.

Table 4.2 Numbers and live-weight biomass of soil flora and fauna commonly found in surface soils (from Brady, 1990).

Organism	Number per m^2	Biomass ($g\ m^{-2}$)
Microflora		
Bacteria	10^{13}–10^{14}	45–450
Actinomycetes	10^{12}–10^{13}	45–450
Fungi	10^{10}–10^{11}	110–1100
Algae	10^{9}–10^{10}	5.6–56
Microfauna		
Protozoa	10^{9}–10^{10}	1.7–17
Nematoda	10^{6}–10^{7}	1.1–11
Other fauna	10^{3}–10^{5}	1.7–17
Earthworms	30–300	11–110

60–80% of soil metabolism attributable to the microflora. In some areas, ants and termites are important, and the activities of termites and their associated fungi in some tropical soils play a major part in soil regeneration.

Bacteria are single-celled organisms whose capacity for rapid reproduction means that they can adjust their activities quickly in response to changes in their environment. Their length rarely exceeds 4–5 μm and the smaller ones are about the same size as a clay particle (2 μm). The bacteria exist in clumps or colonies, and different types use different food sources. Most soil bacteria are heterotrophic, meaning that their energy and carbon source come directly from organic matter. These bacteria, together with fungi and actinomycetes (filamentous relatives of bacteria), undertake the general breakdown of organic matter in soils. A smaller group of bacteria are metabolically self sufficient, obtaining their energy from inorganic substances such as ammonium, sulphur and iron, and most of their carbon from carbon dioxide (as plants do). This group plays a vital role in controlling the availability of nutrients, especially nitrogen, to garden plants (see later).

Another process in soils that depends largely on bacteria is nitrogen fixation. In this process, nitrogen gas in the soil atmosphere is 'fixed' to form ammonia, which is then incorporated into organic compounds. The process occurs both in bacteria living in soil (about 5–20 kgN ha^{-1} per year in the UK), and also, most effectively, in bacteria living inside nodules associated with roots of legumes such as peas and beans (50–200 kgN ha^{-1} per year can be fixed by legume crops in this way).

Actinomycetes, like bacteria, are unicellular, but are filamentous and frequently branched. They are similar to bacteria in size and, like bacteria, their main food source is soil organic matter which they break down. They appear to be more capable than many bacteria of breaking down some of the more resistant compounds produced by plants. However, while bacteria can survive across a wide range of pH levels, the optimum for actinomycetes is between 6.0 and 7.5.

Most soil fungi are multicellular organisms forming long hyphae normally 5–20 μm in diameter, though a few species of non-cellular slime moulds and unicellular yeasts are also found. The soil fungi can use a wide variety of carbon substrates as a source of energy. Some fungi can utilise only simple carbohydrates, alcohols and organic acids, while others can decompose polymers such as cellulose and lignin. Some are parasites of plants (see Chapter 11), while others (for example, mycorrhizas) have a symbiotic relationship with their host (see Chapter 1). Overall, fungi are the most versatile group of microorganisms and, because of their role in humus formation and aggregate stabilisation, play a significant role in all soils.

Much has been written on the importance of earthworms. They ingest leaf and root material along with mineral particles, which are subjected to digestive enzymes and grinding processes within the animal. The mass of material passing through the body of a worm daily may equal its own body mass, and the casts produced have significantly greater nutrient content and structural stability than the bulk soil. Earthworms must have organic matter as a food source, and prefer a well-aerated but moist habitat. Most thrive when the pH of the soil is between 6 and 7, although a few species tolerate lower levels of pH. Many native species of earthworms are killed by the disruption of their food sources and burrows that results from cultivation. Where earthworms are not present (in acidic, dry or very wet conditions) the accumulated litter is broken down only very slowly.

From the description above it is clear that inputs of new organic materials are essential to maintaining an active population of living soil organisms. It should also be clear that adding organic materials to soils will not necessarily increase the organic matter content of the soil; much may simply decompose to carbon dioxide, which is lost to the atmosphere, and inorganic ions, which may be lost by drainage.

MANAGING SOILS IN THE GARDEN

Soils rarely provide ideal conditions for the growth of garden plants, and the objective of soil management is to try to create better conditions. Maintenance and improvement of soil fertility involves managing a suite of biological, chemical and physical properties and processes to achieve the desired end.

Land under natural vegetation has an inherent fertility, but in a garden, different aspects of fertility will be altered to different degrees. For example, aeration, which depends mainly on transmission pores, can be decreased by compaction or increased by careful cultivation, but water availability, which depends on storage pores, cannot easily be changed by any practice.

What do plants want from the soil?

It is plant roots that interact directly with the soil to influence growth. The following requirements of plants can be supplied by soil:

- anchorage for roots to stabilise the plant and enable the leaves to intercept sunlight
- water
- air – especially oxygen to allow roots to respire
- nutrients
- a buffer against adverse changes of temperature and pH.

As a general rule, it is rare for roots to occupy more than 5% of the total soil volume, even in the upper 10–15 cm where they are generally most abundant. The volume occupied decreases very rapidly with depth, and is often no more than 0.01% at 50 cm. This means that only a small fraction of the soil within the rooting zone is in direct contact with roots. The implication of this for horticulturists is that managing soil fertility is essentially about providing a medium in which:

- roots can grow and proliferate to form a network capable of harvesting water and nutrients
- water and nutrients can be stored in forms that are available as required by the plant
- water, nutrients and air are able to move to the limited areas where a soil/root interface occurs
- biological activity is manipulated to favour beneficial organisms and reduce the incidence of pathogens.

ROOTS: ACTIVITY AND GROWTH

Roots have several functions related to the requirements of plants for all the substances that they assimilate: they absorb water and nutrients from the soil and transport these from the site of absorption to the shoot; they form the site of synthesis for several plant hormones which may act in either the root or the shoot or both; they may act as storage organs as, for example, in root vegetables such as carrot; and they serve to anchor the plant.

Roots are difficult to study because they cannot easily be seen and, until recently, it has been inevitable that any study in soil has disturbed the very medium in which they are living. For these reasons, much less is known about the functioning of roots than leaves, and most of what has been learned has involved young roots, often growing in solutions. In many cases, root function has been inferred from observations of form and structure, rather than observed from direct evidence.

A typical young root is shown in Fig. 4.6. It comprises a *root cap* behind which are the *meristem* (i.e. dividing) cells, which later differentiate into cells forming the *epidermis*, the *cortex* and the *stele*. During this differentiation the cells increase in volume and elongate to push the root through the soil. Behind this zone of elongation, the cells of the stele themselves differentiate to form the *endodermis*, *phloem* and *xylem*. In roots the stele often forms a solid core, which better enables the root to withstand tension and compression; in stems it is often a hollow cylinder. The xylem is essentially a pipe that transports water and nutrients upwards to the shoot, while the phloem consists of living cells that allow bi-directional movement of assimilated materials. The endodermis gradually becomes thickened with deposited *suberin* (corky material) as the plant ages. The suberisation of the roots strengthens them, but also prevents the movement of some nutrients to the xylem and thence to the shoot.

The outer cells of the root cap may exude materials which coat the root tip with mucilage (see Fig. 4.7 left). This mucilage is very noticeable in some plant species (for example, grasses); it ensures good contact between the root and the soil, and aids the transfer of water and nutrients to the plant. So effective is this material that it is impossible to detach soil particles from some roots, and they exist within a *rhizosheath* (Fig. 4.7 right). The root hairs (often in association with mucilage) also act to increase contact between the root and the soil particles. In addition, because they effectively extend the surface of the root outwards, they are able quickly to mine a zone of soil that would otherwise take some time to deplete. This process is particularly important in the acquisition of nutrients, such as phosphate, that otherwise move slowly through the soil. Mycorrhizal fungi (see Chapter 1) may serve a similar purpose.

However, most roots of most plants do not

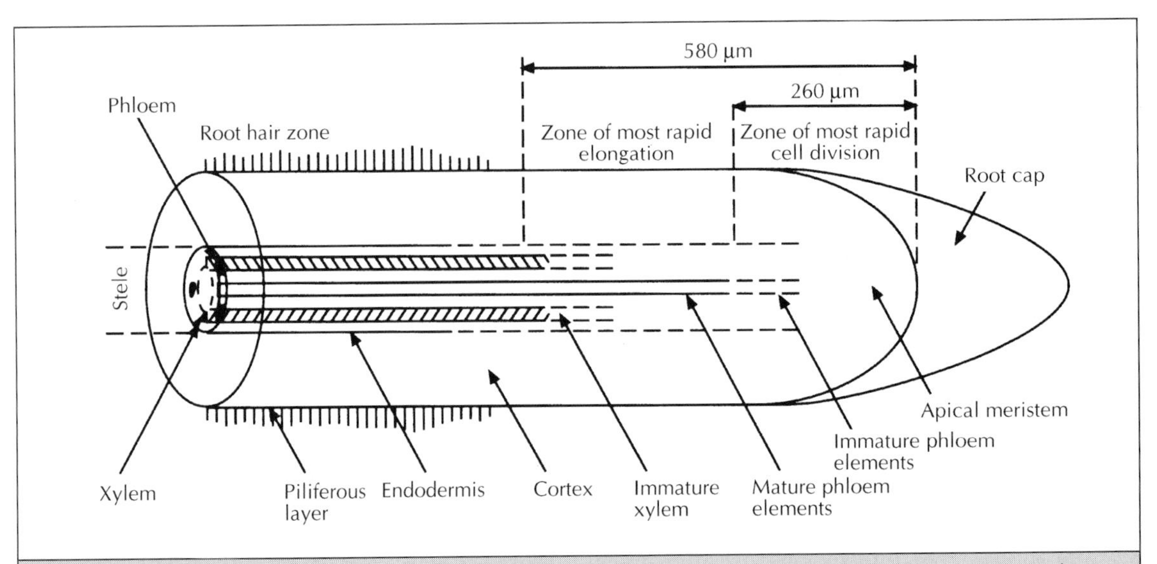

Fig. 4.6 Diagrammatic representation of a longitudinal section through a young root showing the spatial arrangement of the root tissues.

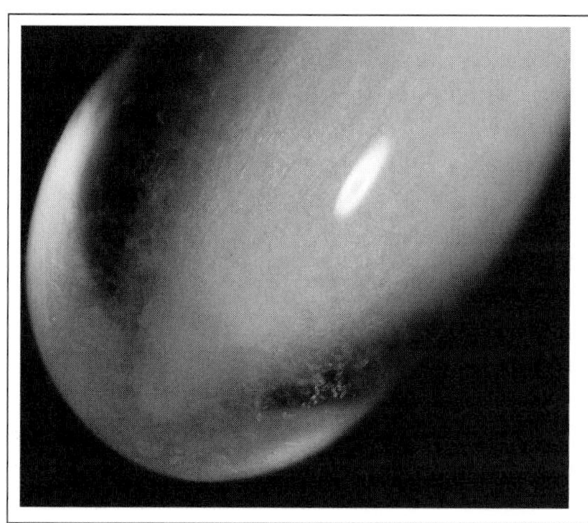

Fig. 4.7 (*left*) Mucilage production at the root tip of a maize root; and (*right*) the binding of soil particles to a root by mucilage and root hairs to produce a 'rhizosheath' that is very difficult to remove. The production of rhizosheaths is common in many species especially as the soil dries.

look or function like young roots. Root hairs are ephemeral and generally last for only a few days. As roots age, the epidermis and cortex are attacked by other organisms, become desiccated during dry periods, and eventually disappear. Older roots of annual species may have the endodermis as the outer layer for a significant part of the growing period.

Water can be taken up along the whole length of the root system if it is available in the soil and if contact between the root and soil is good. The water directly surrounding a root system is only a small fraction of the total needed to supply the amount which evaporates (transpires) during the course of a sunny day. So the liquid water must flow from some distance away to reach the root surface – often as much as 10 mm in a day. This cannot happen if there is an air gap between the soil and the root.

Two different situations have to be managed by horticulturists. The first is on sandy/silty soils (often alluvial deposits) with high organic matter content, where 'puffiness' may lead to poor soil-root contact. This can be overcome by firming the soil, manually or with a roller. The second situation is more difficult to manage because it occurs where there are a few cracks in the subsoil and the roots are limited to following them. As the soil dries, the roots may lose contact with the soil, and hang in the air between the cracks.

The activity of roots in taking up nutrients is rather more complex than that of taking up water, because of the role played by the endodermis. Older books on this subject often suggest that nutrient uptake is restricted to an absorbing zone close to the root tip. This is now known to be incorrect, at least for certain ions in plant species that have been tested. Ions such as nitrate, ammonium, phosphate and potassium can be taken up along the whole length of a root and transported to the xylem for onward transportation to the shoot. For calcium and iron, however, while limited uptake can occur along the length of the root, uptake is greatest nearest the root tip, and translocation to shoots is confined to this zone. Suberisation of the endodermis restricts movement of these ions, so that only young roots contribute to uptake and acquisition by the shoot. The reason for this differing behaviour by different nutrients lies in the pathways that they use to move from cell to cell within the root. Nitrate, ammonium, phosphate and potassium move mainly inside the cell while calcium and iron move initially via the cell walls whose route is blocked by the suberised endodermis.

Little research has been done on the roots of

garden plants, and there are only few data available on the roots of vegetables, mostly in soils with sandy texture. Pulling up a plant to examine the roots often gives a false impression of the depth and size of the root system. The roots of most garden vegetables will normally grow to at least 50 cm (see Table 4.3), and the roots beneath a lawn will often penetrate to 1.5 m if the soil is deep enough. Beneath each square metre of ground there will be several kilometres of roots with about 40 km m^{-2} in the case of grass. In fact, the values shown in Table 4.3 are something of an underestimate, because the methods used to obtain them ignore the very fine roots and root hairs.

The weight of a root system is also a substantial component of the total plant weight, although its relative size generally decreases with time. For example, in many crops of peas and beans (*Vicia faba*), up to 40% of the total plant weight may consist of roots up to flowering. After flowering, as the pods start to grow, root growth occurs at a much slower rate, and the roots typically form only 15–20% of the total plant weight. In the final stages of pod filling, the plant allocates most of its resources to the pod and only a small amount to the roots, so the root system begins to degenerate.

In general, most of the root length is located in the top 15 cm of soil, and the quantity decreases very rapidly (often logarithmically) with depth. The zone of maximum root length coincides with the most chemically and biologically fertile zone of the soil. Water, however, is more evenly distributed in the soil profile, so that the depth of rooting and the roots growing in the subsoil are important for the acquisition of water. Drought-tolerant plants, for example, often have a deep root system.

There are many soil conditions that can restrict root growth. In horticultural practice, the most important that can be managed include:

- mechanical impedance such as occurs in compact layers
- a shortage of oxygen normally due to water-logging or poor drainage
- dry soil
- low nutrient supply
- a pH that is too low.

CULTIVATING THE SOIL AND ESSENTIALS FOR PLANTING

Cultivation of soil has several functions. It is done to:

- obtain a seed-bed
- kill weeds
- remedy damage done by previous traffic or cultivation
- increase the permeability of the surface soil or subsoil to water
- incorporate crop residues and manures
- provide a hospitable medium for root growth.

Table 4.3 The maximum depth of rooting and length of root systems of some vegetable crops grown on a silt loom (from publications at Horticulture Research International, Wellesbourne, UK).

Crop	Maximum rooting depth (m)	Total length (km m^{-2})
Broad bean	0.8	1.76
Cauliflower	0.8	11.9
Lettuce	0.6	2.37
Onion	0.6	1.83
Parsnip	0.8	5.65
Pea	0.7	5.46
Turnip	>0.8	15.42
Grass	1.5	>40

In the particular case of double digging, this is done to increase the depth of topsoil leading to an increased store of readily accessible nutrients and water through the encouragement of a more extensive root system. Given the constraints of time and labour and the availability of fertilisers, this form of cultivation is practised much less than in the past. Many gardeners use digging as their primary means of cultivation, but apart from its virtue as a form of exercise and a good reason for being out in the fresh air on a bright spring morning, it is vastly overrated. Digging is a good method of weed control and is essential for many perennial weeds, although selective use of herbicides and many ground-cover plants can substantially obviate the need. However, digging alone almost always destroys soil structure, and it should rarely be done on clayey soils where the weather (frost in winter and drying in summer) will regenerate the structure naturally. On sandy and silty soils, some digging may be beneficial but, except for vegetable production, a light cultivation with a fork, hoe or rake will often suffice.

The essential requirements of a good seedbed are that the tilth produced by cultivation should be sufficiently fine and firm to allow the seed to be placed at a uniform depth, and to have sufficient contact with the soil to allow it to take up water. The soil around the seed must have sufficient pore space to maintain good aeration, while the soil above and below the seed must be sufficiently loose to allow both the shoot to emerge into the air and the roots to elongate. For rapid growth, there should be a supply of essential nutrients close to the seed, and weeds should be absent, so that there is no competition for soil resources. On clayey soils it is difficult to achieve such conditions in a timely manner, but on sandy and silty soils it is practicable. Many silty soils have poor structural ability when wet, so that the surface aggregates collapse after wetting and then form a hard cap on drying. This can substantially reduce the number of seedlings that emerge, unless the cap is disrupted by a light hoeing.

Root growth occurs as the cells behind the root tip elongate longitudinally and radially to push the root tip forward. The roots of most plants have a diameter greater than 0.1 mm, so they must either grow through transmission pores or deform the soil to make a pore of sufficient size for elongation to proceed. Growth of the root will only occur if the pressure exerted within the elongating cells is sufficient to overcome the constraints imposed by cell walls and the soil matrix. The resistance of the soil to compression depends on its composition and the packing of soil particles, and increases as the soil dries.

One function of cultivation is to create transmission pores, or to loosen the soil sufficiently to allow the roots to elongate and create such pores.

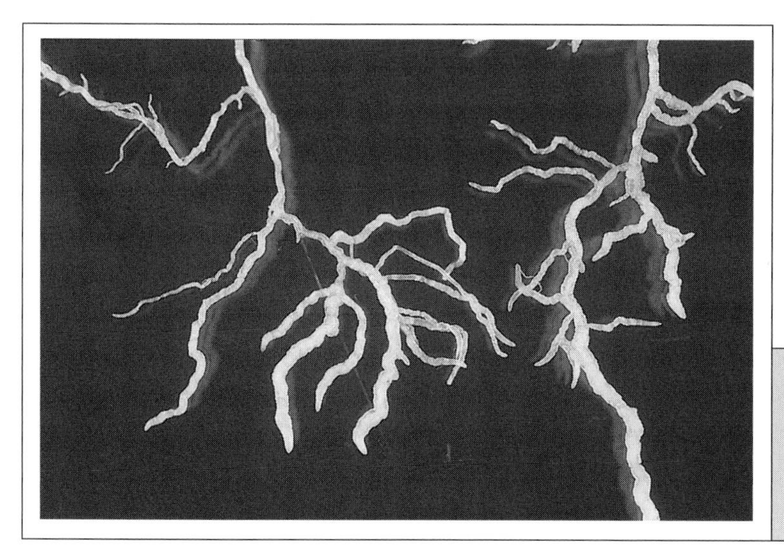

Fig. 4.8 Effects of compaction on the growth of wheat roots. Note the thicker and more distorted roots with sharp bends. These roots were grown in a sand with a pore space of only 30%.

However, if cultivation is done in the wrong conditions or with the wrong implement, it can exacerbate the problem of poor root growth. An obvious sign of compaction is that the root diameter increases immediately behind the root tip (Fig. 4.8). This feature is an adaptive mechanism that reduces soil strength in front of the growing tip, thereby allowing some root elongation. In extreme conditions, as in Fig. 4.8, the root becomes distorted in the attempt to find gaps between soil particles, and the root tip may eventually die.

In gardens, there are two frequent situations that may lead to soil conditions that impede root growth. The first is due to the impact of machinery and feet. Building work, laying pipes and walking on the top of exposed subsoil can all lead to the formation of compacted layers that will reduce root growth. Essentially, these layers should be broken up with a spade or fork before planting. In addition, the use of powered cultivators in wet conditions can cause smearing of the soil at the depth of cultivation, which produces a compact layer and blocks large pores. In agriculture this is known as a 'plough pan'. Figure 4.9 shows the results of such a pan on a crop of leeks; this crop was unmarketable.

The second set of problems arises when transplanting. Some soils (especially silty and clayey soils) smear when wet, so that the use of a dibber or trowel can produce a lining to a hole that confines roots. Figure 4.10 shows the roots of a strawberry plant confined to the dibber hole. A related problem can occur when transplanting a container-grown plant. Frequently such plants form a root mat against the inside of the container, which needs to be teased apart and the roots spread out in a larger hole if normal root growth is to be resumed. This is particularly important in drought-prone areas, because the volume of water available within the root mat is typically sufficient only for a day or two.

Why is it so important to get the physical aspects of soil fertility right? Besides the obvious effects of constraints on the size of the root system, and therefore the root surface available to intercept water and nutrients, there are less obvious effects. There is now ample experimental evidence that when a root experiences a constraint to growth, it rapidly produces a signal (a plant hormone) which is transmitted to the shoot, and this reduces the rate of leaf elongation. In this way, reduced expansion of the root system is matched by reduced canopy expansion and the sizes of the root and shoot systems remain in equilibrium. Big shoots, therefore, normally cannot be produced if the root system is constrained.

Fig. 4.9 Effects of a cultivation pan (produced by cultivating at the same depth for many years at the topsoil/subsoil interface) on the growth of leeks. The crop was not marketable.

Fig. 4.10 Roots of a strawberry plant restricted by 'dibbing in'.

MANAGING SOIL NUTRIENTS: MANURES, COMPOSTS AND FERTILISERS

Nutrient requirements, functions and symptoms of deficiency

Plants require some fourteen essential mineral nutrients to produce healthy growth and to reproduce. The elements considered essential for higher plants are as follows.

- Macronutrients
  - □ nitrogen
  - □ phosphorus
  - □ potassium
  - □ calcium
  - □ magnesium
  - □ sulphur
- Micronutrients
  - □ iron
  - □ manganese
  - □ copper
  - □ zinc
  - □ boron
  - □ chlorine
  - □ molybdenum
  - □ nickel
- Beneficial elements
  - □ sodium
  - □ silicon
  - □ cobalt.

The *macronutrients* and *micronutrients* (sometimes called trace elements) are all essential nutrients although the known requirement for chlorine and nickel is, as yet, restricted to a limited number of plant species. The distinction between macronutrients and micronutrients is somewhat arbitrary but is based on their typical concentrations in plant dry matter. The concentration of macronutrients is normally at least ten times that of the micronutrients and often the difference is much greater. For example, the concentration of nitrogen will be typically one million times greater than that of molybdenum and one thousand times greater than that of manganese.

Some elements are not essential to all plants but may be beneficial to some. For example, cobalt is essential for biological nitrogen fixation by bacteria, including those in the nodules of leguminous plants such as peas and beans. Sodium is not essential for most plants except the salt bush *Atriplex vesicaria*; it is beneficial to the growth of sugar beet. Similarly, silicon is essential for some grasses and also for *Equisetum arvense* (see Chapter 1, Table 1.3) but for most plants it is beneficial rather than essential.

Other, non-essential elements may be present in plants (e.g. vanadium) but it is anticipated that, as analytical techniques continue to improve, the list of essential micronutrients will increase.

Each mineral nutrient may perform several functions within the plant, functioning as a part of an organic structure, a catalyst of enzyme reactions, a charged carrier to maintain electrochemical balance, or a regulator of osmotic pressure. Generally the micronutrients, because

Table 4.4 Symptoms of nutrient deficiency.

Nutrient	Visible symptoms
Nitrogen	Uniform pale colour with yellowing especially of lower leaves. Feeble growth and lack of branching.
Phosphorus	Reduced growth especially soon after emergence. Often there are no other symptoms though some species show purple coloration of older leaves.
Potassium	Margins of older leaves scorched and curled up or down.
Sulphur	New leaves a uniform golden yellow or cupped and deformed. Foliage frequently stiff and erect.
Calcium	Cupping and burning of leaf tips with blackening of young leaves.
Magnesium	Yellowing between the veins on older leaves to give a mottled appearance.
Iron	Yellowing between the veins of older leaves and the production of nearly white young leaves.
Manganese	Yellowing between veins of new leaves but may quickly spread to others.
Zinc	Specific to individual species.
Copper	Very dependent on species. New leaves become greyish-green, yellow and sometimes white.
Boron	Brittle tissues which crack easily. Death of growing point and production of side shoots.
Molybdenum	Rare except in cauliflowers where new leaves become progressively twisted and reduced until only the midrib appears ('whiptail').
Chlorine	Wilting.

of their low concentration, play little part in either osmotic regulation or in the maintenance of electrochemical balance. They mainly assist in the activation of enzyme reactions. Conversely, potassium and chlorine (the only nutrients that are not part of organic structures) are very important in osmotic regulation and for electrochemical balance.

Nitrogen is an essential constituent of many organic components of plants including proteins, nucleic acids, chlorophyll and hormones. Similarly phosphorus in plants is present as part of many organic compounds which play a very important part in the very large number of enzyme reactions that depend on phosphorylation.

For the horticulturist, though, probably the most important aspect of plant nutrition is being able to recognise symptoms of deficiency (and in some cases, toxicity), and to remedy them. Often the growth of plants will be reduced considerably before deficiency symptoms first appear so any measures to correct a shortage of nutrients will be beneficial only to subsequent plantings. While generalisations can be made about the deficiency symptoms of the macronutrients, the symptoms are often specific to the plant species because some species are more susceptible than others. For micronutrients the symptoms are almost entirely plant specific. Table 4.4 gives some general deficiency symptoms. A gardener suspecting nutrient deficiency as a cause of poor growth would be best advised to consult a specialist textbook and to keep some plant leaves for laboratory analysis. In the UK, deficiencies of macronutrients are more common than those of micronutrients and if nitrogen, phosphorus and potassium are provided then there are usually sufficient amounts of other nutrients in soils to allow good growth. However, this is not the case elsewhere especially on sandy soils and those where leaching of nutrients by excess rain occurs.

Toxicity can occur if there are excess amounts of manganese, boron, aluminium or chlorine in the soil solution. Chlorine toxicity can occur when water with a high salt concentration is used for irrigation but the others are associated with soils of low pH (less than 5) and can be remedied by the application of lime.

All of the essential nutrients are obtained predominantly from the soil, though deposition on to leaves can lead to uptake of some nutrients in polluted industrial areas. The nutrients are obtained in the following ways:

- Potassium, calcium and magnesium are taken up as the positively charged ions K^+, Ca^{2+} and Mg^{2+} from the soil solution
- Nitrogen, phosphorus and sulphur are taken up as the negatively charged ions NO_3^-, $H_2PO_4^-$ and SO_4^{2-} from the soil solution. Nitrogen is also taken up as ammonium ions (NH_4^+) from the soil solution and is obtained from fixation of atmospheric nitrogen gas in plants with symbiotic nitrogen-fixing bacteria
- Other elements (such as copper, Cu^{2+}) are obtained as ions from the soil solution.

Two essential features emerge from this. First, plants take up most of their nutrients from the soil solution, and not from the solid soil materials. Second, the nutrients are taken up predominantly as inorganic ions, not as organic ions or attached to organic matter. A further feature of nutrient uptake by plants is that it is generally highly selective. Membranes in root cells are selective about what they allow into the plant. Some ions (for example, phosphate) are present in very much greater amounts inside the plant than would be anticipated from their concentration in soil solution, while others (for example, calcium) are present in much smaller amounts. In turn, this selectivity at the root surface means that some ions (for example calcium and magnesium) tend to accumulate there as a consequence of the mass flow of soil solution to meet the demand of water for transpiration, while other ions (such as phosphate and nitrate) are depleted, resulting in a concentration gradient which permits diffusion towards the root surface.

Sources of major nutrients in soils

Because potassium, phosphorus and nitrogen are the major macronutrients, this section concentrates on the processes that control their replenishment in the soil solution.

Earlier sections of this chapter showed that the clay minerals and the organic matter in the soil have negatively charged surfaces. This property is very important in maintaining chemical fertility, especially for the retention of positively charged nutrient ions such as potassium (K^+). Potassium ions are held on the negatively charged surfaces of the clay minerals and replenish the soil solution as K^+ is removed from the solution by plant roots. Most soils contain quite large quantities of potassium in minerals such as feldspar, but this potassium is unavailable to plants. It only becomes available as it is weathered from the crystalline minerals and is held on the negatively charged surface of clay minerals. For K^+, then, the replenishment of the soil solution is essentially an inorganic, chemical process.

Phosphate in soil solution is present as two negatively charged ions, $H_2PO_4^-$ and HPO_4^{2-}, over the usual range of soil pH values, so it is not held by the negatively charged surfaces of clay minerals. The main feature of soil phosphate is its very low solubility, because these ions bind strongly to iron and aluminium sesquioxides and calcite to give very low concentrations in soil solutions. Typically, the concentration of phosphate in solution is about 10–100 less than that of K^+ and nitrate. As a result, phosphate supply is a major limitation to plant growth worldwide, and plants have evolved many mechanisms (including the release of organic acids and solubilising enzymes) to help in its acquisition.

Phosphorus (P) is also present in soil organic matter and this is released to the soil solution by microbial decomposition. In some tropical regions, the organic forms of P are a major contributor to plant P uptake but, in most places, although the total amounts of inorganic and organic P are similar in a soil, the inorganic minerals dominate in providing plant-available P. The processes leading to the release of organic P to the soil solution are similar to those for nitrogen (see below).

In contrast to both K and P, the processes

controlling plant availability of nitrogen (N) are biologically mediated, rather than dominated by inorganic chemistry. Although soil minerals contain some N, it is the soil organic matter that dominates the supply of N in the soil. This N is part of a very dynamic system that moves through a variety of forms as a result of the activities of plants and microorganisms. Plants and microbes take up N from the soil solution predominantly as either ammonium (NH_4^+) or nitrate (NO_3^-) and convert it, together with any nitrogen fixed from the atmosphere, into organic nitrogen. When the organisms die, the organic materials returned to the soil are, in turn, decomposed, and the organic nitrogen converted back to ammonium (in a process known as *mineralisation*), which is then normally rapidly followed by nitrification to nitrate by specific bacteria (Fig. 4.11). The nitrate produced is available for plant uptake, but if the microorganisms cannot satisfy their needs for nitrogen from the organic compounds on which they are 'feeding', they will use this nitrate themselves, and re-convert it to organic nitrogen. This process is known as *immobilisation*. Microbes and plants, then, compete for the inorganic nitrogen (NH_4^+ and NO_3^-), depending on the nitrogen-richness of the soil organic materials available to the microbial population.

This dynamic set of processes, mediated by the soil microorganisms, is at the heart of nitrogen management in the garden. Understanding something of the nitrogen requirements of the microorganisms will assist in effective management of nitrogen for plants.

The organic matter in soils (which includes the microorganisms) has a conservative carbon: nitrogen ratio of about 10:1 to 12:1 (the range worldwide is 8–15:1). The carbon:nitrogen ratio in plant materials returned to soil is typically 20:1 to 30:1 in green plants and farmyard manure, about 100:1 in straw, and as high as 400:1 in sawdust. In microorganisms the ratio is more stable and much lower at between 4:1 and 9:1. It follows that there will be keen competition between microorganisms for soil nitrogen when residues rich in nitrogen are added to soils. For gardeners, this has two practical consequences:

- Most compost heaps contain plant materials both rich in nitrogen (early spring grass cuttings) and low in nitrogen (autumn prunings), so mixing will promote more even decomposition. Overall the carbon:nitrogen ratio will be higher than 10:1, so a compost enhancer (essentially, a source of nitrogen) will promote decomposition
- Addition of fresh organic materials to soil (such as farmyard manure) will, in the short term, reduce the availability of nitrogen to

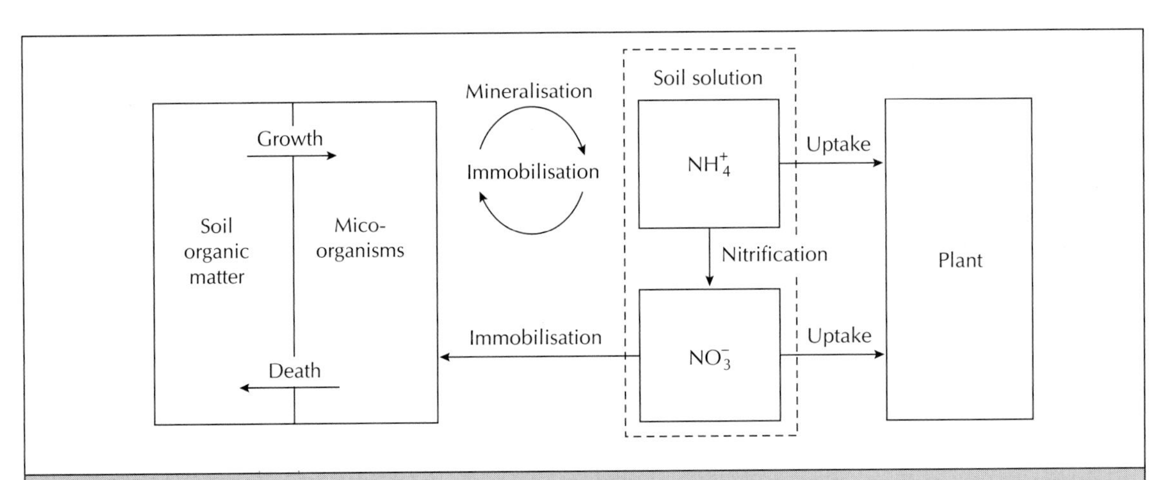

Fig. 4.11 Diagrammatic representation of the cycling of nitrogen in soils. Most plants take up their nitrogen as inorganic nitrate and ammonium from the soil solution, which is replenished by the breakdown of organic matter by microorganisms.

plants, as nitrate is immobilised by the population of microorganisms.

A final practical point is that because the activities of soil microorganisms dominate the mineralisation process to produce nitrogen that is available to plants, it is impossible at present to predict the rates and amounts of inorganic nitrogen that will be produced. There is no soil test available to assist with this prediction.

Maximum rates of mineralisation are achieved during:

- warming of moist soils in the spring
- re-wetting of warm soils in the autumn (particularly if cultivated)
- additions of fresh plant material.

Soil organic matter also contains P and sulphur (S) in a combination of about 100C: 10N: 1 P and S. So, as nitrogen is mineralised, smaller quantities of P and S are also made available for plant and microbial uptake.

Fertilisers, manures and composts

The purpose of adding fertilisers, manures and composts is primarily to provide a supply of essential nutrients for plants; a subsidiary purpose is to maintain the biological activity of the soil. Most garden soils in the UK and Northern Europe contain ample resources of micro-nutrients and attention can be focused on N, P and K. In other parts of the world, micro-nutrients are often a more important issue, and specialised products exist to remedy the deficiencies.

Fertilisers

A *fertiliser* is defined as any material that provides a concentrated source of one or more essential nutrients. Fertilisers are often divided into 'organic' or 'inorganic' products in an attempt to differentiate them by source. 'Organic' indicates that the material is of plant or animal origin (for example, bone meal or hoof and horn), while 'inorganic' refers to materials which have been chemically synthesised (such as ammonium nitrate) or are the result of processing mineral deposits (for example, superphosphates). Fertilisers are available as *straight fertilisers*, which supply only one or two nutrients (see Table 4.5), and as *compound fertilisers* which provide a more complete range of the elements essential for plant growth (especially N, P and K). An example of an inorganic compound fertiliser is 'Growmore', which supplies mainly nitrogen, phosphorus and potassium. In contrast, 'fish, blood and bone' may be considered to be an organic compound fertiliser, being made up of fish meal, dried blood and bone meal, and designed to supply a broad range of major nutrients and micro-nutrients in an organic form.

Table 4.5 The composition of some common inorganic fertilisers.

Name	N, P or K concentration (%)	Nutrients supplied
Nitrogen fertilisers		
Ammonium sulphate	21	Ammonium plus some sulphur
Ammonium nitrate	32–34.5	Ammonium plus nitrate
Calcium ammonium nitrate (CAN)	25	Ammonium, nitrate plus some calcium
Urea	46	Converted to ammonium
Phosphorus fertilisers		
Single superphosphate	8–9	Phosphate, calcium plus sulphur
Triple superphosphate	20	Phosphate plus calcium
Diammonium phosphate (DAP)	23	Phosphate plus ammonium
Potassium fertilisers		
Potassium chloride (muriate of potash)	50	Potassium plus chlorine
Potassium sulphate	42	Potassium plus sulphur

Fertilisers may also be differentiated by the rate at which the nutrients they contain are made available for plant uptake. Although there are exceptions, the straight inorganic fertilisers tend to be highly soluble and the nutrients they contain are in forms which are readily available to plants, once applied to a moist soil. In contrast, many organic fertilisers contain a large proportion of their nutrients bound up as complex organic compounds, which are made available for uptake only slowly. Common exceptions to these generalisations are fertilisers such as dried blood, which is rapidly mineralised, rock dusts, which are finely ground insoluble mineral deposits that require chemical weathering to render their nutrients available, and *controlled-release fertilisers*, which release their inorganic nutrient load over an extended period, depending on temperature.

The objective of fertiliser application is to supplement the available nutrient reserves in a soil in order to ensure that optimum nutrient levels are available for plant growth at the right time. Exactly which nutrients need to be supplied, and at what time of year, is dependent upon a variety of factors including the plants being grown, the type of soil they are growing in, the prevailing climate and the existing nutrient reserves.

Much research has been done on commercial crops with the aim of providing accurate recommendations about what fertiliser to supply, how much to use and when to use it. These recommendations are extremely valuable to professional growers, since under-application of nutrients can lead to reduced yields and products of poor quality, while over-application leads to excessive wastage of fertiliser, which in turn leads to lost profits. For the individual gardener, fertiliser usage is small by comparison, but it still makes sense to employ good judgement when using these chemical compounds.

Professional growers depend upon regular soil and plant analysis to ensure they are applying exactly the right amounts of fertilisers, but the cost of laboratory analysis means that most gardeners will only do this occasionally. Perhaps the best time for having a soil analysis carried out is when buying a new area of land whose cultivation history is unknown. Another situation is where heavy vegetable cropping has taken place over a prolonged period, and nutrients and organic matter may be depleted. In practice, rather than spending money on soil testing, many gardeners make 'insurance dressings' of fertiliser just in case they are needed. This is understandable but wasteful, and can lead to toxic or unbalanced levels of nutrients developing in the soil.

When purchasing fertilisers, it is a good idea to read all of the small print on the label, both to ensure that the product contains the correct nutrients, and that it represents value for money. By law, the container must state the composition of the contents. Table 4.5 gives the N, P and K content of most common single fertilisers. Unfortunately, the law obliges manufacturers to specify the composition not as N, P and K but as N, P_2O_5 and K_2O. This is an unhelpful historical anomaly dating back to the times when chemists carried out analyses in terms of the mass of an oxide. The oxides can be converted to P by multiplying P_2O_5 by 0.44, and to K by multiplying K_2O by 0.83. So, for example, 'Growmore' is labelled as 7:7:7 and contains 7% N: 7% P_2O_5: 7% K_2O or 7% N: 3% P and 5.8% K.

Because many gardens have received inputs of P and K over prolonged periods, building up the soil's reserves of bound P and exchangeable K, there is generally much less need for P and K fertilisers now than in the 1950s and 1960s. Results of tests on garden soils sent in by members of the Royal Horticultural Society during 2000 showed that of the 289 samples supplied, 72% contained sufficient P to produce most vegetable crops without additions of P fertiliser. Only 30 samples contained low amounts of P. Relatively more emphasis should be placed on supplying adequate nitrogen at the appropriate times.

Manures

Whereas fertilisers are basically a source of mineral nutrients only, manures and composts have additional attributes that contribute to the fertility of a soil. Traditionally, the term *manure* referred to any material applied to soil to improve fertility. Today we tend to use it when referring to animal dung and urine. Both these waste products are valuable sources of major nutrients and trace elements, so that manures are useful fertiliser materials. However, because most manures are mixed with carbon-rich bedding materials such as straw and wood shavings, they can also supply significant amounts of organic matter to the soil. The relative value of a particular manure depends on:

- the animal species from which it was obtained
- the amount and type of bedding material mixed with the dung
- and the degree of decomposition that has taken place before use.

As described earlier, the use of fresh straw-rich manure can reduce the amount of plant-available nitrogen due to immobilisation.

In terms of their value as fertilisers, manures are a valuable source of N, P and K, as well as magnesium, sulphur and micronutrients. The quantities of these in manure vary with both the species of animal and the bedding material used (see Table 4.6). Much research has been done on nutrient contributions from cattle, pig and poultry manures, because these are the types most frequently used by farmers. However, the manure most frequently used by gardeners is horse manure, together with the waste products of domestic pets such as rabbits and pigeons. Where nitrogen-rich manures are applied, such as those derived from poultry, or where there is little bedding material present to dilute the manure, there is a risk that ammonia gas may be released following application. Ammonia is toxic to plants and can cause leaf scorch and root death, especially in seedlings and young plants. Ammonia loss from the soil is most likely to occur when the weather is warm and dry and when the manure has been left on the surface rather than being incorporated. It is more prevalent in sandy, alkaline soils and those low in humus.

Because of the complex microbial processes of mineralisation and immobilisation described earlier, only a portion of the applied manure is of value as fertiliser in the season of application. However, a large reserve of nutrients can gradually be established in the soil if manuring is introduced as an annual gardening activity.

Composts

The term *compost* has two meanings for the gardener; it can refer either to a medium in which container plants are grown (a 'potting compost'), or to the product of the composting process. The second meaning is employed here. Although any organic material can be composted, garden compost is usually a mixture of composted garden and kitchen waste, sometimes supplemented with animal manure. Compost supplies a low level of nutrients, and its main value is as a source of humus (stabilised organic matter) which can help to improve soil structure, support soil life and recycle nutrients.

Composting is the rapid aerobic decomposition of organic matter by microorganisms at elevated temperatures. It differs from natural decay processes both in terms of the speed of decay and the heat that is generated. Whereas leaves on a woodland floor can take months to decompose, decay will occur in an active compost heap in a matter of weeks.

For a compost heap to function effectively, a large volume of undecomposed raw material has to be accumulated before the heap is built, and it needs to be a good mixture of both soft, nitrogen-

Table 4.6 The composition of some natural fertilisers and animal manures (note that many natural fertilisers do not contain potassium).

Material	%N	%P	%K
Dried blood	12–14		
Bone meal	6–10	8	
Fish meal	7–14	4–7	
Wood ash	0.1	0.13	0.8
Farmyard manure (cattle)	0.9	0.18	0.5
Farmyard manure (pigs)	0.6	0.26	0.33

rich materials such as grass clippings and kitchen scraps (ideally 25–50%) with drier, carbon-rich materials such as woody prunings, wood chippings and straw (50–75%). When the compost heap is built, the large volume of organic material acts as an insulating layer, allowing heat generated by microorganisms contained within to build up. Accumulating a large quantity of fresh material at the start provides an ample supply of 'food' material for the microorganisms to metabolise, and this allows a massive flush of microbial growth over a short time, resulting in the characteristic rise in temperature. For the reasons described earlier, the C:N ratio of the materials to be composted is important.

The heat generated within a compost heap has two effects. Up to about 45°C, the increasing temperature speeds up the biochemical reactions taking place and kills off undesirable pathogens and weed seeds. The degree of pasteurisation increases as the temperatures continue to rise, but at temperatures greater than 55°C the microorganisms start to decline. Although some microorganisms will function at much higher temperatures, the rate of composting slows down. For this reason, commercial composting heaps are normally maintained at a maximum temperature of 55–65°C. The compost should also be turned, to prevent it becoming either too wet or too dry.

Making compost in gardens is often a small-scale process, using lower temperatures and involving the gradual addition of unselected materials over a long time. For a small-scale garden compost heap, it is more realistic to expect a batch of compost to be ready in eight to twelve months, rather than the three to four months anticipated in a commercial operation. Having said this, garden composting is still a very effective means of recycling garden and kitchen waste materials, diverting these away from the domestic waste stream and boosting soil fertility at no cost other than that of time. The resulting material is an excellent soil conditioner and mulch, although it will usually still contain viable weed seeds; decomposition, particularly of tough vegetable matter, may remain incomplete.

Composting bins and wormeries are available to gardeners for the production of small quantities of high-grade compost but these are too small and too expensive for many. The following guidelines may be useful for improving the quality of compost produced in a typical compost heap:

- accumulate as much mixed raw material as possible for adding to the heap at one time
- build the heap over bare soil and include some old compost with the raw materials to ensure rapid colonisation by the necessary microorganisms
- mix all the materials together and chop up or shred large pieces to speed up decomposition
- Protect the heap from the rain with a lid or cover to avoid waterlogging, especially over the winter months
- turn the heap once, or if possible twice, early in the composting process.

MANAGING SOIL pH

Most soils will become acidic if they are exposed to rainwater for sufficient time, and drainage occurs. Acidity, as strictly defined, depends on the concentration of hydrogen ions in a solution. The cation exchange capacity of clay minerals and organic materials, allows hydrogen ions to be adsorbed and to exchange with other ions in the soil solution. The *pH of a soil* is a measure of the hydrogen ion concentration in the soil solution. The scale used is logarithmic, so that a difference of one unit of pH reflects a ten-fold difference in concentration. By definition, the pH of pure water is 7 and is neutral, while acid solutions have a pH below 7 and alkaline solutions have a pH greater than 7.

Acidification is a natural process; it occurs because water in contact with atmospheric carbon dioxide produces a solution of dilute carbonic acid with a pH of 5.6. In heavily polluted air the pH may be much lower (typically 4.4 over eastern Britain in the early 1990s), and even in unpolluted air rain picks up small quantities of naturally occurring acids, resulting in a pH of about 5. As the rain enters the soil it is affected by numerous other chemical processes, acting both as sources of and sinks for acidity. The presence of sinks for acidity, notably the presence of large quantities of basic minerals and exchangeable calcium in many soils, has the effect that the pH of the soil does not immediately fall to that of the rainwater, but is *buffered* (i.e. kept at a near constant value) for a period of time. Generally, clay soils and soils with appreciable amounts of soil organic matter are better 'buffered' than sandy soils.

Acidification of soils and the management of pH are important, because as soils grow more acidic, there are several associated changes in their properties.

- The amounts of exchangeable calcium and magnesium decrease. These are essential nutrients
- The amount of exchangeable aluminium increases. Aluminium is toxic to plant roots
- The negative charge on humus decreases and the positive charge on sesquioxides increases
- The availability of many plant nutrients (for example, phosphate) is reduced, while that of manganese may be increased to toxic levels
- The activity of many soil organisms is reduced, resulting in lower availability of N, P and S and accumulation of undecomposed organic matter.

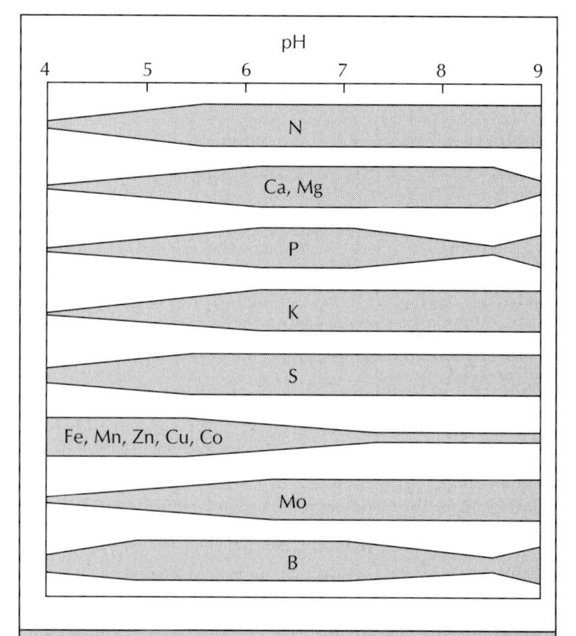

Fig. 4.12 Relationships between the availability of plant nutrients and soil pH. The widest portion of the bands indicates the most ready availability of nutrients to plants. Overall, it is evident that a pH range of about 6 to 7 will ensure the best availability of nutrients.

Figure 4.12 shows the effects of pH on the availability of the essential nutrients. These considerations mean that the optimum pH for many common plants growing in soils is about 6.5 and in peats about 6. Maintenance of the correct pH wherever possible is therefore essential if the optimum benefit is to be obtained from any applications of fertiliser or manure.

Plants vary considerably in their tolerance of acid or alkaline conditions, and some diseases are also affected by soil pH. A common example of a pH-related nutritional problem is lime-induced chlorosis, which develops under alkaline soil conditions when the trace elements iron and manganese are rendered insoluble. Susceptible plants, such as acid-loving rhododendrons and azaleas (see Chapter 5 for other examples), are unable to absorb these nutrients in sufficient quantities for their needs, resulting in a characteristic yellowing of the foliage between the veins, particularly at the growing tip. In this case, the most effective treatment is to amend the soil with a compound such as elemental sulphur which is converted to an acid, thereby reducing the pH and increasing the solubility of iron and manganese. As an alternative, in the early stages a 'chelated' fertiliser mix may be applied; this supplies the nutrients combined with a complex organic molecule known as a *chelate*, rendering them soluble despite the pH of the soil itself.

Where soils are too acid, the chief means of managing soil pH is through the application of lime (calcium carbonate, $CaCO_3$). The amount of lime needed depends on the soil pH, the buffer capacity of the soil and the target pH. Strictly speaking, a laboratory test is required for accurate results, but Table 4.7 provides useful guidance based on extensive experience of field trials in England and Wales. It shows, for example, that

Table 4.7 The amounts of lime to add (kg m^{-2}) to bring the soil to the optimum pH (6.5 for most soils but 6 for peat) for the growth of most vegetables and flowers. The pH shown on the left is the measured pH of the soil before the addition of lime (from MAFF publications).

Measured pH	Sandy soils	Loamy soils	Clayey soils	Peats
6.5	0	0	0	0
6.2	0.3	0.4	0.4	0
6.0	0.4	0.5	0.6	0
5.8	0.5	0.6	0.7	0
5.5	0.7	0.8	1.0	0.8
5.2	0.9	1.1	1.2	1.3
5.0	1.0	1.2	1.4	1.6
4.8	1.1	1.3	1.5	1.9
4.5	1.3	1.5	1.8	2.4
4.2	1.5	1.8	2.0	2.9
4.0	1.6	1.9	2.2	3.2

if the measured pH of a sandy soil is 4.8, then 1.1 kg m^{-2} should be added to bring the pH to the optimum value of 6.5. In the east of England, where annual drainage is around 200 mm, losses of lime by leaching will be smaller than in the southwest, where drainage of 1000 mm may occur. Liming, then, needs to be more frequent in western England, or plants that are more tolerant of acid must be grown.

A final point is that ammonium fertilisers such as ammonium sulphate and ammonium nitrate produce hydrogen ions as they are nitrified by microorganisms to nitrate. For every 1 kg of fertiliser nitrogen applied as ammonium, about 4 kg lime is required to neutralise the acidity produced. This is not to say that lime should be added at the same time as fertiliser (it should not, because there is a strong possibility that the ammonium will be converted to ammonia gas and be lost to the atmosphere) but rather that the natural processes of acidification will be enhanced when ammonium fertilisers are used.

MANAGING WATER

An earlier section in this chapter showed that soils contain a network of interconnected pores in which the largest pores do not retain water, the smallest pores retain water too strongly for the water to be available to plants, and only the medium-sized (50–0.2 µm in diameter) pores retain water available to plants (see Table 4.1). This wide range of pore sizes is essential for a large number of biological processes that occur in soils, and includes the following associations.

- animal burrows – rapid movement of water during periods of flooding ('mole' drains are common in clay soils)
- earthworm channels – freely drained allowing rapid movement of water when water-filled, and an easy supply of oxygen for aerobic respiration
- root channels – the pores in which they grow are also freely drained, and therefore well aerated
- storage pores – water is held in close association with particle surfaces, allowing rapid replenishment of nutrients when uptake occurs. Water movement can occur over a wide range of water contents
- bacteria and other microorganisms – live in storage pores; most require a water film to function normally, as well as a supply of oxygen.

A well-structured soil contains a distribution of pore sizes that allows for these functions and processes. Water management in soils is essentially about ensuring that the largest pores can operate to remove excess water during wet periods, and that the storage pores are re-filled with water during dry periods.

Excess water can be removed from gardens, and waterlogging eliminated, by the installation of a drainage system. The necessity for such a system is largely confined to those clayey soils, where swelling of clay minerals occurs on wetting to reduce the volume of transmission pores. The solution is to install a system of pipes or channels which provide a stable system of pores to direct the water away from the site. The outlet of the drainage system must, however, be above the level of any water table or water in an open watercourse.

Occasional waterlogging can also occur on other types of soils, although this is frequently the result of past mismanagement of the site. The solution is to create an interconnecting system of pores. For example, during construction work it often happens that the use of machines or repeated treading on an area smears over the transmission pores, causing them to close. If topsoil is then placed on top of this, waterlogging will often ensue, because the interconnectedness of the transmission pores is disrupted at the interface. Sometimes this problem resolves itself (roots dry the soil in summer, re-forming a network of cracks), but light cultivation of the interface with a fork to recreate some large pores is a more reliable and faster method of amelioration.

Controlling excess water is important not only because of the direct influence of the water on the aeration status of the soil, but also because it has an effect on the speed with which soils warm up. Growth of plants in early spring is often related to temperature. Clayey soils in temperate climates warm up more slowly than loams in spring, because their high water content and high thermal capacity mean that for a given input of energy from the sun, the temperature increase is less. Clayey soils are sometimes colloquially described as 'cold' for this reason.

The water available to plants in a freely drained soil is determined by the volume of the storage pores. Many measurements of this volume have been taken in many soils, so that it is possible to make generalisations about the likely storage capacity of soils with different textures (see Table 4.8). For many loam soils the amount of water available to plants (the *Available Water Capacity*) is about 20–22 mm for every 100 mm depth of soil. In sands and loamy sands, it is about half this value, and in clays, silty clays and sandy loams, it is roughly three-quarters of this value (see Table 4.8).

The frequency of watering in the garden depends on the size of the store and how quickly it is depleted. As shown above, the size of the store is governed principally by soil texture, and also by the depth of rooting. So a plant with its roots confined to the topsoil (usually about 250 mm deep) has only half the amount of water available to it compared with a plant that can root to a

Table 4.8 The amount of water available in different soils, the number of days taken to deplete soil water to a point where growth of vegetables and flowers might be adversely affected, and the amounts of water required to replenish the soil store. The values in column 3 were calculated assuming that the rooting depth was 0.5 m, the potential rate of evaporation is 3.5 mm per day, and that only 0.25 of AWC can be depleted before growth is adversely affected.

Soil texture	Available water (mm water per 100 mm depth of soil)	Days to deplete stored soil water	Volume of water required to replenish soil water store (litres per square metre)
Sand	10	3.5	12.5
Loamy sand	12	4.0	15.0
Sandy loam	14	5.0	17.5
Other loams	22	7.5	26.0
Clay and silty clay	16	5.5	20.0

depth of 500 mm (half a metre). This is one reason why horticulturists should pay attention to the subsoil. The time for which the store will last depends on the evaporation rate of the atmosphere. In the UK, this varies from about 1–2 mm per day on a bright sunny day in April to about 4–5 mm per day on a bright sunny day in June and July. Typically, of course, the weather is mixed and days are rarely wholly sunny, so that a reasonable long-term approximation of summer conditions is 3.5 mm per day. So a plant with a rooting depth of 500 mm growing on loam in a UK summer has sufficient water for 31 days ($[22 \times 5]/3.5$) before it will wilt permanently. In reality, the plant will start to suffer reduced growth long before all of the available water is exploited. Plants like potatoes and many vegetables and flowers are adversely affected once about a quarter of the available water has been used, while grass growth is reduced when about half of the available water is used.

Table 4.8 shows that plants will need to be watered about every three to seven days to prevent water shortage restricting growth, depending on the type of soil. It also shows the volume of water that needs to be added to replenish the store of water once it has become depleted. The quantities are substantial, and often underestimated by amateur gardeners. Essentially, watering should be conducted infrequently in large amounts, rather than frequently in small amounts. This is because small amounts of water stay in the soil close to the surface, and can easily be lost by evaporation without ever passing through the plant. A large amount of water is more effective, as long as:

- The quantity is not unreasonably excessive; adding water beyond the maximum storage capacity of the soil (Field Capacity) wastes both water and the nutrients leached as it drains into the water table.
- The rate of application must be slow enough for the water to infiltrate, not run off the area.

Besides replenishing the store by watering, there are other ways of influencing soil water content. Where soils are bare, applying a mulch to the surface will reduce the rate of water loss from the soil. The mulch needs to be coarse enough and thick enough to ensure that the hydraulic continuity of the soil pore system is broken between the soil and the atmosphere. Mulches of coarse organic material some 5–10 mm thick can assist the endeavour but, unless they have a low C:N ratio (for example, bark), will be incorporated into the soil by organisms within 12 months, and thus lose their effectiveness. Additions of organic materials to soil can increase the size of the soil water store. This results from the twin effects of organic materials containing large amounts of storage pores and the aggregation of soil materials to create storage pores. In studies at an agricultural scale, it has been found that it requires the application of fairly large quantities of material over sustained periods in order for there to be any measurable impact at a particular site. For example, at Horticulture Research International, Wellesbourne, UK, applications of 50 tonnes ha^{-1} of farmyard manure applied twice a year for five years on a sandy loam soil increased the Available Water Capacity of the upper 150 mm by only 5 mm (the equivalent of about one day's worth of evaporation in summer). Given the quantities of manure available to most gardeners, the effects on water storage *per se* are likely to be insignificant if the manure is applied broadly. Limited benefits can be gained by burying large volumes of organic material close to specific plants at the time of planting, although the benefits will disappear with time as the organic materials are decomposed.

Of course, an alternative means of managing soil water is to select plants that will survive and flower even if summer drought occurs. With the current awareness of possible changes in climate and generally drier summers in southern Britain, horticulturists and the water supply industry are both promoting this approach. A wide range of flowering plants is available (see Chapter 5) but so far there have been few attempts to introduce more drought-tolerant vegetables.

FURTHER READING

Brady, N.C. (1990) *The Nature and Properties of Soils*. 10th edn, MacMillan Publishing Co., New York.

Marschner, H. (1995) *Mineral Nutrition of Higher Plants*, 2nd edn. Academic Press, London.
Rowell, D.L. (1994) *Soil Science: Methods and Applications*. Longman Scientific & Technical, Harlow.
Scaife, A. & Turner, M. (1983) *Diagnosis of Mineral Disorders in Plants. Volume 2: Vegetables*. The Stationery Office, London.
Wild, A. (1993) *Soils and the Environment: An Introduction*. Cambridge University Press, Cambridge.

5

Choosing a Site

- Plant adaptations to their environment
- Water: the need for water and the role of the stomata
- Stomatal opening and how it is controlled by light and carbon dioxide
- Water stress and the role of abscisic acid
- Mechanism of water loss from the leaves
- Strategies for conserving water
- Succulence and crassulacean acid metabolism (CAM)
- Water conservation by C-4 plants
- Excess water – waterlogging
- Aquatic plants
- Soil pH: calcicole and calcifuge plants
- Light: shady situations; how plants detect shade; the role of phytochrome
- Shade adaptors and shade avoiders
- Choosing plants for particular situations

Because they are unable to move and find new and better habitats, plants have evolved a number of ways to survive in a variety of different conditions. This makes it possible to choose plants that can grow well in the environment that they occupy. Selecting the plant for the site is thus an important factor in growing healthy plants with the minimum of attention. To some extent the site can be improved (for example, by making peat beds for acid-loving plants, or installing irrigation systems in dry habitats) but the better and more environmentally-friendly option is to live with what you have, and to try to match the plant to the conditions in which it is to be grown. Among the most important local factors are the availability of water, pH of the soil, the amount of light and the degree of exposure to wind. These are discussed in this chapter. Seasonal factors such as day-length and temperature are discussed in Chapter 9.

WATER

How water is lost from the leaf

The basic features of water movement in the plant and its loss through transpiration are discussed in Chapter 1. Here, it is only necessary to restate the central problem. Plants require a supply of carbon dioxide for photosynthesis. This means the provision of entry pores (the *stomata*) for the gas and it follows that water vapour is lost to the atmosphere through these open channels. The plant must, therefore, be able to maintain an adequate supply of carbon dioxide to the photosynthesising cells in the leaf while preventing a net loss of water. In this context, it is important to note that the amount of water contained in the majority of land plants is small, compared with the potential loss of water by transpiration on a dry, sunny day. If water is lost from the leaves at a greater rate than it can be supplied from the roots, *turgor* is lost and wilting occurs. When the guard cells around the stomata lose water and become flaccid, their physical structure means than the outer surfaces move together and close the pore. Obviously this reduces the loss of water vapour from the leaf, especially where the cuticle is thick and impervious to water vapour, and restores the water balance as long as there is available water present in the soil (see Chapter 4). The guard cells then absorb water again and re-open.

The way in which the opening and closing of stomata is controlled has been found to be quite complex and involves both physical and biochemical factors. This is not surprising, since the regulation of water loss by stomatal movements is a major factor in determining the survival of the plant. If the stomata remain open, the plant can wilt and may die from dehydration. If the stomata remain closed, there is no photosynthesis and the plant will ultimately die from starvation.

Stomata open because the guard cells absorb water. Many studies have shown that they absorb water from the surrounding cells by *osmosis,* because there is an increase in the amount of dissolved substances present inside the guard cells. It is now clear that the main increase is from potassium ions that move into the guard cells from the surrounding cells. Fundamentally, therefore, the opening and closing of stomata depends on the movement of potassium ions into and out of the guard cells.

The stomata of most plants open in the light, allowing the entry of carbon dioxide for photosynthesis, and close in darkness to reduce water loss. They are mainly sensitive to blue light, which results in a larger and faster response than when they are exposed to red light. Exposure to light causes potassium ions to build up inside the guard cells. A reduction in carbon dioxide in the air also increases the content of potassium in the guard cells. Thus, stomata open in the light in part as a direct response to light and, in part, because the concentration of carbon dioxide within the leaf is reduced by photosynthesis. Darkness and an increase in the amount of carbon dioxide have the opposite effect.

Water stress

An insufficient supply of water to the roots, or a too-rapid loss from the leaves, leads to water stress, a common problem in horticulture. When leaves are subjected to water stress, the hormone abscisic acid (ABA), begins to build up in their tissues. The application of ABA results in closing of the stomata. On this basis, it was originally thought that ABA is produced in the leaves when they experience a water deficit, resulting in stomatal closure and a reduction in transpiration. More recently, though, it has been found that ABA can control water loss before any measurable water stress occurs in the leaves. It appears that, when they begin to encounter dryness in the soil, the root tips generate a signal which is transported to the leaves and results in partial closing of the stomata and a reduction in leaf expansion, both of which improve the water economy of the plant. The general consensus is that this signal from the roots is ABA and it is probably true to say that the regulatory role of ABA is essential for the survival of land plants in all but the least stressful environments. This is supported by the discovery of the so-called 'wilty' mutants, which are deficient in ABA and are always partly wilted, unless this hormone is applied artificially.

It seems unlikely, however, that the ABA signal itself is the only story and sensitivity to the hormone also needs to be taken into account. In other words, how do the stomata react to the signal arriving from the roots? This may well be important for future breeding strategies to improve drought resistance, since simply increasing the amount of ABA in the plant will not necessarily work.

There is now compelling evidence that ABA from the roots plays a central role in the control of stomatal opening in the field. The ABA signal is produced as the soil begins to dry out, thus allowing the plant to take corrective action to improve the water economy of the leaves before wilting occurs. This is a good example of the ability of plants to 'detect what is going to happen next', thus allowing them to take action before the major stress arrives. Similarly, the perception of a day-length signal allows plants to prepare for the subsequent stress of low temperature (see Chapter 9) and another example (from shading) is given below. Once water stress occurs and plants wilt, recovery may take some time even when the water balance is restored. This is an important fact in relation to watering and irrigation practice. Many cell functions are depressed when plants are subject to water stress, leading to an often severe reduction in growth and yield (Fig. 5.1).

Does the ABA signal from drying soils have any practical application? It has been found that, if only half of the root system is allowed to dry out while the remaining half is plentifully supplied with water, the stomata still close, presumably in response to the ABA signal from the roots located in the drying soil. This observation has already found some practical use on irrigated vines, where water is supplied to alternate sides in turn. Irrigation is rotated on a two- to three-week cycle so

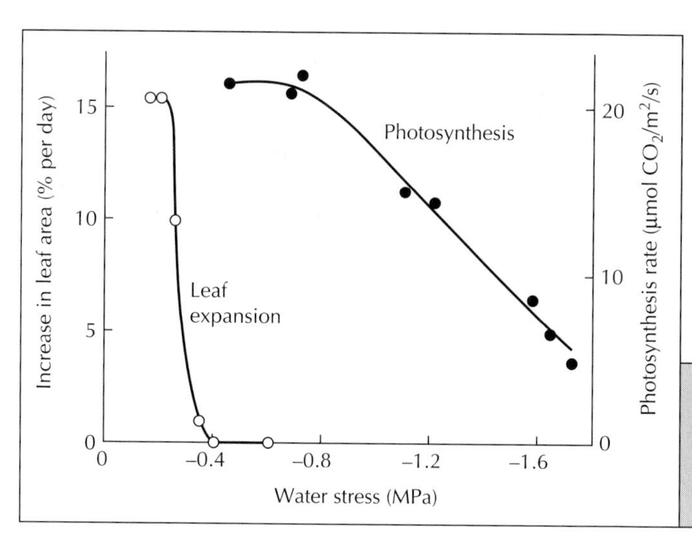

Fig. 5.1 Water stress affects the rates of both leaf expansion and photosynthesis. Leaf expansion is often more sensitive to water stress than photosynthesis, as shown here for sunflower (*Helianthus annuus*).

that first one side and then the other is drying out (Fig. 5.2). The reduces the amount of water used and also results in more and better quality fruit by reducing the excessive leaf growth found in a normally-irrigated crop. The possibility of using this half-root watering system is currently being explored for other crops, such as tomatoes.

Drought tolerance

Natural habitats often suffer from a shortage of water. This may occur in the form of a seasonal drought as in Mediterranean regions, where the annual rainfall (between 350 and 900 mm) occurs almost exclusively in winter, so that the native plants of these regions experience a summer drought. Water is more permanently scarce in arid regions of low rainfall. Many strategies, both biochemical and physical, have evolved to allow plants to survive under such conditions and an understanding of these enables the gardener to choose plants that are more tolerant of, or are actively resistant to periods of water shortage. Some of these strategies are summarised in Table 5.1.

Avoidance

Some plants survive by avoiding or escaping the dry conditions. A good example of drought avoidance in nature is that of the desert ephemerals and annuals (for example many South African composites such as Namaqualand daisies, *Dimorphotheca sinuata*, Plate 11) which escape the dry conditions as seeds. These only germinate when there has been sufficient rainfall for them to grow, flower and set seed and the resultant synchronous flowering of deserts after rain is one of the most beautiful sights in the natural world. In such plants, germination often depends on the leaching of inhibitors from the seed (see Chapter 6).

Other plants may become dormant during periods of drought, frequently surviving by producing underground storage organs (see Chapter 9) which lose only small amounts of water into the surrounding soil (e.g. crown anemone, *Anemone coronaria*). Many grasses (e.g. *Poa bulbosa*) also survive drought by dormancy and leaf die-back.

Succulent plants that store water in their leaves or stems can also be considered as drought avoiders, since during periods of drought they survive on their stored water, which is replenished when conditions allow. Succulents have a mainly shallow root system which dies during periods of water stress, leaving only the supporting roots surviving. When the rains come, new absorbing roots form and refill the cells with water.

Some plants of arid regions avoid drought by having deep roots that reach the permanent water table. Examples of these include alfalfa (*Medi-*

Fig. 5.2 The half-root irrigation method is used on vines in Australia. The irrigation lines are used to supply water to alternate sides on a 2–3 week cycle so that first the roots on one side and then the other are drying out. (Photo, courtesy of W. Davies, University of Lancaster.)

Table 5.1 Some strategies for the adaptation of plants to water stress (Modified from Table 26.5 in *Plant Physiology* by F.B. Salisbury and C.W. Ross, 1992, Wadsworth Publishing Co.).

Strategy				
Drought escapers	**Water spenders**	**Water collectors**	**Water savers**	**Tolerance of dehydration**
seeds or dormant structures	deep roots	succulents with CAM photosynthesis	sunken stomata small leaves hairs shedding leaves reflective leaves grey/blue leaves	seeds lichens creosote bush

cago sativa) and mesquite (*Prosopis glandulosa*) where the roots can extend 7–10 m. Mesquite in particular is very deep rooted and has become a pest in the semi-arid regions of SW United States, where its profligate water use has prevented the re-establishment of agronomically valuable grasses. The success of this strategy depends, of course, on there being sufficient water for root growth at some stage.

Although it is not directly relevant to arid-zone plants, it is worth mentioning here the recommendation that if irrigation is undertaken, it must provide enough water to wet the soil to deeper layers, and not just the surface. Roots are only able to grow in moist soil and when only the surface is moist, they will mainly be present near the surface. As the surface layers of the soil dry out most rapidly, plants will suffer if the roots are only close to the surface.

Toleration

The second major strategy that plants have evolved for dry conditions is to be able to tolerate them. True toleraters are those plants that can withstand a considerable degree of water stress in their cells. For example the creosote plant (*Larra divaricata*), a desert shrub, can lose 70% of its fresh weight before the leaves die, whereas most leaves will withstand only 25–50% loss. Several mosses and some ferns also belong here. The 'resurrection plant' (*Selaginella lepidophylla*), for example, can dry out almost completely and yet becomes metabolically active again on re-hydration.

Other plants can tolerate dry conditions by reducing their water loss into the atmosphere. These plants can be considered as 'water savers' (Table 5.1) and they conserve water in a number of different ways, using both biochemical and physical strategies.

Physical strategies that conserve water

Water is lost from the leaves because of a gradient in water vapour content between the internal spaces of the leaf and the surrounding air (see Chapters 1 and 7). Anything that affects this gradient will, therefore, alter the rate of loss of water vapour from the leaves. The air in the sub-stomatal cavity is normally saturated with water vapour (100% relative humidity) and so water vapour diffuses into the surrounding air when this has a lower water vapour content. As discussed in more detail in Chapter 7, water vapour is lost even into a saturated atmosphere when the leaf is at a higher temperature than the surrounding air. When water vapour diffuses into still air, a *boundary layer* with a higher content of water vapour forms between the leaf surface and the surrounding air. Based on these facts, we can understand many of the physical strategies that are designed to reduce the rate of water loss.

The presence of hairs helps to maintain the boundary layer and so increases the distance between the leaf cells and the external atmosphere. This increase in the length of the *diffusion pathway* means that it takes longer for water vapour to reach the external atmosphere and so reduces the rate at which water vapour molecules are lost. Hairs also reduce the drying effect of wind, which tends to blow away the protective boundary layer and increase the rate of water loss. The holm oak (*Quercus ilex*), a characteristic species of the Mediterranean flora, has small leaves (see below) with dense hairs on the undersurface where most of the stomata are. Other drought-tolerant plants such as *Lavandula lanata* and *Phlomis fruticosa* have soft leaves, often with hairs on both surfaces.

The stomata of many drought-resistant plants are sunk below the main surface of the leaf. Examples include oleander (*Nerium oleander*) and many pines (e.g. *Pinus strobus*). This increases the length of the diffusion pathway from inside the leaf to the outside air and reduces the rate of water loss. The diffusion pathway is also lengthened in leaves that roll under in response to drought (for example in *Rhododendron*) because most of the stomata are on the lower surface of the leaf.

Perhaps the most obvious way of decreasing water loss during periods of drought is to reduce the leaf area by shedding leaves. Drought deciduous plants are found both in arid zones and in regions with seasonal dry periods such as those with hot dry summers. For example, *Lotus scoparius*, a shrub native to California, sheds its leaves in summer and remains dormant until the arrival of the winter rains. This is an extreme example and many plants shed some of their leaves, especially the older ones, in response to water stress. This also occurs in species that are not drought tolerant and often results in unsightly plants with bare basal stems and only a few leaves remaining at the tip of the shoot, a phenomenon that is well known to gardeners. It is thought that the shedding of older leaves in response to water stress is due to an increase in ethylene production by the plant.

Water loss is also decreased by any factor that reduces the heat load on the leaf (see Chapter 7). The obvious one is shade, and shading plants in hot weather or when water uptake is reduced (as in cuttings or newly potted plants) is a common garden practice. Although shading can be disadvantageous because of the importance of light for photosynthesis, water stress and stomatal closure themselves severely depress the rate of photosynthesis (Fig. 5.1). Leaf surfaces that reflect at least some of the incoming radiation also act to reduce the heat load on the leaf. Hence many drought-tolerant plants have shiny leaves (*Quercus ilex*, *Ceanothus* spp.), or grey foliage (*Helictotrichon sempervirens*) that may also be covered with reflective hairs (*Lavandula lanata*). Leaf orientation can also reduce the heat load and some plants (for example soya bean, *Glycine max*) can change the leaf angle away from the sun when the plants are water stressed.

Wind also affects the leaf temperature because the transfer of heat by convection from the leaf surface to the air is most rapid when the boundary layer is thin. In general this layer is thinnest for small leaves and high wind velocities, so that smaller leaves have temperatures closer to the surrounding air temperature than do larger leaves, especially if there is a wind. That small leaves are a factor in drought tolerance is clear from the characteristic small leaves of many of the shrubs of the Mediterranean maquis (Plate 12) and the fynbos of the South African Cape. Well-known examples include Spanish broom (*Spartium junceum*) and many species of *Erica*. Where leaf area reduction is carried to extremes, as in some desert cacti, the stems are often green and able to carry out photosynthesis.

Although wind helps to prevent a rise in leaf temperature and thus can decrease the rate of water loss, the overall effect of high wind is usually to *increase* the rate of water loss, especially when the heat load is low. This is obviously a major factor in exposed sites. The drying effects of wind are particularly evident where the cuticle is thin and does not, itself, offer a significant barrier to water loss. Plants with thick cuticles are better able to withstand the effects of wind on exposed sites. These include many broad-leaved evergreens, such as Portuguese laurel (*Prunus lusitanica*). Shelter from drying winds is obviously of the greatest importance for expanding foliage that has not yet developed a protective cuticle.

The usefulness of these different physical strategies for conserving water depends to a large extent on the conditions in which the plant is growing. When water is short, it is advantageous to the plant to conserve water and maintain a relatively low leaf temperature. Many drought-tolerant plants have small leaves and thin boundary layers, resulting in efficient heat transfer and leaf temperatures close to the air temperature. The rate of transpiration is further reduced by features that increase the length of the diffusion pathway for water vapour, such as sunken stomata and hairs, and by factors such as grey leaves that reflect much of the sun's radiation. When water is plentiful, however, high rates of transpiration cool the leaf by evaporation. Some drought-tolerant plants utilise this strategy by having long roots that can reach the water table and large leaves to maximise the rate of water loss. This strategy is only successful in air that has a relative humidity low enough for transpiration to continue, even when the leaf is several degrees below the ambient temperature.

Biochemical strategies that conserve water

Crassulacean acid metabolism

Many drought-tolerant species have thick, succulent leaves or stems with a low surface to volume ratio. This, together with their thick cuticle, results in low rates of transpiration when the stomata are closed. Keeping the stomata closed during the daytime, when rates of water loss would be high, poses the problem of obtaining sufficient carbon dioxide for photosynthesis. Many succulents have developed a biochemical strategy that allows them to open their stomata at night and still allows them to carry out photosynthesis. This mechanism was first investigated in members of the family Crassulaceae and is, therefore, called *crassulacean acid metabolism* (CAM for short). Its occurrence is, however, much more widespread than this and CAM has been found in at least 26 Angiosperm families and in some ferns. CAM plants normally inhabit areas where water is scarce or difficult to get at, such as arid and semi-arid regions, and salt marshes. It is also found in some epiphytic orchids that must obtain their water from the air.

In contrast to normal photosynthesis, CAM plants open their stomata at night in order to take up carbon dioxide, which is temporarily stored in the form of an organic acid, malic acid. This accumulates in the vacuoles of the leaf cells and increases their osmotic concentration, which in turn allows the plants to absorb and store water, leading to the observed succulence. During daylight when the stomata are closed, the malic acid moves out of the vacuoles and is broken down to release carbon dioxide within the cells. Utilising light energy, the carbon dioxide is then re-fixed into the usual products of photosynthesis, sugars and starch.

CAM is confined to certain plants and is determined by the plant's genes. However, many CAM plants (for example *Clusia rosea*, a tree that begins life as an epiphyte in the rain forest) can switch to normal photosynthesis under favourable conditions, when the stomata remain open longer during daylight hours. When CAM is occurring in these succulent plants, there is little water loss during the day because of thick cuticles and closed stomata, but the absence of cooling by transpiration can lead to high cell temperatures which the plants must be able to tolerate.

C-4 plants

Another variation of photosynthesis which leads to water conservation is that carried out by C-4 plants. This is similar to CAM in many ways, except that, in C-4, both processes occur in daylight. In contrast to C-3 plants, where the initial product of photosynthesis is a molecule with 3 carbon atoms (see Chapter 1), C-4 plants produce an organic acid as the first step. This organic acid, *oxaloacetic acid*, contains 4 carbon atoms, which is why this type of photosynthesis is called C-4.

As with CAM plants, carbon dioxide in C-4 plants is initially fixed into an organic acid, which is stored in the leaf cells and is later broken down to release carbon dioxide inside the leaf. The C-4 pathway is an adaptation to environments with high light intensities and periods of water stress, and is commonly found in plants from desert and sub-tropical regions. The incorporation of carbon dioxide into oxaloacetic acid occurs at high rates when the light intensity and temperature are high, the stomata are open and carbon dioxide is available. When water stress occurs and stomata close, the release of carbon dioxide from the stored oxaloacetic acid enables the plants to maintain a high concentration of carbon dioxide inside the leaf cells and continue to carry out normal C-3 photosynthesis.

Many, but not all C-4 plants exceed most C-3 plants in their productivity, especially under high light and temperature conditions (see Chapter 10). The C-4 pathway is not automatically better than C-3 photosynthesis; indeed, C-4 photosynthesis is less efficient and uses more light energy to fix carbon dioxide. It is, however, particularly advantageous in high light, high temperature environments, where water is limiting.

C-4 plants are found in several groups of tropical grasses (for example in sugar cane and maize) and sedges, as well as in some dicotyledons. The North American prairie grasses that have recently become very popular in British landscapes are C-4 plants. These include, for example, little bluestem (*Schizachyrium scoparium*), Indian grass (*Sorghastrum nutans*) and *Sporobolus heterolepis*. They grow best in high summer and are drought tolerant. Several families (e.g. Chenopodiaceae and Amaranthaceae) contain both C-3 and C-4 species, suggesting that the C-4 photosynthesis pathway has arisen independently in different plant groups and is a recent evolutionary development.

EXCESS WATER

While lack of water can be the major environmental stress in many habitats, excess water can also pose many problems. Hairy leaves become waterlogged in heavy rain and may rot, while most plants cannot tolerate waterlogged soils due to their lack of oxygen.

Roots normally obtain their oxygen from gas-filled pores in the surrounding soil. When the soil is well drained and has a good structure (Chapter 4) these pores permit the diffusion of gaseous oxygen to a depth of several metres so that, in such soils, the oxygen concentration is similar to that in humid air. However, in flooded soils, water fills the pores and oxygen must diffuse through liquid. This occurs much more slowly than in air. If the temperature is low and the plants are dormant, oxygen depletion in the soil is slow and may be relatively harmless. However, at temperatures above about 20°C, oxygen consumption by roots and soil microorganisms can totally deplete oxygen from the bulk of the soil water in as little as 24 hours. Growth and survival are then severely depressed in many species as, for example, in garden peas (*Pisum sativum*) where yield may be halved by flooding the soil for 24 hours.

Specialised plants, such as marsh and bog plants and paddy rice, can survive for long periods in flooded soils. As with drought resistance, we find that several different strategies have evolved, although most involve some means of enabling oxygen to reach the roots rather than the ability to tolerate low oxygen tensions within the cell. In most cases, survival is aided by the development of *aerenchyma*, a tissue consisting of long files of interconnected gas-filled spaces. This allows for the diffusion of gaseous oxygen from the air, through the plant and into the roots. Wetland vegetation often also has heavily *suberised* roots (i.e. they are covered with an impermeable corky layer), thus preventing the loss of oxygen from the roots into the soil.

Wetland species that grow at the margins of ponds and rivers, or in marshy conditions, are usually dependent on the presence of aerenchyma for their survival. The dormant rhizomes of salt-marsh bullrush (*Scirpus maritimus*) and cat-tail (*Typha* spp.) overwinter in anaerobic mud at the edges of lakes but, when the leaves begin to expand, survival of the plant is dependent on the diffusion of oxygen through the gas-filled spaces of the aerenchyma tissue. Similarly, although the seeds of paddy rice can germinate in low oxygen concentration, the roots are not tolerant of anaerobic conditions and their survival depends on the emergence of the *coleoptile* above the surface of the water. This acts as a kind of 'snorkel' that allows oxygen to diffuse from the air to the roots. In other plants, such as the swamp cypress (*Taxodium distichum*) and mangroves, portions of the roots (known as 'knees' in *Taxodium*) remain above the water or soil surface and allow oxygen to diffuse down to the submerged parts.

In flood-tolerant plants such as willow (*Salix* spp.) and sunflower (*Helianthus* spp.), very little aerenchyma is present in plants growing in well-aerated soils. Aerenchyma develops in response to low oxygen concentrations in the soil. This is an interesting process that involves the gaseous hormone, ethylene. A low concentration of oxygen induces the production of ACC (a precursor of ethylene) in the roots, but the absence of oxygen blocks the final step in the pathway, the conversion of ACC to ethylene. However, the ACC is transported from the roots to the aerial parts of the plant, where the increased oxygen level results in the production of ethylene and the formation of aerenchyma.

The production of ethylene by the shoots of flooded plants has effects other than to induce the development of aerenchyma. In tomato (*Lycopersicon esculentum*), the formation of ethylene results in *epinasty* (rolling under of the leaves) within 6–12 hours after flooding. In tomato, maize (*Zea mays*) and sunflower (*Helianthus annuus*), flooding leads to swelling at the base of the stem, probably due to ethylene production. It is thought that this *hypertrophy* may increase the porosity in this region of the stem and so enhance aeration. However, although ethylene is a key hormone in the ability of some species to resist flooding, prolonged exposure to increased levels of ethylene leads to yellowing, senescence and ultimately death in sensitive species.

Waterlogging causes the formation of adventitious roots in some resistant species. These can replace old roots that have died off in response to flooding and so increase the possibility of survival. For example, in *Rumex palustris* (a wetland species), waterlogging resulted in the development of 49 adventitious roots within four days, whereas *Rumex thyrsiflorus* (a dry-land species) produced only eight.

Death due to flooding of the soil is mainly caused by the lack of oxygen for normal aerobic respiration by the roots. However, other factors are also important. Anaerobic soil microorganisms produce toxic compounds such as hydrogen sulphide (a respiratory poison) and butyric acid, which are also partly the cause of the characteristic unpleasant odour of waterlogged soils. Moreover, the increased production of ethylene gas in the aerial parts of the plants, which occurs as a result of the low oxygen concentration around the roots, results in senescence and yellowing of the leaves leading to their early death.

AQUATIC PLANTS

Well-managed ponds (i.e. those that have a thriving population of oxygenating submerged aquatics such as *Elodea canadensis*) have more available oxygen than flooded soils but, because gases diffuse more slowly through water than through air, submerged leaves may have difficulties in obtaining sufficient carbon dioxide.

There are several adaptations that increase the ability of plants to live in an aquatic environment. Some of these are immediately obvious. They include, for example, long petioles that allow leaves to be in an aerial environment while their roots remain in the soil at the bottom of the pond, floating leaves that have most of their stomata on the upper surface (land plants have most of their stomata on the lower surface in order to conserve water) and submerged leaves that are finely divided, with a high surface to volume ratio allowing more rapid diffusion of carbon dioxide into the leaf cells. Several aquatic plants have entire aerial leaves and finely divided submerged ones (e.g. water crowfoot, *Ranunculus aquatilis*). The presence of aerenchyma (see under 'excess water' above), which allows gases to diffuse rapidly through the plant, is also of major importance in aquatic species.

SOIL pH

Many groups of plants have evolved on soils with particular characteristics and are, consequently, better suited to some soils than others. Important groups for gardeners are those plants that have adapted to growing on acid soils (the *calcifuges*), and those that prefer chalky, or calcareous soils (the *calcicoles*). Table 5.2 gives some examples.

Calcifuges have a preference for ammonium as their source of nitrogen, whereas calcicoles utilise nitrate for preference. Calcicoles are often very efficient in taking up phosphorus and, in some cases, this is because the roots are heavily infected with mycorrhizal fungi.

The constraints to growth imposed by high soil pH include high concentrations of bicarbonate and calcium in the soil solution and a low availability of iron and zinc. At low pH, the constraints on plant growth may include a high concentration of aluminium in the soil solution (see Chapter 4).

Table 5.2 Plants suitable for growing on acid or chalky soils.

Plants that prefer acid soils (less than pH 7)*	**Plants that tolerate chalky soils** (more than pH 7)**
Corydalis spp.	Border carnations and pinks (*Dianthus* spp.)
Gentian (*Gentiana septemfida*)	Pasque flower (*Pulsatilla vulgaris*)
Meconopsis spp.	Poppy (*Papaver* spp.)
Rhodohypoxis spp.	Sage (*Salvia* spp.)
Bearberry (*Arctostaphylos* spp.)	Pincushion flower (*Scabiosa caucasica*)
Bottlebrush (*Callistemon* spp.)	Bearded rhizomatous iris (*Iris* spp.)
Ling (*Calluna vulgaris*)	*Clematis* spp.
Camellia spp.	Honeysuckle (*Lonicera* spp.)
Clethra spp.	Paper-bark maple (*Acer griseum*)
Corylopsis spp.	Horse chestnut (*Aesculus* spp.)
Lantern tree (*Crinodendron hookerianum*)	Judas tree (*Cercis siliquastrum*)
St Dabeoc's heath (*Daboecia cantabrica*)	Sun rose (*Cistus* spp.)
Chilean fire bush (*Embothrium coccineum*)	*Escallonia* spp.
Enkianthus spp.	Beech (*Fagus sylvatica*)
Heather (*Erica* spp., except *E.* x *darleyensis* and *E. carnea*, which are lime tolerant)	*Forsythia* spp.
Fothergilla spp.	Holly (*Ilex* spp.)
Gaultheria spp.	Juniper (*Juniperus* spp.)
Halesia spp.	Batchelor's button (*Kerria japonica*)
Witch hazel (*Hamamelis* spp.)	Privet (*Ligustrum* spp.)
Calico bush (*Kalmia* spp.)	Daisy bush (*Olearia* spp.)
Tea tree (*Leptospermum* spp.)	*Photina* × *fraseri* 'Red Robin' (evergreen)
Leucothoë spp.	Bird cherry (*Prunus avium*)
Magnolia spp.	Sumac (*Rhus* spp.)
Photina beauverdiana, P. villosa (deciduous)	Mountain ash (*Sorbus* spp.)
Pieris spp.	Lilac (*Syringa vulgaris*)
Rhododendrons and azaleas (*Rhododendron* spp.)	*Viburnum* spp.
Blueberry (*Vaccinium corymbosum*)	*Weigela* spp.
Cranberry (*Vaccinium vitis-idaea*)	

* Although they cannot tolerate chalky conditions, many of the plants listed will grow successfully in neutral soil.
** These plants will grow well in chalky soil but most also grow well in neutral or acid conditions. However, acid soils can adversely affect growth in some plants (e.g. below pH 6.0 for beans, pH 5.9 for peas and pH 4.9 for potato).

The low availability of iron in chalky soils can lead to iron deficiency in the plant, especially in calcifuge species such as *Rhododendron*. This results in lack of chlorophyll, especially in the young leaves, a condition known as 'lime-induced chlorosis'. It is known that calcicoles possess adaptive mechanisms that enable them to cope with the high concentrations of calcium in calcareous soils but these seem to be complex and are poorly understood. In some calcicole species a high concentration of calcium within the plant tissue is avoided (i.e. calcium is excluded by the root cells), while in others it is tolerated. High levels of calcium-binding proteins are found in the cytoplasm of calcicole species and this compartmentation of calcium at the cellular level is probably a key process in their tolerance of high calcium concentrations. Calcifuge species, in contrast, have only low levels of calcium binding proteins.

Although it is preferable not to try to grow acid-loving plants such as rhododendrons on chalky soils, the application of elemental sulphur would have some which benefit in lowering the soil pH. Alternatively foliar sprays or soil dressings of a chelated form of iron (which is taken up by the plant irrespective of the soil pH) can be given and may correct lime-induced chlorosis. The treatments need to be repeated at intervals. Where soils are too acid, the chief means of managing soil pH is through the application of lime (calcium carbonate). Chapter 4 gives more detailed information on how to manage the soil pH.

LIGHT

All higher green plants are dependent on receiving sufficient energy from sunlight for the manufacture of organic compounds. Gardens often have areas of shade either from trees or from surrounding buildings and it is important to choose plants that will grow well under these conditions. 'Sun-loving' plants, which will only grow well in areas with plenty of light, will often tolerate some degree of water shortage; shade-tolerant plants may well grow happily in full sun, provided that there is plenty of available water.

Plants have evolved a range of mechanisms to enable them to adjust to the prevailing light conditions. For example, in the shade, leaves are often larger and thinner, a response designed to maximise the collection of light in a situation where light is severely limiting for photosynthesis. Internal changes, such as an increase in the amount of light-harvesting chlorophyll (see Chapter 1), also take place under low light conditions and increase the ability of the plant to absorb light. Plants in the shade can also divert their energy into stem growth, which enables the plant to compete for available light and outgrow its neighbours to reach better light conditions. Such competitive elongation would, of course, be of no value to plants of the woodland floor and so we have the *shade avoiders*, such as many arable weeds, which grow taller in response to shade, and the *shade tolerators* which don't, although they have other responses which improve the competitiveness of the plant under shady conditions, such as larger and thinner leaves. Photosynthesis in the shade avoiders is usually relatively inefficient under poor light conditions, but they have the ability to re-direct their growth into stems at the first sign of shading.

Shade

How are plants able to detect whether or not they are in a shady habitat? Under natural conditions, the most important shade is from leaf canopies, and the shade-avoidance response depends on the fact that the light quality is profoundly altered when it passes through leaves. In order to understand this we need first to look at the spectrum of sunlight and how leaves affect the light passing through them.

Sunlight contains red (R) and far-red (FR) wavelengths in approximately equal amounts and the effect of leaves (mainly due to the absorption of light by chlorophyll) is to remove the red wavelengths (Fig. 5.3). This means that under the leaf canopy the relative amounts of red and far-red light are changed. Shade avoiders respond to canopy shade because they can detect changes in the ratio of R to FR light through the pigment *phytochrome*. The major characteristics of this pigment are discussed in Chapter 6, the most important from the point of canopy shade being the red/far-red reversible reaction. Red light produces the biologically active form, Pfr, while far-red light converts Pfr back to the biologically inactive form, Pr. Thus both forms are present in light that contains a mixture of red and far-red wavelengths, as does sunlight. Furthermore, the amount of Pfr that is present depends on the relative amounts of red and far-red wavelengths in the light (the R/FR ratio). For example, in sunlight, which has a R/FR ratio of 1.15, about 55% of the phytochrome is present as Pfr.

With increasing amounts of canopy shade, the proportion of red light decreases and so the amount of Pfr also decreases. Since Pfr (the form which actually controls the response) *inhibits* stem elongation, the result is that stem length increases as the proportion of red light decreases with increasing canopy shade. (Fig. 5.4).

Early experiments demonstrated that the red/far-red reversible reaction of phytochrome controls many responses of plants to light, including germination (see Chapter 6) and the changes that

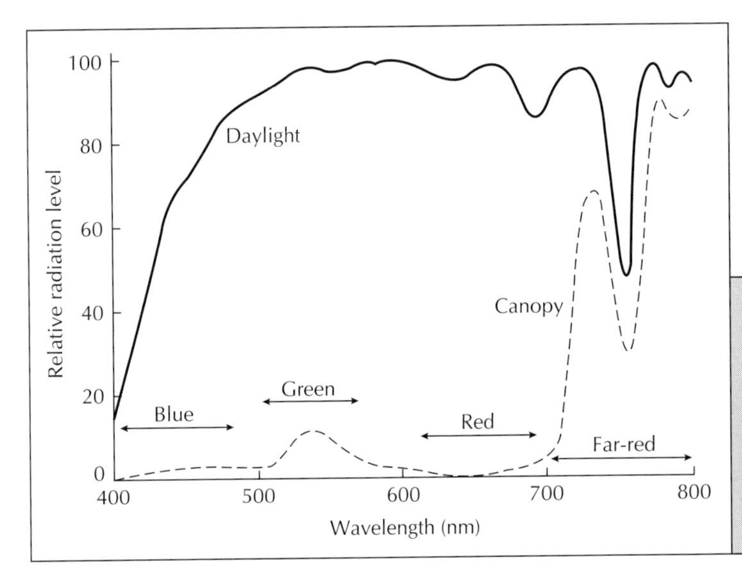

Fig. 5.3 A leaf canopy changes the ratio of red to far-red light beneath it. Red light is largely absorbed by the leaves but a high proportion of far-red light is transmitted through them. There is, therefore, a marked relative enrichment in far-red light under the leaf canopy and an almost complete lack of red and blue light. (Adapted from Smith, 1994.)

Fig. 5.4 The effect of the ratio of red (R) to far-red (FR) light on elongation growth in the potato. The plant (right) growing in a *low* R/FR ratio (equivalent to deep shade) is markedly elongated. The plant (left) growing in a *high* R/FR ratio (light from fluorescent lamps containing little FR) is considerably shorter than the one (centre) growing in a *medium* R/FR ratio (equivalent to sunlight, in which the R/FR ratio is approximately 1.0). (Photo, courtesy of Brian Thomas, Horticulture Research International.)

occur when dark-grown plants are initially exposed to light (collectively known as *de-etiolation* (Fig. 5.5). It was observed that the responses brought about by a short exposure to red light could be prevented when this was immediately followed by a short exposure to far-red. However, the relevance to natural conditions was not immediately apparent since plants are not normally exposed to brief sequences of R and FR light and some scientists even doubted whether such a pigment actually existed. Indeed, in some early papers it was suggested that phytochrome might be a 'pigment of the imagination' and it was also dubbed a 'red/far-red herring'. It was several years before phytochrome was finally purified and shown to be a protein. Today (due largely to work at Leicester University) we understand that this red/far-red reversible reaction of the pigment is a sophisticated way of measuring the degree of shading by a plant canopy. As the canopy becomes thicker and the amount of light decreases, the R/FR ratio of the light also decreases. Thus, the thicker the leaf canopy and the denser the shade, the longer the stems.

In addition to the canopy effect, the way in which leaves absorb light results in an increased sideways reflection of FR wavelengths towards neighbouring plants (Plate 13). By sensing the FR light reflected from their neighbours, some species are able to detect the presence of a neighbour at a

Fig. 5.5 The effect of darkness (etiolation) and light on seedling growth. Seedlings of mung beans (*Vigna radiata*) grown in darkness (left) are tall, with apical hooks, small leaves and no green or red pigmentation. Exposure to light (right) reduces elongation, unfolds the apical hook and increases leaf expansion; it also results in the formation of green (chlorophylls) and red (anthocyanins) pigments. Many of these responses depend on exposure to red light (see text).

distance of 20–30 cm and thus obtain an early warning of the competition for light that they are likely to encounter in the future. This enables them to adjust their growth patterns before the shading actually occurs and is another good example of the plant's ability to 'detect what is going to happen next'.

Shade avoiders and shade tolerators

Not all plants respond to the ratio of R/FR light in this way. The ones that do so are considered to be *shade avoiders* since this response enables them to compete with their neighbours by out-growing them. Many arable weeds fall into this category. Such a response is often undesirable in the garden, since rapid stem growth results in 'leggy' plants, with weak stems. For shady conditions, therefore, the *shade tolerators* are more desirable. These are often plants from woodland habitats where an elongation response would be of no value. As described above, shade-tolerant plants often have mechanisms that improve their ability to collect light under shaded conditions. Energy is directed into leaf expansion to produce a larger light-collecting area and more light-harvesting chlorophyll develops. This too is a response to light but appears to be largely controlled by a pigment system that responds primarily to the amount of *blue* light received by the plant, rather than to the R/FR ratio.

Shade from neighbouring buildings reduces the amount of light available for photosynthesis but, unlike shade from trees, it does not alter the R/FR ratio. Nevertheless, many plants show typical shade responses, probably detecting the shade conditions through the blue-light sensitive pigment system.

CHOOSING PLANTS FOR PARTICULAR CONDITIONS

Water, wind and light are major factors in determining survival of plants in a particular habitat and plants have evolved many strategies that enable them to tolerate, or avoid unfavourable conditions of water stress, excess water, wind and/or shade. Some of these are readily observable. For example, grey reflective leaves, hairy leaves, succulence, thick cuticles, or small leaves are usually indicative of a greater resistance to water stress, while low-growing plants with small leaves are likely to be more tolerant of wind. Other strategies (such as an increase in the amount of light harvesting chlorophyll during acclimation to shade) are not immediately apparent. Although the plant's natural habitat gives some indication of the conditions in which it

Table 5.3 Some plants for particular habitats.

Dry and sunny		
Agave (*Agave* spp.) Wormwood (*Artemesia* spp.) Sun rose (*Cistus* spp.) Broom (*Cytisus* spp.) Eryngium (*Eryngium* spp.) Wallflower (*Erysimum* 'Bowles' Mauve') Treasure flower (*Gazania* spp.)	Broom (*Genista* spp.) Halimium (*Halimium* spp.) Hyssop (*Hyssopus* spp.) Shrubby veronica (*Hebe pinguifolia*) Rhizomatous iris (*Iris* spp.) Lavender (*Lavandula* spp.) Catmint (*Nepeta* spp.)	Marjoram (*Origanum* spp.) Shrubby potentilla (*Potentilla* spp.) Rosemary (*Rosmarinus* spp.) Stonecrop (*Sedum* spp.) Spanish broom (*Spartium junceum*) Thyme (*Thymus* spp.) Yucca (*Yucca* spp.)
Dry and shady		
Spotted laurel (*Aucuba japonica*) Box (*Buxus* spp.) Cyclamen (*Cyclamen coum*) Mezereon (*Daphne mezereum*) Wood spurge (*Euphorbia amygdaloides* var. *robbiae*) Herb Robert (*Geranium robertianum*)	Ivy (*Hedera helix*) Dead nettle (*Lamium* spp.) Mahonia (*Mahonia aquifolium*) Portuguese laurel (*Prunus lusitanica*) Sweet box (*Sarcococca* spp.) Periwinkle (*Vinca* spp.)	Many spring-flowering bulbs
Wet and shady		
Chokeberry (*Aronia arbutifolia*) Astilbe (*Astilbe* spp.) Dogwood (*Cornus alba*) Aconite (*Eranthis hyemalis*) Dog's tooth violet (*Erythronium* spp.) Hepatica (*Hepatica* spp.) Summer snowflake (*Leucojum aestivum*)	Cardinal flower (*Lobelia cardinalis*) Bog myrtle (*Myrica gale*) Willow (*Salix daphnoides*) Spiraea (*Spiraea* × *vanhouttei*) Wood lily (*Trillium* spp.) Blueberry (*Vaccinium corymbosum*) Guelder rose (*Viburnum opulus*)	Ferns – including Hart's tongue (*Asplenium scolopendrium*) Lady fern (*Athyrium filix-femina*) Buckler fern (*Dryopteris affinis*) Shuttlecock fern (*Matteuccia struthiopteris*)

will thrive, the choice of plants that are most suitable for any particular situation often has to depend on knowledge from experiments and trials. Many gardening books give lists of such plants and some examples are given in Table 5.3.

FURTHER READING

Barnes, P. (1999) Ingenious strategies. *The Garden*, **124**, 374–9.

Bidwell, R.G.S. (1979) *Plant Physiology*. 2nd edn. MacMillan Publishing Co, London.

Davies, W. J., Tardieu, F. & Trejo, C.L. (1994) How do chemical signals work in plants that grow in drying soils. *Plant Physiol.* **104**, 309–14.

Grant-Downton, R. (1998) Dry Lazarus. *The Garden*, **123**, 657–9.

Mansfield, T.A. (1994) Some aspects of stomatal physiology relevant to plants cultured *in vitro*. In: *Physiology, Growth and Development of Plants in Culture*. P.J. Lumsden, J.R. Nicholas & E.J. Davies (eds). Kluwer Academic Publishers, Dordrecht.

Prasad, M.N.V. (Ed.). (1997) *Plant Ecophysiology*. John Wiley and Sons Inc, New York.

Salisbury, F.B. & Ross, C.W. (1992) *Plant Physiology* 4th edn. Wadsworth Publishing Company, Belmont, Ca.

Smith, H. (1994) Sensing the light environment: the functions of the phytochrome family. In; *Photomorphogenesis in Plants*. 2nd edn. R.E.Kendrick and G.H.M. Kronenberg (eds). Kluwer Academic Publishers, Dordrecht.

Vergine, G. & Jefferson-Brown, M. (1997) *Tough Plants for Tough Places*. David and Charles, Newton Abbot.

Vince-Prue, D. (1992) Phytochrome: a remarkable light sensor in plants. *The Plantsman*, **13**, 203–14.

6

Raising Plants from Seed

- Seed structure and development: pollination and the formation of the embryo
- Apomixis
- Seed reserves
- Seed maturity and ripening
- Germination: water uptake; metabolic events
- Dormancy: types of dormancy
- Special requirements for germination
- The role of phytochrome and how it acts to control germination
- Stratification and the promoting of germination by low temperature
- Dormancy breaking by high temperature and alternating temperatures
- Fire
- After-ripening in dry storage
- Chemical treatments
- Factors influencing seed vigour
- Methods for ensuring high-vigour
- Primed seeds and pelleted seeds
- Seed storage: optimum conditions and loss of viability
- New developments; genetic manipulation and terminator gene technology
- Gene banks

Flowering plants and their relatives usually reproduce sexually, developing seeds that are dispersed by wind, water, or by biological agents such as insects and birds. In the strict botanical sense a seed is a fertile and ripened ovule that contains an embryonic plant, usually supplied with stored food and surrounded by a protective coat called a *testa*. Seeds are produce by two groups of plants, the Gymnospermae (conifers and their relatives) and the Angiospermae (the flowering plants).

SEED STRUCTURE

A major difference between the seeds of the gymnosperms and those of the flowering plants is that the former do not have an ovary, so that the seeds are naked. Indeed, the term gymnosperm means naked seed. In contrast, the ovules of flowering plants are enclosed in an ovary which later develops into the fruit (Fig. 6.1). The ovary contains one or more ovules, depending on the species. For example, one-seeded fruits include lettuce (*Lactuca sativa*), sunflower (*Helianthus annuus*) and ash (*Fraxinus excelsior*). In contrast, peanut (*Arachis hypogaea*) contains two, or occasionally three seeds while peas (*Pisum sativum*) and beans (*Phaseolus vulgaris*) contain five or more.

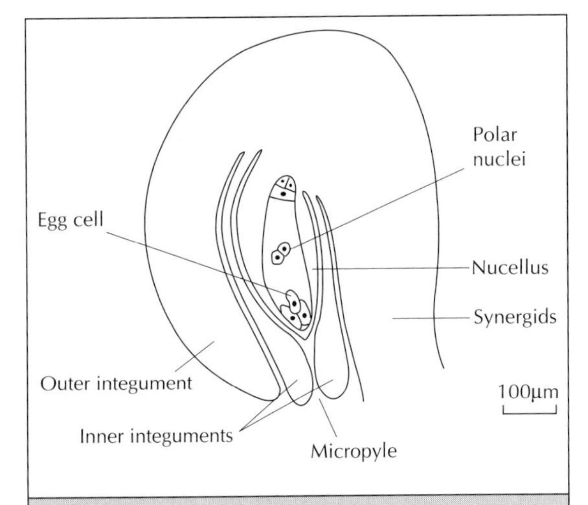

Fig. 6.1 Longitudinal section through a typical angiosperm ovule. (Re-drawn from Bryant, J.A. (1985) *Seed Physiology*, Edward Arnold.)

Each ovule contains an egg cell and a central cell with two nuclei, called the *polar nuclei*. Both of these cells are *haploid*, that is their nuclei contain only a single set of chromosomes having been produced by *meiosis*, a type of nuclear and cell division that occurs prior to sexual reproduction and results in cells containing nuclei with only one set of chromosomes (see Chapter 3). In contrast, the other cells of most plants contain two sets of chromosomes and are, therefore, said to be *diploid*. Pollination occurs in plants when pollen from the anthers (the male parts of the flower) is deposited on the stigma (the female parts). This process is usually carried out by intermediary agents such as insects, wind, water or even gravity. The pollen grain on the stigma germinates by sending down a pollen tube, which grows down the style into the embryo sac. The two pollen nuclei (which are also haploid) enter the embryo sac. One fuses with the egg cell to form the *zygote*, which divides to produce the young embryo. Each embryo consists of a *radicle* (embryonic root), a *plumule* (embryonic shoot), and a *hypocotyl* connecting the radicle with the plumule. The other pollen nucleus fuses with one of the polar cells to form a storage tissue, called the *endosperm*. In dicotyledons, the endosperm is short lived and its storage function is taken over by the two cotyledons. These are the leaves of the embryo and are sometimes known as seed leaves, for example in peas (*Pisum sativum*) and beans (*Phaseolus vulgaris*).

Although the production of a seed is normally initiated by a sexual process (the fertilisation of the egg cell), a non-sexual process called *apomixis* occurs in some angiosperms. In this instance, the egg cell is formed from one of the *diploid* cells (with two sets of chromosomes) of the *nucellus*. The nucellus provides tissue for the origin and nurture of the embryo sac as it develops (very similar in function to the placenta in human embryology). Apomixis is common in several grasses, such as annual meadow grass (*Poa annua*), Kentucky blue grass (*Poa pratensis*) and dandelion (*Taraxacum officinale*). In citrus fruits, several apomictic embryos can develop together with a normal, sexually produced embryo, leading to *poly-embryony* (several embryos in the same seed). However, poly-embryony can also arise by division of the sexually derived embryo, as in many orchids.

Seeds vary enormously in their size and shape. Orchid seeds, for example, are microscopic while the double coconut (*Lodoicea maldivica*) can weigh up to 10 kilograms. Factors which determine the size and shape of seeds are the form of the ovary, the size of the embryo, the amount of endosperm, what other tissues are associated with the seed and the conditions under which the parent plant is growing during the formation of the seed. Polymorphic seeds (i.e. seeds of more than one shape) occur in certain species such as cocklebur (*Xanthium strumarium*) and the differently shaped seeds often have different requirements for germination.

Food reserves in seeds

Most seeds contain enough stored food to support germination and early growth until the seedling is able to photosynthesise. The main storage materials accumulating in the seed storage tissue are carbohydrates (simple and complex sugars), proteins and fats. Seeds also contain mineral reserves, especially phosphates. Table 6.1 gives examples of the storage materials in certain seeds.

However, the smallest seeds do not have any reserves, either because the seedlings are parasitic and live off their hosts, for example broomrapes (*Orobanche* spp.) and witchweed (*Striga* spp.) or, like epiphytic orchids, they form symbiotic associations with mycorrrhizal fungi that provide nutrients during germination. Seeds of orchids are extremely minute. Each consists of a tiny embryo surrounded by a single layer of protective cells. They are so small that the food reserves in the embryo are inadequate by themselves for the early development of a new plant. In nature most orchid seeds begin life in partnership with a symbiotic fungus. The fungal hyphae, which are present in the soil or in the bark of the host tree, invade the seed and enter the cells of the embryo. The orchid soon begins to digest this fungal tissue and obtain nutrients from it, thus using the fungus as an intermediary in obtaining nutrients from decaying material in the soil. In horticultural practice, orchid seeds are often germinated on artificial nutrient media such as those used for tissue culture (see Chapter 7).

In larger seeds, the food is stored in special tissues. In the gymnosperms and some angiosperms, a tissue called the *perisperm*, derived from the remains of the nucellus, provides the main store of nutrients. It is found in certain plants in which the endosperm does not completely replace the nucellus, e.g. Caryophyllaceae and coffee (*Coffea* spp.). In angiosperms it is more usual for either the endosperm (e.g. in onions and their relatives) or swollen cotyledons (e.g. in legumes such as French bean, *Phaseolus vulgaris*) to be used as food stores. Other tissues can also act as food stores, for example an enlarged hypocotyl in Brazil nut (*Bertholletia excelsa*), while more than one tissue is sometimes involved, as in beet (*Beta vulgaris*) where there is both a well-developed endosperm and a perisperm.

Seed ripening

The embryos of all seed plants, whether produced sexually or apomictically, develop to a particular

Table 6.1 The major storage materials of certain seeds.

Species	Percentage in air-dry seed			
	Sugars	Proteins	Fats	Starch
Zea mays (maize)	1–4	10	5	50–70
Pisum sativum (pea)	4–6	20	2	30–40
Helianthus annuus (sunflower)	2	25	45–50	0
Papaver somniferum (opium poppy)	25	20	34	0

stage which is characteristic of the species. When the embryo reaches full size, the deposition of food reserves is completed, physiological maturity is attained and ripening or dehydration begins. In hot summers the moisture content of the seed may be reduced to 10%, although a water content of 15% is usual in an average summer. Once this stage of desiccation in reached, the seed is usually shed from the parent plant. By this stage, it is physiologically inert and will remain in a dormant state until the conditions are right for germination.

GERMINATION

Germination is the term used to describe the physiological and physical changes immediately prior to and including the first visible signs of growth. The dry seed is inert until sufficient water enters the cells to allow metabolic reactions to begin. Once in the soil, water becomes available and enters the dry seed by the process of *imbibition*. This is a physical process, for both dead and living cells can imbibe water. Water is taken up through openings in the seed coat and causes the seed to swell. This ruptures the seed coat, allowing the entry of gases and the uptake of more water. As the cells become hydrated, enzymes are activated and begin to break down the storage tissue and transfer nutrients to the embryo. Some of the metabolic activity that occurs during early germination is directed at repairing or replacing cell membranes (which are selectively permeable and control the exchange of materials through the cell, see Chapter 1) and organelles (sub-cellular structures that are found throughout a cell and are the sites for respiration, photosynthesis etc., see Chapter 1) that were damaged during the drying process. However, most activity is directed towards embryo growth.

Water

A key factor for the earliest stages of germination is the availability of water. Dry seeds absorb water in three phases: an initial, extremely rapid water uptake; a second, lag phase, in which little or no water enters the seed; and a third phase of water uptake, less rapid than the first and associated with embryo growth and seedling emergence. Very dry seeds must be allowed to absorb water gradually, as rapid entry into dry tissue can cause severe damage and even, in some cases, lead to death of the embryo. Because of this, the common practice of soaking seeds before sowing in order to improve germination is all too often detrimental and results in poorer germination than when the seeds are sown dry. The temperature at which imbibition (water uptake) occurs is also important; the lower the temperature the greater the damage.

Temperature

Once the cells are sufficiently hydrated, metabolic processes become important and require a suitable temperature and a supply of oxygen. Most seeds will germinate fully in air, although in some cases, notably cocklebur (*Xanthium strumarium*) and certain cereals, a higher percentage germination is achieved by increasing the oxygen level.

Most seeds contain large reserves of nutrients such as carbohydrates, proteins and fats. These large molecules have to be broken down by enzymes before they can be assimilated by the embryo and support the growing seedling until it is capable of photosynthesis. The metabolic processes that are necessary for the breakdown of stored nutrients and for embryo growth can occur only when the temperature is within a certain range. The optimum temperature is that at which the highest percentage of germination is attained in the shortest possible time. The range is defined by the minimum and maximum temperatures, which are the lowest and highest temperatures at which germination will occur. For example, too high a temperature is often the cause of poor emergence and establishment of summer-sown lettuces (*Lactuca sativa*) in the UK. Both the optimum temperature and the temperature range vary with species, and the latter can be slightly widened by changing the conditions during seed maturation. Some seeds germinate best when exposed to alternating temperatures (see below under *dormancy*).

As germination proceeds, more water enters

and the storage nutrients are gradually assimilated. The embryo increases in size and undergoes a number of cellular changes, including an increase in several enzymes. Finally, the primary root begins to emerge, followed by emergence of the young shoot.

DORMANCY

Not all seeds are able to germinate when supplied with water and maintained at temperatures that are favourable for growth. Such seeds are said to be *dormant*. There are several reasons why a period of seed dormancy confers an advantage to the plant. For example, germination can be prevented until conditions are favourable for seedling growth or germination can be spread out in time so that competition is reduced. The conditions needed to break dormancy usually relate to the natural environment and confer a selective advantage to the species. For gardeners, however, seed dormancy is often a disadvantage leading to delayed and erratic germination. Indeed many tried and tested gardening techniques to improve germination are no more than dormancy-breaking mechanisms.

Types of dormancy

There are three types of dormancy: these are *innate* dormancy; *induced* dormancy; and *enforced* dormancy. In other words:- some seeds are born dormant, some achieve dormancy and some have dormancy thrust upon them.

Enforced dormancy

This occurs when seeds are capable of germination but are prevented by the immediate environmental conditions, such as lack of water or temperatures outside the range that permits germination in the particular species. Germination occurs as soon as the restriction is removed and suitable conditions prevail. Such seeds are not truly dormant and this term is best restricted to a condition internal to the seed which does not allow germination, even though water is available and the temperature is favourable for growth.

Innate dormancy

This is under genetic control, but it is influenced by the environmental conditions during seed maturation. Seeds of the vast majority of species have a period of innate dormancy, which begins immediately the embryo ceases to grow and while it is still attached to the parent plant. Such dormancy prevents the seed from germinating while still on the plant and usually persists for some time after the seed is shed or harvested.

The most usual cause of innate dormancy is that the embryo is insufficiently developed at the time of shedding, a common problem in members of the Apiaceae family (carrot, *Daucus carota*, and parsnip, *Pastinaca sativa*), in several palms, and in magnolia (*Magnolia grandiflora*). In other seeds, chemical inhibitors, such as phenolic compounds and abscisic acid, may be present in the seed coat and have to be removed by leaching before germination can proceed; such inhibitors are found in beet (*Beta vulgaris*) and in certain East Asian species of ash (*Fraxinus* spp.). A further cause of innate dormancy is that certain biochemical changes must occur in the seed before germination is possible. This is known as *after-ripening* and is usually influenced by environmental factors such as temperature and light. Fresh barley (*Hordeum vulgare*) seed, for example, has to be stored for three months at 20°C in order to break its innate dormancy. The need for after-ripening may, in some cases, be associated with the presence of inhibitors of germination which can either be leached out as described above, or may be broken down to inactive forms by chemical reactions within the seed.

Another type of innate dormancy is found in many species in a range of families, including the Papilionaceae (pea and bean family), Malvaceae (mallows), Convolvulaceae (bindweed family), Liliaceae (lilies) and Chenopodiaceae (goosefoot family). Here germination is prevented because the seeds have impermeable seed coats which present a barrier to the uptake of water. Under natural conditions the impermeable barrier in such *hard seeds* is removed by conditions which result in cracking or abrasion of the seed coat, such as repeated wetting and drying or exposure to particular temperatures. Digestion by animals also results in loss of the hard seed coat.

Induced dormancy
This occurs when seeds that are capable of germination are exposed to conditions that lead to the dormant state. These include particular temperatures and sometimes prolonged burial. As with innate dormancy, germination cannot take place until the seeds have been exposed to conditions that bring about a further change within the seed. Many of the environmental conditions that are effective in breaking innate dormancy are also effective in breaking induced dormancy and, in many cases, it is not easy for the gardener to distinguish between these two types of dormancy.

Fig. 6.2 The effect of sowing depth on the germination of light-requiring cress seeds. The seeds were completely covered with soil (left), sown on the surface (right) or covered with the soil surface lightly scratched to expose some of the seeds (centre).

SPECIAL REQUIREMENTS FOR GERMINATION

A number of special treatments are effective in breaking seed dormancy or are required for rapid germination. It is not really possible to distinguish between these two situations and any viable seed that fails to germinate when given an adequate supply of water and maintained at temperatures favourable for growth can be considered as being dormant. The factors that are effective in breaking dormancy and permitting germination confer a selective advantage on the species and are related to the conditions experienced in its natural habitat. These factors include light, temperature, fire, water and digestion by animals.

Light

Many seeds will only germinate when exposed to light (Fig. 6.2). However, they need to be fully, or partly hydrated before they become responsive to the light stimulus. A requirement for light has obvious advantages for the plant as it prevents the germination of deeply buried seeds until they are uncovered by cultivation or the activities of animals. Many of these species are *ruderals*. These are plants with a short life cycle which are poor competitors in a perennial sward but are able to grow, flower and set seed rapidly if the ground is disturbed and the competition thus reduced. In the garden environment, the majority of these are annual weeds. Garden plants with seeds that normally require exposure to light before they can germinate include cress (*Lepidium sativum*, Fig. 6.2), poppy (*Papaver* spp.), birch (*Betula* spp.), tobacco (*Nicotiana tabacum*), some cultivars of lettuce (*Lactuca sativa*) and several grasses. These seeds must be sown at the soil surface. Although it is not common, germination is actually inhibited by light in a few plants; these include pansy (*Viola* spp.), love-lies-bleeding (*Amaranthus* spp.) and love-in-a-mist (*Nigella damascena*). Such seeds must be well covered with soil or black plastic to exclude light. Some further examples of the responses of seeds to light are given in Table 6.2.

Light quality
The kind of light to which the seed is exposed is also important and most light-requiring seeds are responsive to red light, although a small number will germinate only in blue light (e.g. *Nemophila* spp.). As sunlight contains both red and blue

Table 6.2 Some examples of the responses of seeds to light (From data of Kinzel, 1926).

Germination favoured by light	Germination favoured by darkness	Germination indifferent to light or darkness
Adonis vernalis	*Aithanthus glandulosa*	*Anemone nemorosa*
Alisma plantago	*Aloe variegata*	*Bryonia alba*
Bellis perennis	*Cistus radiatus*	*Cytisus nigricans*
Capparis spinosa	*Delphinium elatum*	*Datura stramonium*
Colchicum autumnale	*Ephedra helvetica*	*Hyacinthus candicans*
Erodium cicutarium	*Euonymus japonicus*	*Juncus tenagea*
Fagus sylvatica	*Forsythia suspensa*	*Linaria cymbalaria*
Genista tinctoria	*Gladiolus communis*	*Origanum majorana*
Helianthemum chamaecistus	*Hedera helix*	*Sorghum halepense*
Iris pseudacorus	*Linnaea borealis*	*Theobroma cacao*
Juncus tenuis	*Mirabilis jalapa*	*Tragopogon pratensis*
Lactuca scariola	*Nigella damascena*	*Vesicaria viscosa*
Magnolia grandiflora	*Phacelia tanacetifolia*	
Nasturtium officinale	*Ranunculus crenatus*	
Oenothera biennis	*Tamus communis*	
Panicum capillare	*Tulipa gesneriana*	
Reseda lutea	*Yucca aloifolia*	
Salvia pratense		
Sueda maritima		
Tamarix germanica		
Taraxacum officinale		
Veronica arvensis		

wavelengths, this difference is not something that normally impinges on horticulture. However, it was from studies of the responses of lettuce (*Lactuca sativa*) seeds to different wavelengths of light that the light-sensitive pigment, *phytochrome* was discovered.

The discovery of phytochrome stemmed from an observation in 1935 that the germination of lettuce seeds was suppressed by light of wavelengths in the part of the spectrum between 700 and 760 nm, now called far-red light (see Chapter 8 for the spectrum of light and its physical properties). Nearly twenty years later, in 1952, a classic scientific paper established that the germination of lettuce seeds was stimulated by exposure to a few minutes of red light (between 600 and 700 nm), and that this was prevented when the red light was immediately followed by a brief exposure to light in the far-red region. After a series of alternate exposures to red and far-red, the response depended on the last exposure in the sequence. When the sequence ended in red, most of the seeds subsequently germinated in darkness but, when the sequence ended in far-red, many of the seeds failed to germinate (Table 6.3).

The authors concluded that the response must be mediated by a single light-sensitive pigment, which they called *phytochrome* (the word phytochrome simply means ‘plant pigment’). They also concluded that phytochrome could exist in two forms, and that these could be converted from one to the other by exposure to the appropriate wavelength of light in the following reactions:

$$\text{Pr} \xrightarrow{\text{red light}} \text{Pfr}$$
$$\text{Pr} \xleftarrow{\text{far-red light}} \text{Pfr}$$

The Pr form of phytochrome absorbs red light and is converted to Pfr. In the reverse reaction, the Pfr form absorbs far-red light and is converted to Pr. Since the germination response required exposure to red light, it was further deduced that Pfr (formed by red light) is the biologically active form and that phytochrome must be synthesised in darkness in the inactive form as follows:

Table 6.3 Exposing seeds of lettuce 'Grand Rapids' to red light promotes germination, while exposure to far-red light decreases germination.

Light treatment	Percentage of seeds germinating
Darkness	9
1 minute of **red light**	98
1 minute red, followed by 4 minutes of **far-red light**	54
1 minute red, 4 minutes far-red, followed by 1 minute **red light**	100
1 minute red, 4 minutes far-red, 1 minute red followed by 4 minutes **far-red light**	43

When the sequence of light treatments ended in **red**, germination was increased.
When the sequence of light treatments ended in **far-red**, germination was decreased.

$$\text{darkness} \longrightarrow \text{Pr} \underset{\text{far red}}{\overset{\text{red}}{\rightleftharpoons}} \text{Pfr} \longrightarrow \text{response}$$

Phytochrome was purified in 1966 and shown to be a protein associated with a light-absorbing region, called a chromophore. Much more recently it has been found that plants contain more than one kind of phytochrome and that these different molecules may have distinct functions. Phytochrome is now known to control a wide range of plant responses to light in addition to germination, such as the elongation of stems, branching and the expansion of leaves, as well as responses to day-length (see Chapters 5, 8 and 9).

To return to germination, we now understand the reasons for the evolution of the photo-reversible phytochrome system. Initially, scientists could not understand why such a pigment should have developed, as plants are not exposed to sequences of red and far-red light under natural conditions. However, red light is strongly absorbed by leaves, while far-red light passes through them. Since far-red light strongly inhibits germination by converting phytochrome to the inactive form, germination is prevented under a leaf canopy close to the parent plant. Thus, the presence of the inactive form in the dark prevents the germination of buried seeds and the reversible reaction, which converts phytochrome to the inactive Pr form in far-red light, prevents germination in heavy shade where the seedlings would not flourish. As discussed in Chapter 5, the reversible property of phytochrome is also an important factor in the survival strategy of plants that are shaded by their neighbours.

Light requirements for germination

Not all seeds require light in order to germinate. Nor indeed is the light requirement always satisfied by a brief exposure. In some cases, quite long durations of light, up to several hours each day, are necessary. This would allow them to ignore sun flecks passing between the overhead leaves and ensure that germination takes place only where there is a gap in the canopy with sufficient light to allow the new seedling to become established.

Measurements of the penetration of light into soils have shown that they are very opaque and only in the top few mm is there enough light for germination. For example, in field mustard (*Sinapis arvensis*) germination was reduced to zero when seeds were buried 10 mm deep, and to 20% at only 5 mm. For light-requiring seeds, therefore, it is important to sow them at, or very close to the surface of the soil (Fig. 6.2). This applies to most turf grasses, leading to the well-known problem of birds eating the seeds that are at or near the soil surface.

Large seeds often germinate in the dark and the seedlings are able to reach the surface before their food reserves are exhausted. In contrast, a high proportion of small seeds need light and so only germinate when they are close to the soil surface; this allows them to emerge into the light and establish photosynthesis before their small reserves are depleted. Most soils contain a large reservoir of seeds that are potentially capable of germination but are prevented from doing so by lack of light until the soil is disturbed. Gardeners know only too well that cultivating the soil often leads to an explosion of germinating weeds. However, under natural conditions, the need for soil disturbance results in germination being spread over time and increases the chances of survival of the population as a whole.

Temperature

Low temperature

In many temperate and cool climate species, the hydrated seed must be exposed to a period of chilling before germination can occur. The effective temperatures are similar to those that break winter dormancy in the buds of woody species and bring about vernalisation in biennial plants (see Chapter 9); the optimum temperature is usually about 5°C. Under natural conditions, a chilling requirement delays germination until the spring thus allowing the longest possible time for annual plants to complete their life cycle, and for seedlings of perennial plants to become established before being exposed to low winter temperatures.

A chilling requirement is the underlying reason for the horticultural technique of stratification, which involves layering the seeds in moist soil and maintaining them at low temperatures. The optimum temperatures (3–5°C) are in the range found in most domestic refrigerators. Stratification at low temperatures probably acts by changing the balance between growth inhibitors and growth promoters in favour of the latter. For example, in seeds of Norway maple (*Acer platanoides*), the amount of abscisic acid (a growth inhibitor) decreased during stratification, while the amount of gibberellins (growth promoters) increased. A decrease in abscisic acid has also been observed on seeds of ash (*Fraxinus americana*), walnut (*Juglans regia*) and hazel (*Corylus avellana*).

High temperature

A requirement for high temperature to break dormancy is rather rare in temperate species. The best example is bluebell (*Hyacinthoides non-scripta*), where the seeds are shed in early summer and fail to germinate unless they are exposed to high temperatures during the summer; germination will then occur in the autumn. This is a deciduous woodland species, so germination and growth must occur in autumn, winter and early spring when the leaves are off the trees and light can reach the seedlings.

Contrary to popular belief, a requirement for exposure to high temperature does not underlie the use of boiling water to break dormancy. This treatment is effective because it softens or cracks the seed coat, allowing water to enter. Although popular, especially for parsley seed (*Petroselinum crispum*), it is not an advisable technique as the seed may be damaged or even killed.

Alternating temperatures

Many seeds will not germinate until they have been exposed to particular alternating temperature regimes. Alternations between low and high temperatures during each day promote germination in a number of species; these include evening primrose (*Oenothera biennis*), tobacco (*Nicotiana tabacum*), poppy (*Papaver* spp.) and many common weeds such as dock (*Rumex crispus*) and rough meadow grass (*Poa trivialis*). However, alternating temperatures are often only effective when combined with other factors such as light, as in celery (*Apium graveolens*) and love-lies-bleeding (*Amaranthus* spp.). The stimulation of germination by alternating temperatures may result from mechanical changes in the seed or may possibly be due to specific temperature requirements of sequential reactions that take place during germination.

Seasonal alternations of temperature are necessary for the germination of some seeds. An example is that of *epicotyl dormancy* (the epicotyl is the portion of the stem of an embryo or seedling above the cotyledons) in the Guelder rose (*Viburnum opulus*) where the root emerges when the seeds are maintained in warm temperatures (20–30°C) but the shoot fails to grow until the well-developed root system has been exposed to cold. Epicotyl dormancy also occurs in *Actaea* spp., *Allium burdickii*, *Asarum canadense*, *Fritillaria ussuriensis*, *Hydrophyllum*, *Lilium*, *Paeonia* and *Viburnum* spp. The radicle of the seed emerges in autumn (i.e. at warm temperatures between 10 and 30°C, depending upon species) while the shoot apex exhibits dormancy. This is similar to bud dormancy (see Chapter 9) and requires cold temperatures (0–5°C) in order to break dormancy and allow the shoot to grow. Cold stratification for three to seven months over the winter is the usual way of breaking the dormancy of the shoot, but soaking in a solution of the growth hormone, gibberellin, may also be effective.

Epicotyl dormancy allows the plant to establish a root system following germination in the autumn, but prevents the emergence of the shoot until the following spring in order to avoid damage from the cold. It also enables germination of seeds over several years and ensures a reserve of ungerminated seeds in the soil.

Another example is that of *double-dormancy* which occurs in *Trillium* and the so-called 'two-year' lilies. Here the embryo has not completed its development at the time of shedding and the seeds have first to be exposed to warm temperatures to allow the embryo to develop. This must be followed by a chilling treatment in order for the hypocotyl and roots to be formed. A second exposure to warm temperature allows these to grow until, once a certain stage of development is reached, a second exposure to cold allows the shoot to emerge.

Under natural conditions, seeds with epicotyl or double-dormancy fail to develop into seedlings in one year and are, therefore, called biennial seeds. Some doubt exists concerning the possible ecological advantage of this type of dormancy. Species exhibiting such phenomena will apparently only produce fully developed seedlings in nature after two or more seasons. However, as soon as growth has started, they lose the protection which the 'uncommitted' state provides against the elements.

Hard seed coats

These inhibit germination by preventing the uptake of water into the dry seed. Under natural conditions, hard seed coats are rendered permeable to water by the action of microorganisms in the soil or by passing through the digestive tract of an animal. However, the gardener has a number of short-cuts available. These include scarification with coarse sandpaper or chipping, as commonly practised with sweet pea seeds (*Lathyrus odoratus*) and treatment with alcohol, which is particularly effective on members of the family Caesalpiniaceae. The dormancy-breaking action of abrasion, alcohol or sulphuric acid is directly related to an increase in the permeability of the seed coats to water, although they probably also induce other changes such as permeability to gases, sensitivity to light or temperature, and even a reduction in the amount of inhibitory substances present.

Fire

Several trees and shrubs of sub-tropical and semi-arid regions have seeds that fail to germinate until they have been exposed to fire. Under natural conditions, such a response prevents the seeds from germinating until a fire has burned down the overhead vegetation and so reduces the competition for light, nutrients and water. Several species take advantage of this reduction in competition and are known to be rapid colonisers of burnt ground. The heat of the fire also may be necessary for the dehiscence of the seed from the follicles, as in *Banksia ericifolia*. Fire can also destroy accumulated inhibitors in the soil, removing a possible cause of failure to germinate.

The charred remains of chaparral vegetation appear to stimulate germination in yellow bells (*Emmeranthe penduliflora*), a species found in arid habitats of southwestern North America, suggesting that gases in the smoke or substances generated on burnt sites play a role in the post-fire germination of chaparral annuals. Indeed smoke derived from the burning of plant material has been found to stimulate germination in over 170 species native to Mediterranean-type ecosystems (where there is an ever-present threat) on three continents, representing 37 families and 88 genera and a range of plant types. Within the garden context it has been found that smoke-impregnated disks can stimulate germination and such disks are available for use as dormancy-breaking treatments.

After-ripening in dry storage

When seeds are stored after harvest in an air-dry state, they gradually lose their innate dormancy and become able to germinate (e.g. barley, *Hordeum* spp.; sand dropseed, *Sporobolus cryptandrus*; and sedge, *Cyperus rotundus*). In commercial practice it is extremely common for dormancy to be broken in this way, although it should be used with caution because viability can be adversely affected if the temperature or moisture content of the after-ripening environment increases (see below under seed storage). It is likely that at least some of the changes which take place during dry storage are a result of changes in the seed coats which reduce their tensile strength or possibly their impermeability to water or respiratory gases. In other cases it is likely that changes occur in the embryo itself or the immediate chemical environment of the embryo which result in an increase in the potential of the embryo to exert growth forces against the mechanical restraints of the seed coats.

The need for after-ripening may also exert a protective role against precocious or accidental germination early in the season of maturation when conditions may not be favourable to seedling development (e.g. cotton, *Gossypium hirsutum*; garden balsam, *Impatiens balsamina*; lettuce, *Lactuca sativa* 'Grand Rapids'; and evening primrose, *Oenothera odorata*).

Chemical treatments

A number of chemicals are known to break dormancy and stimulate germination. Among these are plant growth regulators, both naturally-occurring hormones and synthetic growth promoters. Seeds may contain different amounts of hormones, accounting in part both for their varying depths of dormancy (e.g. apple, *Malus domestica*; blackberry, *Rubus idaeus*; peach, *Prunus persica*; ash, *Fraxinus excelsior*; and rowan, *Sorbus aucuparia*) and for the differences in the way that they respond to hormonal treatments. Many seeds respond to the application of gibberellins (e.g. maize, *Zea mays*; and barley, *Hordeum* spp.), although a mixture of gibberellins often gives a better response than treatment with a single gibberellin such as GA_3 (gibberellic acid).

Other chemicals that have been reported to stimulate germination in one or more species include hydrogen peroxide, sodium hypochlorite, thiourea, mercaptoethanol, nitrates, nitrites, acids, cyanide, azide, hydroxylamine, chloroform, ether and ethylene gas (Table 6.4). In general, their action is not understood, although it is interesting that some of them (for example, thiourea) may also be effective in breaking bud dormancy (see Chapter 9). Many of these are hazardous chemicals and are not recommended for general use.

Table 6.4 Miscellaneous chemicals that promote germination.

Chemical	Species in which germination is promoted
Hydrogen peroxide	*Daphne* spp.; carrot, *Daucus carota*; beetroot, *Beta vulgaris*
Sodium hypochlorite	celery, *Apium graveolens* var. *dulce*
Thiourea	lettuce, *Lactuca sativa*
Mercaptoethanol	barley, *Hordeum vulgare*
Nitrates	tobacco, *Nicotiana tabacum*
Nitrites	rice, *Oryza sativa; barley, Hordeum vulgare*; shepherd's purse, *Capsella bursa-pastoris*
Acids	sedge, *Cyperus rotundus*; pineapple, *Ananas comosus*; grape, *Vitis vinifera*
Cyanide	lettuce, *Lactuca sativa* 'Grand Rapids'
Azide	wild oat, *Avena fatua*; guava, *Psidium guajava*
Hydroxylamine	tobacco, *Nicotiana tabacum*
Chloroform	cashew, *Anacardium occidentale*
Ether	upland cotton, *Gossypium hirsutum*
Ethylene gas	lettuce, *Lactuca sativa*; peanut, *Arachis hypogaea*; clover, *Trifolium subterraneum*

In conclusion, it is important to note that treatments that are said to promote germination often interact, and a combination of more than one treatment may be necessary in order to break dormancy. A classical demonstration of this can be seen in dock (*Rumex crispus*), where the effects of light or darkness, alternating or constant temperature and treatment with potassium nitrate were examined. Only the combination of alternating temperatures, light, and potassium nitrate completely broke dormancy, none of the other treatments, singly or in combination produced 100% germination. Recent work on *Daphne longilobata* has demonstrated the same phenomenon. Here alternating temperatures, hydrogen peroxide and gibberellic acid produced nearly 100% germination.

SEED VIGOUR

Vigour is a measure of seed quality and relates to the ability of the seed to germinate and establish under a wide range of environmental conditions. High-vigour seeds will germinate rapidly to produce a uniform stand, even in adverse conditions, while low-vigour seeds germinate erratically, producing a variable, patchy stand. Seed vigour affects various aspects of performance including the rate and uniformity of germination, the rate and uniformity of seedling emergence and growth, the ability of seedlings to emerge under unfavourable environmental conditions and the potential for seed storage.

Changes in growing techniques have made growers and gardeners more concerned with the quality of seeds. New technologies, such as F1 seeds (i.e. first filial; the first offspring of a cross between two distinct cultivars which have previously been self-fertilised or *selfed*), precision drilling, seed pelleting and priming, plant modules, and once-over harvests all rely on high-vigour seeds capable of producing a uniform stand. A new development is that supermarkets now demand crops of specific size. This is achieved by establishing a particular spacing of the plants which, in turn, depends heavily on high-vigour seeds so that germination is uniform over the entire crop.

As might be expected, seed vigour is influenced by the genetic constitution of the plant. For example, seeds of lettuce (*Lactuca sativa*), onion (*Allium cepa*), parsnip (*Pastinaca sativa*) and peanut (*Arachis hypogaea*) typically have low vigour, while cereals and most pea seeds have high vigour. However, wrinkled and round-seeded peas differ in their vigour; round peas have starchy seeds and are usually much more vigorous than their wrinkle-seeded counterparts.

In addition to their genetic constitution, a number of factors influence the quality and vigour of seeds. To maintain high vigour, it is important that the mother plant receives an adequate supply of nutrients, especially nitrogen and phosphorus, and that the seeds are allowed to grow to maturity before harvesting them. Although small seeds are themselves not necessarily of poor quality, they may be immature and, therefore, of low vigour. Other procedures that may be adopted to produce and maintain high vigour in seed lots include the use of protective seed dressings against pests and diseases and the use of new technologies such as pelleting and priming.

Priming involves the controlled hydration of seeds. Enough water is allowed to enter to enable pre-germination metabolic events to occur but this is not sufficient for radicle emergence. Thus the seed lot is made more uniform with all the seeds at the brink of germination, resulting in rapid and uniform germination when the seeds are sown. Primed seeds of leek, carrot, tomato and some flowers are easily obtainable. Pelleted seed is encapsulated in an inert coat (usually clay) to which pesticides can be added. The process makes irregular shaped seeds like celery, carrots, parsnips and lettuce round and allows these seeds to flow easily through seed drills, producing more even stands.

Weather conditions during seed ripening can adversely affect seed vigour. Excessive rainfall can lead to an increase in fungal infection while cycles of wet and dry periods can damage the seeds as they swell and contract under the changing moisture conditions. Damaged seed allows further ingress of pathogens and also attack by insects, resulting in seeds with low vigour.

For the amateur gardener who relies on home-saved seed for the next crop, the best way to ensure high-vigour seed which stores well is to ensure that the mother plant has adequate nutrition so that it matures fully on the parent plant. Once the seed is ripe, it should be harvested, dried quickly at a relatively cool temperature and stored in an air-tight container in a refrigerator until needed. When the seed is removed from the refrigerator, it should be allowed to equilibrate with the air temperature before the container is opened. This ensures that the warm, ambient air (which has a higher humidity than the air in the container) does not rapidly enter the seed possibly causing membrane or cellular damage. This technique also ensures that the seed moisture content does not suddenly increase with the sudden rush of air. A high seed moisture content would result in more rapid loss of viability and seed quality.

Finally, the most important factor influencing the vigour and viability of seeds is the conditions under which they are stored. Poor storage conditions can lead to a rapid loss of vigour and seed viability, even in high-vigour seeds. Low-vigour seeds do not store well and, consequently, the storage conditions are particularly important if vigour and viability are to be maintained.

SEED STORAGE

Seeds are stored for a variety of reasons:

- to maintain seed stocks for growing from one season to the next
- to keep stocks for sale
- to maintain breeding lines and
- to conserve genetic material.

Food and grain seeds are stored before processing and consumption.

Seed vigour and seed longevity are closely related and, other things being equal, high-vigour seeds will retain their viability for longer than low-vigour seeds. Thus the factors described above which increase vigour will also normally increase longevity. Similarly, the genetic constitution affects both vigour and storage potential. Seeds of onions and lettuce have low vigour and store less well than, for example, the high-vigour, starchy seeds of round-seeded peas. However, the conditions under which the seeds are dried and stored are also extremely important in determining longevity.

Since all seeds lose vigour and viability during storage, it is important to establish the optimum storage conditions for different types of seeds. In 1973, Roberts divided seeds into two broad categories, depending on their storage behaviour. Seeds which can be dried to low moisture contents and stored at −18°C for long periods are termed *orthodox* seeds, while those that are killed if their moisture content is reduced below some relatively high level (12–31%) are termed *recalcitrant* seeds. The latter group includes many tropical species, especially tropical fruits such as mangoes (*Mangifera indica*), coconuts (*Cocos nucifera*) and jackfruit (*Artocarpus heterophyllus*), as well as several large-seeded temperate species such as oak (*Quercus robur*), horsechestnut (*Aesculus hippocastanum*) and sweet chestnut (*Castanea sativa*). More recently an intermediate group has been identified, where drying below 10–12% moisture results in early loss of viability; coffee (*Coffea arabica*) belongs in this group.

Damaged seeds do not store well and it is important to minimise damage during and after harvesting. A common cause of post-harvest damage with consequent loss of viability is drying the seeds at too high a temperature. There is no real 'safe' temperature at which to dry seeds, but seed banks minimise deterioration and thereby safeguard longevity by drying at 15°C and 15% moisture content of the air.

Seed moisture content and storage temperature
Generally, the lower the moisture content and the lower the temperature the greater the seed longevity, as under these conditions the seeds are maintained in a state of suspended animation. Thus the wetter and warmer the seeds are in storage, the faster they lose viability and vigour. Therefore orthodox seeds should be stored under dry and cool conditions. In order to maintain the viability of orthodox seeds for as long as possible, the ideal conditions are hermetic storage at a temperature below −18°C (the lowest limit of a domestic freezer) and a moisture content below 5%. In these conditions, metabolic processes are reduced to a minimum level. Predictions of storage longevity under these conditions are given in Table 6.5.

In summary, the best way to store seeds is to keep them cool and dry. A popular and effective method is to keep them in an air-tight container (preferably with a desiccant) in a refrigerator.

Table 6.5 Estimate of probable regeneration intervals* for seeds stored at −20°C and 5% moisture content. From (Roberts, E.H. and Ellis, R.H. (1977). Prediction of seed longevity at sub-zero temperatures and genetic resources conservation. *Nature*, **368**, 431–3.)

Plant	Cultivar	Probable regeneration interval (years)
Barley	'Proctor'	70
Rice	'Norin'	300
Wheat	'Atle'	78
Broad bean	'Claudia Superaquadulce'	270
Pea	'Meteor'	1090
Onion	'White Portugal'	28
Lettuce	'Grand Rapids'	11

* The regeneration interval is the predicted time for viability to fall to 95% of its initial value. Thus, it will take the barley cultivar 'Proctor' 70 years to fall from 99% viability to 94% viability, if stored under optimum conditions.

However, it is important that the container is hermetically sealed, otherwise the seeds will equilibrate with the moisture content of the refrigerator. This will hasten their deterioration even though they are being stored at a low temperature.

Loss of seed viability

Surprisingly, scientists do not know what causes loss of viability and ultimately the death of seeds. Aseptic seeds are known to last longer than non-sterile seeds, but the former still lose viability. Loss of viability also occurs at relative humidities below 65% when storage fungi are inactive and so it is concluded that fungi accelerate the loss of viability in stored seeds, but do not cause it. Other suggestions include the accumulation of inhibitory chemicals, the depletion of essential food reserves, or the breakdown of proteins and nucleic acids, but so far there is little evidence to support them. However, it has been demonstrated that the loss of viability is accompanied by genetic deterioration. Gardeners who grow peas, particularly from home-saved seeds, will be familiar with the odd albino pea (without chlorophyll) which emerges. This is a direct result of a mutation which occurs during seed ageing. Background ionising radiation was once thought to be a cause of such genetic lesions, but natural levels are too low to bring about this type of damage. It is evident that irreversible damage to the genome occurs as the seeds age, but it is not clear whether this is a cause of the loss of viability or an effect.

NEW DEVELOPMENTS IN SEED TECHNOLOGY

Terminator gene technology

This refers to the technology of genetic manipulation which prevents the seeds of genetically modified crops from germinating. A promoter sequence (a DNA sequence which activates or switches on a specific gene) is attached to a gene that prevents germination and this is then inserted into the seed of the target species. The genetically modified seed is itself capable of germination but the plants that develop from the seed are sterile because the germination-inhibiting gene is activated during seed maturation.

This technology has caused widespread concern because it undermines a common premise of food production, namely the ability to produce self-saved seed. However, this is also true of the well-established practice of producing F1 hybrid seeds, which are the product of crosses between inbred parental lines. Seed produced on these plants do not breed true and so self-saved seed is not possible. Nevertheless, public outrage has halted, for the time being, any further developments in terminator gene technology. No new cultivars containing terminator genes are at or near trialling, or registration, so it is unlikely that such genes will be found in new genetically modified crops. At present not all plants are amenable to sterility engineering but, once perfected, it would be an effective way of preventing 'genetic pollution'.

Gene banks

The need to conserve genetic material for future breeding programmes and to combat the loss of gene pools that is occurring in centres of plant diversity, especially those in developing countries, is now widely recognised. An international network of gene banks has been set up under the auspices of the Food and Agriculture Organization of the United Nations (FAO) and the International Plant Genetic Resources Institute (IPGRI) to prevent this rapid, world-wide depletion of stocks. International gene banks have been established for many crop plants, such as those at the Horticulture Research International, Wellesbourne, UK (for temperate brassicas, radishes and *Allium* spp. and cultivars), the International Rice Research Institute, Philippines (for rice), and the International Center for Agricultural Research in Dry Areas, Syria (for certain legumes and cereals). The gene banks distribute and exchange seed collections for research purposes, as well as conserving *germplasm*. Germplasm is a general term covering the variety of forms, such as seeds, cuttings and tubers, in which a plant's genetic material can be preserved. The easiest and simplest way to store genetic material is to store it in the form of seeds. If care is taken during collection to ensure that a representative sample of genetic diversity is obtained then the seed collections in gene banks are an effective means of conservation.

Seed banks can keep live seeds in a state of suspended animation for hundreds of years. Thus if a plant become extinct in the wild, it will not be lost forever if its seeds are secure in a seed bank. A system of monitoring the accessions is in place so that when viability falls below a pre-determined level (currently 85%), the collection is regenerated to maintain genetic stability. Much of the collecting of crop plants is now complete and efforts are being concentrated on describing and characterising the collections.

One of the largest conservation projects ever undertaken is the Millennium Seed Bank project at Royal Botanic Gardens Kew, Wakehurst Place in West Sussex. This project aims to secure the future of almost all the UK's native flowering plants as well as help safeguard 24 000 species of dry land plants world-wide against extinction (i.e. 10% of all the world's flora). Some 300 of the UK's wild plants are threatened with extinction. The seed bank aims to collect and conserve representative samples of all the 1400 native plants that set seed. The flora of the dry lands will also be conserved as they are amongst the most threatened environments on earth and home to an immense variety of plant life.

The Millennium Seed Bank will be a world resource and, as such, the seeds will be made available to researchers, conservationists and scientific institutions around the world free of charge. Duplicate collections of seed will also be kept in the country of origin. These seeds could be used, for example, to breed plants with resistance to possible future outbreaks of pest and disease, plants more adapted to climate change (e.g. drought, frost) or plants tolerant to saline conditions. They could also be used to restore damaged or even destroyed environments.

FURTHER READING

Attridge, T.H. (1990) *Light and Plant Responses*. Edward Arnold, London.

Atwater, B.R. (1980) Germination, dormancy and morphology of seeds of herbaceous ornamental plants. *Seed Science and Technology* **8**, 523–73.

Baskin, J.M. & Baskin, C.C. (1998) *Seeds: Ecology, Biogeography and Evolution of Dormancy and Germination*. Academic Press, San Diego, CA.

Black, M. & Bewley, J.D. (2000) *Seed Technology and its Biological Basis*. Sheffield Academic Press Ltd, Sheffield.

Bleasdale, J.K.A. (1973) *Plant Physiology in Relation to Horticulture*. MacMillan, London.

Bryant, J.A. (1985) *Seed Physiology*. The Institute of Biology's *Studies in Biology* **165**. Edward Arnold, London.

Ellis, R.H., Hong, T.D. & Roberts, E.H. (1985) *Handbook of Seed Technology for Genebanks Vol. 1: Principles and Methodology*. IBPGR, Rome.

Khan, A.A. (ed.) (1977) *The Physiology and Biochemistry of Seed Dormancy and Germination*. Elsevier, North-Holland.

Kozlowski, T.T. (1972) *Seed Biology. Vols 1 & 2*. Academic Press, Ltd, London.

Mayer, A.M. & Poljakoff-Mayber, A. (1982) *The Germination of Seeds*. Pergamon Press, Oxford.

7

Vegetative Propagation

- General principles: clones
- Regeneration of roots and shoots
- Natural and synthetic auxins, their movement within the plant and their effects on root initiation
- Natural and synthetic cytokins; effect of cytokinin/auxin ratios on shoot and root production; cytokinins and leaf senescence
- Juvenile and adult plants
- Perpetuation of juvenility in cuttings
- Maintaining juvenility and the transition from juvenile to adult
- Avoiding juvenility by propagation from adult plants
- Etiolation: effects on rooting
- Types of cutting and their management: leafy shoot cuttings and the effects of wounding; optimum conditions for rooting; reducing water loss; hardwood cuttings; leaf and leaf/bud cuttings; layering, root cuttings; air layering; division
- Propagation from specialised structures: bulbs, pseudobulbs, corms, rhizomes and tubers
- Grafting: why use grafted material?
- The effects of rootstocks and how they control the growth of the scion
- Compatible and incompatible grafts
- Grafting techniques
- Micro-propagation

Under natural conditions in the wild, the vast majority of flowering plants and conifers reproduce and increase their number by producing seeds (Chapter 6). In the garden context, however, this is often not possible and gardeners have to resort to some form of vegetative means in order to increase their stock of plants. Many garden plants do not breed true from seed and some forms that do breed true may normally produce little or no seed, or the seed may be difficult to germinate. In such cases, plants must be vegetatively propagated. Other reasons for carrying out vegetative propagation include avoiding a juvenile period by propagating from adult tissue (see *juvenility* below), and the use of specialised rootstocks in order to obtain a desirable habit of growth. Many popular garden plants have arisen by selecting a good form, such as good autumn or flower colour, or dwarf habit, and then propagating it vegetatively

The cultivars of most fruit plants and many ornamentals have originated from a single parent, which has been multiplied by vegetative means. Such plants constitute a *clone* and each group is genetically identical. However, they are not necessarily identical in all characteristics, the actual appearance of the plant, its *phenotype,* being the result of the interaction between its genes (its *genotype,* see Chapter 3) and the environment in which it grows. The maintenance of the clone requires, of course, that the number of individuals can only be increased by vegetative propagation.

The life of any clone is theoretically unlimited, since ageing individuals are usually re-invigorated by propagation. For example, the 'Williams Bon Chrétien' pear is known to have been grown for some 200 years and the 'Thompson Seedless' grape may have been cultivated for more than 2000. Deterioration in particular clones frequently occurs, however; the most common reason is infection with one or more viruses. Genetic mutations can also produce off-types that may reduce the value of the clone, although such mutations sometimes lead to improved forms. For example, new colour sports in ornamental plants such as carnations and chrysanthemums often arise in this way. An important consideration during vegetative propagation is to maintain healthy stocks that are free from virus and true to type.

CELL DIFFERENTIATION

Reproduction from cuttings and grafts requires the differentiation of new types of cell. Stem cuttings must produce root cells, leaf cuttings must produce both root and shoot cells, while buds and grafts must generate new vascular tissues in order to form a union between the scion or bud and the stock to which it is attached. Development into different types of cell is under genetic control and most plants cells contain all the genetic information required to produce a new individual. They are, therefore, said to be *totipotent.*

Successful vegetative propagation requires that specific cells in the plant can be induced to express the genes controlling the production of new roots and shoots. The fundamental biology of what triggers the formation of such *adventitious* tissues remains largely unknown, however. It is evident that cells must begin to initiate cell divisions and that these dividing cells must ultimately become organised into one or more new growing points. Pre-formed root initials are present in a number of easy-to-root species such as poplar. These remain dormant until the cutting is made and then emerge as *adventitious* roots. In plants with no pre-formed root initials, new roots are often induced to form by wounding. It is for this reason that the wounded area at the base of a cutting is sometimes increased, by removing an outer sliver of tissue or by cutting up into the cutting from its base, in order to stimulate the formation of new roots.

Plant hormones

Plant hormones have been shown to influence the division, enlargement and differentiation of cells and they clearly play an important part in the processes involved in the rooting of cuttings and the formation of graft unions. They are organic substances that are synthesised in one region of the plant and move to another, where they bring about a physiological response at very low concentrations. In some cases they can also act at the site where they are produced. Several classes of hormone have been identified in plants and they are known to influence a very wide range of responses.

Based on studies of the formation of new plants from *callus tissue* (the undifferentiated mass of cells formed at the site of a wound) or from pieces of tissue consisting of a few of the outermost layers of a stem, two classes of plant hormone (*cytokinins* and *auxins*) have been found to influence the way in which plants develop. In general, it has been found that a high ratio of cytokinin to auxin favours the formation of shoot cells, while a high auxin to cytokinin ratio favours the formation of root cells (Fig. 7.1). The precise mechanisms through which these hormones control the direction of development are still unknown, although it is probable that they modify gene activity in some way. Because of their considerable influence on the production of root and shoot cells, information about the movement and effects of auxins and cytokinins within the plant helps the gardener to understand more about the management of various types of *propagule* (the name given to any part of a plant that is used to start a new plant).

Auxins

This was the first category of plant hormones to be recognised. The major naturally occurring auxin is β indolyl acetic acid (IAA) but, since its discovery in 1936, many synthetic auxins have been developed Of these, the best known and most widely used are α napthalene acetic acid (NAA), 2,4-dichlorophenoxy acetic acid (2,4-D) and 2-methyl-4-chlorophenoxy acetic acid (MCPA). Because they do not occur naturally in plants, they are not true hormones and are, therefore, called *plant growth regulators* (PGRs). Their effects are similar to those of IAA, but their behaviour in the plant is somewhat different. For example, whereas there are biochemical mechanisms for the destruction or inactivation of IAA, many synthetic auxins are not readily destroyed and so remain active in the plant for a long time. This has important consequences since the effect of auxin is strongly dependent on its concentra-

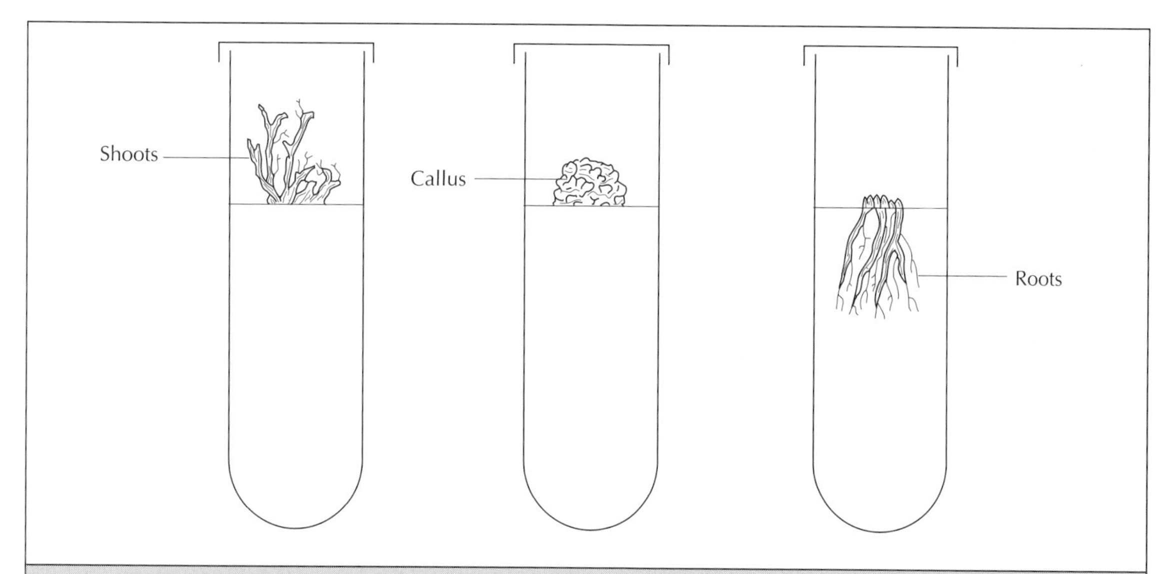

Fig. 7.1 The ratio of auxin to cytokinin affects whether shoots or roots are produced in tissue cultures. A *low* auxin/*high* cytokinin ratio (left) promotes the development of shoots while a *high* auxin/*low* cytokinin ratio (right) promotes rooting. More equal concentrations often lead to the development of undifferentiated callus cells (centre).

tion. At high concentrations, auxins cause excessive growth (called *hypertrophy*) and ultimately death. For this reason, stable synthetic auxins (such as 2,4-D) are often used as weedkillers (see Chapter 11). At lower concentrations, auxins modify tissue differentiation.

An important feature of plants is their *polarity*. This means that they can detect which way is up, even if they are removed from the plant. Cuttings inserted into the soil upside down will still produce roots from the cut surface that was originally towards the base of the plant. This polarity is important for the direction of movement of IAA within the plant; in stems, this auxin moves downwards towards the root end regardless of the orientation of the shoot. Therefore, in stem cuttings, IAA moves downwards from the younger leaves and apex where it is produced and accumulates at the base of the cutting. Because of the concentration effect (high auxin/cytokinin ratio favours root cell differentiation) new roots develop here.

Root formation can often be increased by the application of auxins to the base of the cutting. IAA itself is not normally employed, probably because it is easily converted to an inactive form, especially on exposure to air. The chemicals most commonly used to enhance root formation are, therefore, the synthetic auxin, NAA and the natural auxin, indolyl butyric acid (IBA). The latter is closely related to IAA and has been found to occur naturally in several plants. Since synthetic auxins are not transported in a polar direction like IAA, they must be applied to the base of the cutting where root initials are wanted.

We have already seen that many synthetic auxins are not readily degraded within the plant and the maintenance of high levels of auxins at the base of the cutting can lead to problems once the new root initials have been formed. High auxin concentrations stimulate the production of root initials, but inhibit the elongation of the roots themselves. At low concentrations auxins promote root elongation. For this reason, IBA is the most commonly used rooting hormone in gardening because it is gradually broken down, leading to a more optimal concentration for root elongation once the root initials have been formed. These different responses to auxin depending on its concentration within the plant, illustrate the complexity of hormone action.

Although it is generally accepted that IAA plays a central role in root initiation, many plants are difficult to root and respond poorly to auxin application, suggesting that other factors may also be involved. Seasonal variations in rooting performance are frequently observed and this may, in part, be due to changes in the amount of natural auxin present within the tissues. It is known that, in some cases, hormone levels are influenced by day-length. Both auxin and cytokinin levels can be influenced by day-length (see below), so the auxin/cytokinin ratio, known from experimental studies to be important in cell differentiation patterns, may also vary with season. A few experiments have shown that day-length can influence the rooting behaviour of leafy cuttings (Table 7.1).

Cytokinins

In the 1940s, coconut milk was found to be rich in a compound that increased cell division rates in plant tissues cultured on an artificial growth

Table 7.1 Day-length affects rooting capacity in some plants (Examples taken from Vince-Prue, 1975).

Rooting was increased when the parent plant was grown in long days:
Lorraine begonia, leaf cuttings (*Begonia* × *cheimantha*)
Flowering dogwood (*Cornus florida*)
Canadian poplar (*Populus* × *canadensis*)

Rooting was increased when the parent plant was grown in short days:
Carnation (*Dianthus caryophyllus*)
Japanese holly (*Ilex crenata*)
Weeping willow (*Salix babylonica*)

Rooting was increased when the cuttings were grown in long days:
Abelia (*Abelia* × *grandiflora*)
Canadian poplar (*Populus* × *canadensis* 'Robusta')
Weigela (*Weigela florida*)

Rooting was increased when the cuttings were grown in short days:
Japanese holly (*Ilex crenata*)
Chinese juniper (*Juniperus chinensis*)

medium (*tissue cultures*). Later the active ingredients were identified as *zeatin* and a related compound, *zeatin riboside*. Because of their effect on cell division (*cytokinesis*), these hormones were called *cytokinins*. Zeatin and several other compounds that have since been shown to stimulate cell division are related to the chemical, adenine. Most of these chemicals are not commercially available, with the exception of benzyladenine (BA) which occurs naturally but is not commonly found in plants, and kinetin, a synthetic cytokinin. Cytokinins occur in plants in young organs and root tips, and are probably produced there. Roots are known to be an important source of cytokinins and supply leaves, fruits and seeds via the xylem.

Experimental studies have shown that naturally ocurring cytokinins are essential for cell division in plants and, therefore, play an important role in tissue regeneration. For example, pith from tobacco produces a mass of unspecialised cells (callus) when cultured in the laboratory on a growth medium with equal amounts of cytokinin and auxin. With a high cytokinin to auxin ratio, shoots develop, whereas a low cytokinin to auxin ratio favours the development of roots (Fig. 7.1). Cytokinins are not used in the propagation of cuttings because a high cytokinin/auxin ratio inhibits the formation of root initials. They are used, however, in micropropagation (see below) to stimulate the production of shoots.

Cytokinins from the roots also appear to be involved in the delay of leaf senescence. When leaves are removed from the plant, they rapidly become yellow, especially if kept in darkness. This ageing or *senescence* is greatly delayed, however, if adventitious roots form at the base of the petiole. Moreover, the effect of the roots can be partially replaced by the application of a cytokinin. Thus root formation on cuttings is important not only for water uptake, but also for the supply of essential growth factors to the shoot. Gardeners will be familiar with this phenomenon for, in the absence of roots, leafy cuttings rapidly yellow and die whereas, when roots develop, the leaves remain green and healthy.

Juvenility

Rooting ability is markedly influenced by whether the plant is in its juvenile or adult phase of growth. *Juvenility* is the name usually given to an early phase of growth during which flowering cannot be induced by any treatment. It occurs in all plants but is most readily observed in trees because they have a long juvenile period often lasting for many years, in contrast to many herbaceous plants where the juvenile period may last for only a few days. Consequently juvenility can pose a special problem for tree breeding because of the extremely long period between successive generations.

Size appears to be important in the transition from the juvenile to the mature phase of growth and, in general, conditions that promote growth reduce the duration of the juvenile period. Two possible explanations for the effect of size have been suggested. One is that a large enough plant sends some kind of signal (perhaps a hormone and possibly from the leaves) to the growing point of the plant where the transition from juvenile to adult takes place. The second is that the shoot tip behaves independently of influences from the rest of the plant, perhaps because of some intrinsic ageing process in the dividing cells of this region.

The important point is that the change from juvenile to adult takes place *at the shoot apex* after the plant has reached a certain size. This fact has many implications for propagation, since the base of the plant always remains juvenile and there are many differences between adult and juvenile tissue, other than the fact that only adult plants can flower. Axillary buds retain the maturity level of their point of origin and those that develop from low in the plant are initially juvenile. As they grow, however, individual branches undergo the phase change to maturity at their own shoot tips (Fig. 7.2). The change to the adult phase of growth is more or less permanent until new seedlings of the next generation revert again to the beginning of the juvenile phase. It is important to distinguish between reversion to the juvenile phase (which occurs rarely and with difficulty in adult plants) and the re-invigoration of adult plants by improving their management or by taking cuttings.

Morphology

Plants sometimes have a characteristic morphology during their juvenile phase of growth. One of the best known examples is that of the common

Fig. 7.2 Mature trees may retain their juvenile habit of leaf retention at the base. Buds at the base of the tree remain juvenile and, in some species, retain their juvenile habit of leaf retention. The outer branches are already adult and have shed their leaves.

Fig. 7.3 Adult and juvenile plants of ivy (*Hedera helix*). Flowering plants (above) show adult characteristics of entire leaves and shrubby growth. Juvenile plants (below) have lobed leaves, a climbing habit and adventitious roots.

ivy (Fig. 7.3). The juvenile growth habit is a creeping vine which clings by adventitious roots arising from the stem and has palmately lobed leaves. The adult plant is shrub-like with unlobed, oval leaves. Many experiments have been carried out with ivy because it is relatively easy to handle in the adult phase of growth. Adult plants of ivy can be caused to revert to a juvenile habit of growth by treating them with the hormone, gibberellin (GA, see Chapters 8 and 9). This effect can be prevented by the application of another hormone, abscisic acid (ABA, see Chapter 5) suggesting that a balance of GA and ABA might normally be involved in the transition from one state to the other.

In some plants, the juvenile foliage is considerably more attractive than that of adult plants. A good example is cider gum (*Eucalyptus gunnii*) which has bluish, circular leaves in its juvenile form and far less attractive dark-green, elliptic leaves as an adult. It is grown commercially for its foliage and is also an ornamental garden plant. Since a characteristic of the transition to maturity is that it takes place at the shoot tip while the base of the plant remains juvenile, repeated cutting back is practised both commercially and in the garden to maintain the juvenile foliage.

It is intriguing that cuttings taken from seedlings that are still in the juvenile phase may never undergo the normal phase-change to maturity and, consequently, retain their juvenile foliage and other juvenile characteristics, such as leaf retention (beech, *Fagus sylvatica*), thorniness (wattle, *Acacia* species), and lack of flowering, throughout the life of the plant. This is also seen in some conifers where the juvenile leaf form is fixed when they are propagated from seedling plants (e.g. in the 'Elegans' form of Japanese cedar, *Cryptomeria japonica* Elegans Group). It is suspected that this phenomenon has to do with the hormonal balance in the plant but no detailed studies have been made. From the work with ivy, we know that the application of gibberellins may, in some instances, cause a reversion to the juvenile habit of growth but this still does not explain why the plants do not develop as normal and undergo a phase change to maturity. It is interesting and probably relevant that cuttings taken from horizontally growing branches often retain a prostrate habit of growth, at least for some time. This has been called *topophysis* and is the origin of some prostrate forms.

Rooting capacity

Another difference between juvenile and adult plants is their rooting capacity. Juvenile plants form adventitious roots more readily and, in the adult phase, rooting ability is usually considerably diminished and is sometimes lost. Consequently and especially in plants that are difficult to root, the age of the parent plant can be a very important factor. Several studies have shown that, in some coniferous and deciduous trees known to root only with extreme difficulty, the most important single factor affecting root formation is the age of the tree from which the cuttings are taken. It has been suggested that this may be the reason why some old garden books list species that could be rooted from cuttings but are now considered to be difficult, if not impossible, to propagate in this way. Some of these exotic species were imported as seeds and cuttings were subsequently made from young seedlings which, in their juvenile phase, were easier to root.

The reduction in rooting potential with plant age may be associated with an increased production of chemicals that inhibit rooting. This seems to be the case in one species of gum tree (*Eucalyptus*), where there is a direct relationship between the decrease in rooting ability and the increase, with age, in the production of a root inhibitor at the base of the cutting. There are many cases, however, where there is no obvious correlation between rooting ability and the amounts of promoters and/or inhibitors of rooting present in the tissues, indicating that the decrease with age may, at least in part, be associated with a change in the sensitivity of cells to substances that influence rooting, rather than to the absolute quantities of the substances themselves.

Any treatment that maintains plants in the juvenile phase is of value in preventing the loss of their rooting capacity. One of these treatments is the repeated cutting back of the parent plants, a practice, known as *stooling*, that is employed commercially for the production of hardwood cuttings. In some tree species, leaves towards the base of the tree are retained late into the autumn and often through the winter, indicating the part of the tree that is still juvenile (Fig. 7.2). Cuttings taken from this region are still in their juvenile phase and so would be expected to root more readily. Beech hedges are maintained in the juvenile phase by repeated cutting back and this is the reason why they retain their leaves through the winter. Suitable material for hardwood cuttings of beech and some other species such as oak and some fruit trees is also frequently produced by maintaining a 'juvenile' hedge system.

Flowering

Juvenile plants do not flower and so the question of whether or not the parent plant is flowering does not arise. Cuttings from adult plants may, however, be taken from plants that are either flowering or are vegetative. With easy-to-root species, the presence or absence of flowers makes little difference but, for some difficult species, experiments have shown that cuttings taken from flowering shoots root less well than those from non-flowering shoots. In some cases, removing the flower buds increases the rooting potential but this is not always so. It is prudent, however, to remove any flower buds that are present on a cutting just in case. Although it is evident that the presence of flower buds is often antagonistic to the formation of adventitious roots, the reason is by no means understood. The amounts of several hormones have been shown to change during the transition to flowering but none have been shown to correlate in a clear way with the reduction in rooting potential. Another possibility is that nutrients are diverted into developing flower buds, perhaps reducing their availability at the site of root production.

In woody plants, the juvenile phase with respect to flowering varies from only about a year in some shrubs to as much as 40 years in beech, while juvenile periods of 5 to 20 years are common in trees. Such long juvenile periods pose serious obstacles to tree breeding programmes, as well as leading to unacceptable delays in establishing fruiting trees. Various procedures have been shown to accelerate the transition to maturity. Since we already know that the time at which this occurs is a function of plant size, it is not surprising that most of these procedures involve treatments that increase the rate of seedling growth, such as maintaining them at higher temperatures and/or in long days to prevent dormancy (see Chapter 9).

In horticulture, the problem can sometimes be overcome by taking material for propagation from the adult region of the tree providing that the species is not one in which rooting is inhibited in adult material. In this way the juvenile phase (during which flowering and fruiting do not occur) can be by-passed. The problem does not arise with many fruit tree clones, where generations of vegetative propagation has resulted in 'fixing' the adult phase and the plants are never truly juvenile. For this reason, they are often difficult to root and must be grafted. Recent work has shown that micropropagation can result in some reversion to juvenile characteristics, especially rooting capacity (see under micropropagation, below).

Etiolation

Several trials have shown that keeping plants in darkness (*etiolation,* see Fig. 5.5) is remarkably effective in increasing the production of adventitious roots in stem tissue. This was shown experimentally as long ago as 1864, but the reason is still not fully understood. The inhibitory effect of light on rooting is greatest in red light, suggesting that it is mediated through the pigment, phytochrome (see Chapter 6), which is known to cause many changes in hormones and other biochemical constituents of plants. Complete etiolation would deplete the shoot of carbohydrates and depress rooting, so only the base of the shoot is maintained in the dark, either by insertion into the cutting medium or by mounding up the base of plants to be used as a source of hardwood cuttings. Experimentally, rooting has been increased by covering the entire parent plant for a period.

TYPES OF CUTTING AND THEIR MANAGEMENT

Leafy shoot cuttings

These consist of a portion of leafy shoot, usually including the apical bud. They are obtained from the current year's growth and may be taken during active growth (softwood), when they have become partly woody (semi-ripe) or when they have become fully mature and have set their terminal buds (hardwood). The latter are only from evergreens, since deciduous plants will already have shed their leaves by this stage and are treated differently (see hardwood cuttings below). As far as management is concerned, the difference is between leafy cuttings, where photosynthesis and water relations are important, and leafless hardwood cuttings, where the amount of stored carbohydrate is a major factor.

A large number of dicotyledons can be propagated as leafy cuttings, including both herbaceous and woody plants. Following removal from the parent plant, auxin moving downwards accumulates at the base of the cutting and root cells are formed. As discussed above, rooting can often be improved by dipping the base of the cutting in a solution or powder containing an auxin, usually IBA. Many species have preformed root primordia at the *node* (the point at which the leaf joins the stem), which are arrested until stimulated by auxin. For this reason, cuttings are usually taken just below a node. In many cases, however, the position of the cut is not important, although roots will often still emerge mainly from the node. Where no pre-formed initials are present, cell divisions in the outer layer of the phloem give rise to new roots.

Wounding the basal surface of the cutting can also increase its rooting potential, especially where there are no pre-formed initials. Removal of a few of the lower leaves is normal in order to reduce the incidence of rotting, which is more likely to occur in leaf tissue buried in the rooting medium. Reducing the leaf surface, by removing some of the lower leaves and/or halving large leaves, also helps to reduce water loss from the cutting.

Management of leafy cuttings

The prime requirement in the management of leafy cuttings is to reduce water loss from them. The absence of roots means that water uptake is strictly limited, while water loss by transpiration from the leaves still occurs. Cell division requires metabolism, which depends on oxygen to support respiration, and a source of energy supplied by sugars from photosynthesis. Consequently, for successful rooting, the base of the cutting where cell division is occurring must be in a well-aerated environment, water loss must be reduced to a minimum and photosynthesis should be optimised.

It is important that cuttings are not allowed to wilt before inserting since this leads to prolonged closure of the stomata and a consequent loss of photosynthetic potential by cutting off the supply of carbon dioxide (see Chapter 5 for a discussion of the effects of wilting and the role of abscisic acid). Succulent plants, such as cacti and houseleeks, are an exception to this general rule because they have mechanisms to avoid or minimise water loss (see Chapter 5). Cuttings of succulents are usually left for a short period, up to a few days, in a warm airy place in order to seal the base before they are inserted into the cutting compost.

Temperature

Maintaining the base at a higher temperature than the air (bottom heat) also aids rooting. The higher temperature stimulates metabolism and cell division activity at the site of root formation, while the lower air temperature retards shoot growth until roots develop and water uptake is restored. Maintaining a temperature differential between the aerial part of the cutting (cooler) and the base (warmer) is an important management practice for all types of cutting in order to reduce water loss from shoots until roots have developed sufficiently to maintain the water balance. The optimum temperature is between 18 and 25°C. Too high a base temperature (more than 25°C) has the opposite effect and may inhibit rooting.

Aeration

Adequate aeration at the base of the cutting is ensured by the use of a freely draining rooting medium that retains water but does not become waterlogged. Cuttings rooted in clay pots are usually inserted around the edge since some oxygen can diffuse through the porous clay; however, this does not occur in today's plastic pots. Mineral nutrients are not necessary during the rooting stage itself, since they are not taken up in the absence of roots, but are required later when root growth begins. The most commonly used rooting media, therefore, consist of an inert material such as grit, vermiculite, or perlite (to improve aeration) mixed with soil or, more usually, with a peat- or coir-based compost. Rock-wool blocks are also used successfully.

Water

Leaves lose water to the atmosphere in the form of water vapour diffusing through the stomata (see Chapters 1 and 5). The absolute amount of water vapour present in the air is known as the *vapour pressure* and, like any other gas, water vapour moves along a concentration gradient from a region of higher vapour pressure to a region of lower vapour pressure, i.e. one containing less water vapour. This means that, in order to reduce water loss from the leaves, the cutting must be maintained in an atmosphere of high vapour pressure, i.e. in a high humidity.

The *relative humidity* is a measure of the amount of water vapour in the air compared with the amount that would be present if the air were saturated. When it is saturated, hot air holds more water vapour than cooler air. This means that, even when the air inside and outside the leaf is saturated (i.e. at 100% relative humidity), the amount of water vapour will be greater inside if the temperature is also higher. Leaves in sunlight are often hotter than the surrounding air and, consequently, a leaf can still lose water even when the air around it is saturated with water vapour.

The easiest way to cool the leaf and so reduce water loss is to shade it from direct sunlight. Although this reduces the amount of light for photosynthesis and consequently the supply of sugars for metabolism, photosynthesis would be more drastically reduced by wilting and stomatal closure. The most satisfactory way of cooling the leaf, however, is by spraying it with water. The subsequent evaporation of water, which consumes energy in the form of heat, results in cooling of the leaf surface. Water loss is thereby substantially reduced without the necessity of shading. A number of systems are available which employ an artificial leaf that switches on spray jets each time its surface dries out. Such *intermittent mist* systems are widely used in commercial horticulture and are also available for the amateur gardener.

The usual way of maintaining a high humidity in the atmosphere surrounding the cuttings is to enclose them in a glass or clear polythene chamber (this need be no more than a plastic bag placed over the pot). However, this practice itself can lead to a number of problems. High humidity and stagnant air favour fungal growth while, unless the container is permeable to carbon dioxide, photosynthesis will be reduced if carbon dioxide levels within the chamber fall. Some ventilation is, therefore, desirable. Most plastic bags have a low permeability to water vapour but allow the exchange of carbon dioxide and so they need not be removed until rooting has occurred. Injecting water vapour into the air (known as *fogging*) is another means of maintaining a high humidity round the cuttings.

Seasonal effects

Even under ideal conditions, some species have proved difficult or impossible to root from leafy cuttings for reasons which are not fully understood. Sometimes rooting potential is strongly dependent on season as in some lilac cultivars (e.g. *Syringa vulgaris* 'Madame Lemoine'). Many factors such as stage of growth, light intensity, nutrition and temperature undoubtedly contribute to these seasonal effects. Some species are also sensitive to day-length, perhaps operating through changes in their hormone content. Root initiation in leafy cuttings may be influenced by the day-length experienced by the parent plant or by the day-length treatment given to the cuttings themselves (Table 7.1). However, many of the examples given in Table 7.1 have not been confirmed and amateur gardeners could provide valuable scientific information by carrying out their own experiments.

Hardwood cuttings

These provide one of the easiest and most reliable ways of propagating many trees and shrubs (Table 7.2). They usually consist of pieces of woody material of the current season's growth, about 20 cm long and slightly thicker than a pencil. Hardwood cuttings taken from deciduous plants are leafless but contain sufficient reserves of nutrients to maintain root growth until new leaves emerge in spring. Although timing is not crucial and cuttings can be taken at any time between November and March, provided that the shoot is not growing, the best times are just after leaf fall or just before growth begins in the spring. They are normally trimmed below a node as for leafy cuttings, but the tip is often removed. When long stems are available, several cuttings can be made from a single shoot.

Hardwood cuttings from deciduous trees are leafless, so there is no problem of water conservation as with leafy cuttings and they are not normally enclosed. It is important, however, to avoid desiccation and cuttings in the open should be protected from drying winds and are usually inserted into the soil with only the tip above ground. For trees where a single stem is required the entire cutting is put below the soil in order to achieve a straight stem. Drying out is more of a problem with hardwood cuttings from evergreen plants, and these are often placed in a propagating frame, either out of doors or in a cool greenhouse. As with all cuttings, it is important to firm the soil or compost around the base to avoid air pockets since these interfere with the uptake of water and can also lead to drying out.

Stool beds maintained as a source of hardwood cuttings as, for example, in the production of clonal rootstocks, are frequently earthed up in order to maintain the base of the future cutting in darkness and so increase its rooting potential by etiolation. Juvenility may be maintained by repeatedly cutting back the parent plants.

Hardwood cuttings of some easy-to-root subjects can be inserted out of doors, usually with sharp sand at the base of the cutting to increase aeration and drainage. In this case, rooting will not begin until the soil temperature increases in the spring and the cuttings should not be lifted until the following autumn. The range of plants that can be successfully propagated by hardwood cuttings is increased by giving them the protection of a cold frame or cool greenhouse (Table 7.2). Rooting will occur more rapidly if bottom heat is also provided. When rooting under glass, it is important to ventilate on sunny days to lower the air temperature and prevent premature bud growth before the new root system is established. When rooted over gentle bottom heat and given the protection of a greenhouse, cuttings taken in November will usually be well rooted by the end of February.

Table 7.2 Examples of common garden plants suitable for propagation by hardwood cuttings (From 'Taking the straight and narrow' and 'Simple but effective' by D. Hide (*The Garden*, 1997 (p. 38) and 1998 (p. 731)).

For rooting out of doors: Butterfly bush (*Buddleja davidii*); Forsythia (*Forsythia* × *intermedia*); Privet (*Ligustrum vulgare*); Cherry plum (*Prunus cerasifera*); Elderberry (*Sambucus nigra*); False spiraea (*Sorbaria tomentosa*); Snowberry (*Symphoricarpos* × *doorenbosii*); Weigela (*Weigela florida*).
For rooting in cold frames: Abutilon (*Abutilon* spp.); Barberry (*Berberis* spp.); Caryopteris (*Caryopteris* × *clandonensis*); Flowering quince (*Chaenomeles japonica*, *C. speciosa*); Dogwood (*Cornus* spp.); Deutzia (*Deutzia* spp.); St. John's Wort (*Hypericum* spp.); Kerria (*Kerria japonica*); plus those listed above for rooting out of doors.
For rooting in containers: Actinidia (*Actinidia* spp.); Fig (*Ficus carica*); Virginia creeper (*Parthenocissus quinquefolia*); Tamarisk (*Tamarix* spp.); Vine (*Vitis* spp.); Wisteria (*Wisteria* spp.); plus all those listed above.
Evergreen shrubs suitable for winter propagation: Box (*Buxus* spp.); Escallonia (*Escallonia* spp.); Spindle tree (*Euonymus* spp.); Broom (*Genista* spp.); Ivy (*Hedera helix*); Holly (*Ilex* spp.); Sweet bay (*Laurus nobilis*); Lavender (*Lavandula* spp.); Privet (*Ligustrum* spp.); Cherry laurel (*Prunus laurocerasus*); Portuguese laurel (*Prunus lusitanica*); Rosemary (*Rosmarinus officinalis*); Cotton lavender (*Santolina chamaecyparissus*); Gorse (*Ulex europaeus*); Laurustinus (*Viburnum tinus*)

Leaf and leaf-bud cuttings

Some plants can be regenerated from cuttings consisting of whole leaves, or sections of leaf, or even leaf discs. The latter have been used experimentally to examine the effect of the auxin/cytokinin ratio on the pattern of regeneration in leaves of Lorraine begonia (*Begonia* × *cheimantha*). As found with tissue cultures, the application of cytokinins increased the number of buds that developed, while auxins promoted the formation of roots. Plants frequently propagated by leaf cuttings include members of the Gesneriaceae, for example Cape primrose (*Streptocarpus* spp.) and African violet (*Saintpaulia ionantha*), which grow as rosettes and do not produce suitable stem material.

For this type of propagation to be successful, the specialised leaf cells must be capable of de-differentiating into actively dividing cells, which are then able to undergo *organogenesis* into new meristems that can produce shoots and roots. Leaves also have polarity and must be positioned so that the part of the leaf that was nearest to the parent plant is at the lower end of the cutting. New plantlets then arise at the base of the leaf. Although leaf cuttings are usually inserted vertically in the rooting medium, whole leaves are often simply laid flat on the surface. In this case, large veins on the under surface are partially cut through and new plants arise at these points.

It is important to note that, if chimeras such as the variegated mother-in law's tongue (*Sanseveria trifasciata* 'Laurentii') are propagated from leaf cuttings, the variegation will not be present in the new shoots that are produced. This is because these shoots originate only from the green cells in the core of the leaf (see Chapter 3 for details). A similar problem can occur with root cuttings (see below).

Some members of the family Gesneriaceae (especially *Streptocarpus* spp.) are unusual in that the growing point (*apical meristem*) becomes incorporated into the upper surface of the enlarging cotyledon. In the intact plant, this so-called 'groove meristem' produces the new leaves (termed *phyllomorphs* because they consist of the leaf blade plus its petiole), while roots form at the under surface of the petiole. An organised meristem is, therefore, already present at the base of the phyllomorph when it is detached and enables it to give rise to new daughter plants in this region. The leaves of *Streptocarpus* can also be divided up into several sections. In this case no pre-formed meristems are present and the leaf cells must de-differentiate into new meristems.

Many plants are unable to regenerate new plants from leaves, although they can often produce roots at the base of the leaf. The failure seems to be in the ability to produce a new shoot apex, since many plants unable to regenerate from leaves alone can do so from leaf-bud cuttings, which consist of a piece of stem bearing a single leaf with a bud in its axil. In this case, the lateral bud becomes the new apical shoot meristem. Honeysuckle (*Lonicera* spp.), ivy (*Hedera helix*), *Camellia* spp. and *Clematis* spp. are often propagated by leaf-bud cuttings.

The management of leaf and leaf-bud cuttings is the same as for leafy shoot cuttings.

Root cuttings

These consist of sections of root and are often used when suitable stem material is not available. Roots also have polarity so that it is important that they are inserted into the rooting medium with the proximal end (nearest the parent shoot) pointing upwards, although in many cases it is satisfactory to place the root sections horizontally and so avoid confusion. Since they have no photosynthetic tissue, successful root cuttings must have a supply of stored carbohydrates to support metabolism until new shoots emerge above ground. Consequently plants with thick fleshy roots are usually more successfully propagated by root cuttings than are plants with thin, fibrous roots. An exception is tree poppy (*Romneya coulteri*), where the cutting is made from the thinner, more active part of the root. Not all species can be propagated by root cuttings, but examples of those that propagate well by this method are given in Table 7.3.

Table 7.3 Some common garden plants that can be propagated by root cuttings.

Bear's breeches (*Acanthus mollis*)
Tree of heaven (*Ailanthus altissima*)
Alkanet (*Anchusa azurea*)
Horseradish (*Armoracia rusticana*)
Seakale (*Crambe maritima*)
Oriental poppy (*Papaver orientale*)
Drumstick primula (*Primula denticulata*)
Stag's horn sumach (*Rhus typhina*)
Comfrey (*Symphytum ibericum*)
Mullein (*Verbascum* spp.)

Even where root cuttings are successful, they are not always appropriate, as in the case of *periclinal chimeras* (see Chapter 3), where an outer layer of mutated cells overlays an inner core of the original genotype. In the thornless blackberry, for example, the outer mutated layer contains the gene for thornlessness and plants grown from root cuttings revert to the thorny genotype because the new shoots grow from the inner tissues.

Layering

Layering often occurs naturally when shoots, especially of woody plants such as *Viburnum* spp., touch the surface of the ground. Rooting may then occur at one or more nodes, where preformed root initials are frequently found. The horticultural technique of layering was adapted from the natural layering of wild plants and was already used by the Romans in the propagation of grape vines in the first century BC. It represents an easy and convenient way of propagating, which may be especially useful for plants that are hard to propagate by other methods (Table 7.4).

Table 7.4 Some common garden shrubs that can be propagated by simple layering, taken from 'Bending the rules' by D. Goodwin (*The Garden* 1999, p. 92).

Camellia (*Camellia* spp.); Winter sweet (*Chimonanthus praecox*); Corylopsis (*Corylopsis pauciflora, C. sinensis* and *C. spicata*); Hazel (*Corylus avellana*); Filbert (*Corylus maxima*); Smoke bush (*Cotinus coggygria*); Daphne (*Daphne* spp.); Spindle tree (*Euonymus*, deciduous species); Witchhazel (*Hamamelis* × *intermedia, H. mollis,* and *H. japonica*); Hydrangea (*Hydrangea paniculata* and *H. quercifolia*); Calico bush (*Kalmia latifolia*); Magnolia (*Magnolia liliiflora, M. sieboldii, M.* × *soulangeana,* and *M. stellata*); Pieris (*Pieris* spp.); Cherry (*Prunus*, deciduous shrubby species); Viburnum (*Viburnum* × *bodnantense, V. carlesii, V. farreri* and *V.*

Layering is carried out when the plants are dormant in the early spring or autumn, using wood of the current or previous year's growth that is still pliable enough to bend. In simple layering, a portion of stem near the tip is bent into a u-shape and pegged into a hole about 10 cm deep (Fig. 7.4). This is then filled with a suitable rooting medium and the tip of the shoot is tied to a cane to keep it upright. The likelihood of rooting is increased if the shoot is partially cut through where it is pegged into the ground, resulting in the accumulation of sugars and probably auxin above the cut. Wounding itself also increases the rooting potential, as discussed earlier. Covering the rooting zone with soil to exclude light also helps to promote rooting. and prevent the rooting zone from drying out. The new plant is cut from the parent when well rooted, which usually takes about a year, although rooting may take two years in some cases (e.g. *Rhododendron*).

A variant of simple layering is a technique know as French layering, which can be used to produce more plants from a single stem. Here the stem (of *Wisteria* spp., for example) is pegged down horizontally into a shallow trench and then covered. As shoots begin to grow from the buried axillary buds, they are earthed up to etiolate them and increase the rooting potential. Roots form at the nodes and, when these are well developed, the stem is cut into the rooted portions.

Air layering

This technique is used in some specialised cases, most commonly to improve the appearance of ornamental plants that have become leggy with bare stems, as often occurs in the rubber plant, *Ficus elastica*. It is one of the oldest forms of pro-

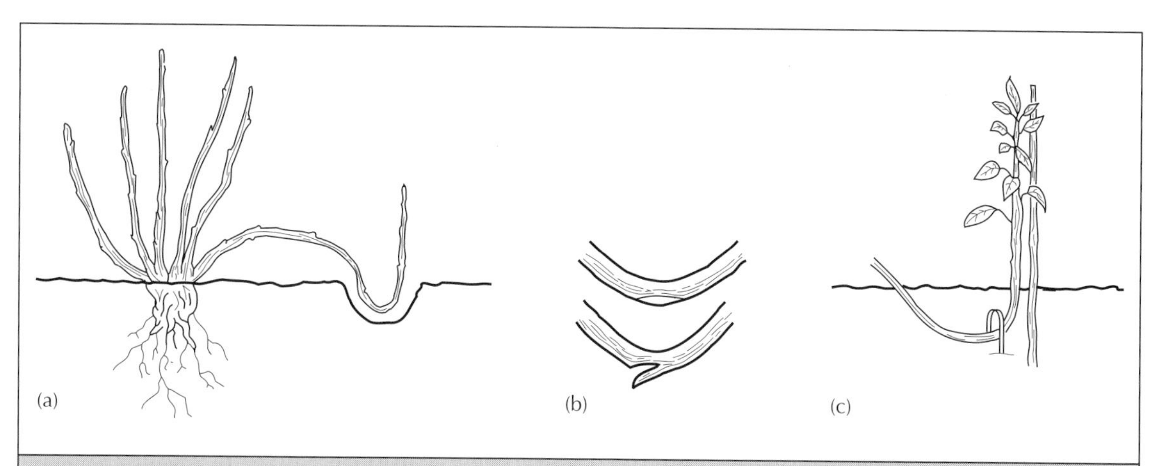

Fig. 7.4 Propagation by simple layering. Shoots are bent over to the ground and the tip is turned up (a). A cut is made at the base of the bent part of the layered shoot (b), which is then covered with soil and pegged into the ground (c). The tip of the shoot is usually fastened to an upright stake. Rooting occurs at, or near, the site of wounding in the buried part of the stem and the rooted plant is later severed from the parent.

pagation, probably originating about 4000 years ago in China and may be successful with both monocotyledons and dicotyledons (Table 7.5).

Table 7.5 Examples of plants that can be propagated by air layering.

Cabbage palm (*Cordyline australis*)
Daphne (*Daphne* spp.)
Dracaena (*Dracaena* spp.)
Rubber plant (*Ficus elastica*)
Swiss cheese plant (*Monstera deliciosa*)
Philodendron (*Philodendron* spp.)

Air layering involves the removal of a layer of outer tissue to girdle the stem and interrupt the downward flow of nutrients, especially sugars, which accumulate above the girdled area; sugars normally move in the phloem from the photosynthesising leaves to the roots. Auxins moving down the plant also accumulate above the ringed area leading to the formation of roots. It is necessary to prevent the girdled area from drying out, usually by covering it with moist sphagnum moss wrapped in polythene. Where possible, a high humidity level should also be maintained around the plant by placing it in a propagating case. When roots have formed, the new plant is severed below the girdled region.

Division

Division is a simple technique that can be applied to plants that produce a mass of closely knit shoots or buds, forming a clump that can be split up into smaller pieces. Each portion will have shoots and/or buds and roots and will, therefore, be capable of independent growth. Although division is most commonly used in the propagation of herbaceous perennials, some woody and semi-woody plants can also be divided (Table 7.6). Most of these plants, if left alone, develop into a crowded patch which often dies out in the centre. In the wild, the outer shoots continue the growth of the plant but, in the garden, the overcrowded or dying centre is unsightly.

The division of herbaceous plants with fibrous crowns and of alpine plants is done either by pulling the mass of shoots and roots apart by hand, or by teasing them apart with two forks placed back to back. Herbaceous plants with tough compacted crowns with fleshy roots and buds are difficult to divide by hand and often need to be cut carefully with a sharp knife. The cut surfaces should be dusted with a fungicide as the fleshy roots are prone to rotting. The semi-woody herbaceous plants that can be divided produce sword-like leaves in dense terminal clusters, each with its own root system. Because of their tough, woody nature, they are difficult to divide by hand and are usually divided by cutting with a knife. The small number of woody plants

Table 7.6 Some common garden plants that can be propagated by division.

(1) Herbaceous plants with fibrous crowns: Yarrow (*Achillea* spp.); Michaelmas daisy (*Aster* spp.); Avens (*Geum* spp.); Leopard's bane (*Doronicum* spp.); Gypsophila (*Gypsophila* spp.); Day lily (*Hemerocallis* spp.); Shasta daisy (*Leucanthemum* × *superbum*); Catchfly (*Lychnis* spp.); Bergamot (*Monarda* spp.); Coneflower (*Rudbeckia* spp.); Scabious (*Scabiosa* spp.); Meadow rue (*Thalictrum* spp.); Globe flower (*Trollius* spp.); Valerian (*Valeriana officinalis*); Speedwell (*Veronica* spp.).
(2) Herbaceous plants with tough, compacted crowns and fleshy buds and roots: Astilbe (*Astilbe* spp.); Hellebore (*Helleborus* spp.); Plantain lily (*Hosta* spp.).
(3) Alpine plants: Aubretia (*Aubrieta* × *cultorum*); Gentian (*Gentiana sino-ornata*); Cranesbill (*Geranium cinereum* var. *subcaulescens*); Primrose (*Primula*, European species); Saxifrage (*Saxifraga* spp., mossy and rosette-forming types); Stonecrop (*Sedum cauticola*); Creeping thyme (*Thymus serpyllum*).
(4) Semi-woody herbaceous plants: Astelia (*Astelia chathamica*); Sedge (*Carex* spp.); Pampas grass (*Cortaderia selloana*); New Zealand flax (*Phormium* spp.); Adam's needle (*Yucca filamentosa*).
(5) Woody plants: Purple chokeberry (*Aronia prunifolia*); Buckeye (*Aesculus parviflora*); Vine maple (*Acer circinatum*); Common quince (*Cydonia oblonga*).

that can be divided produce clumps of stems from suckers arising below ground level.

The best time to divide most subjects is at the start of the dormant season, when the plant dies back. For most alpine plants, division is best done soon after flowering, although autumn-flowering plants should be divided in the spring to minimise over-wintering losses.

Propagation from specialised structures

In specialised structures such as bulbs, corms, tubers and rhizomes, *sympodial* branching is common. This is a pattern of growth in which the apical bud withers at the end of the growing season and growth is resumed in the following year by the lateral bud or buds below the old apex. This characteristic ensures that units of growth, such as daughter bulbs, offsets, cormels or branched rhizomes, already exist and are capable of independent growth once roots are produced. The natural rate of multiplication varies with the species. For example, in daffodils (*Narcissus* spp.), only about one and a half new bulbs are produced each year compared with five in tulips (*Tulipa* spp.). In contrast, a plant of *Gladiolus* can produce a hundred or more cormels.

Most of these plants can be propagated by lifting the plant, separating out the individual offsets, such as daughter bulbs or cormels, and replanting them. This is usually done when the tops have died back and the plants are dormant. Exceptions are snowdrops (*Galanthus* spp.) which must be lifted 'in the green'. If left undivided, the clump can become congested and flower poorly, although a few, such as *Nerine* spp. and some *Sternbergia* spp., flower best when congested and should be divided only to increase stocks.

Bulbs

The main storage parts of a bulb are the fleshy modified leaves (*scales*) that enclose the flower bud and are attached to a base plate, which is the compressed stem from which the roots grow. Some bulbs, such as daffodil and tulip, have tightly packed scales and are called non-scaly or *tunicate* bulbs, because the bulb is enclosed in a protective papery tunic. Others, known as scaly bulbs, such as fritillary (*Fritillaria* spp.) and lily (*Lilium* spp.), have loose, easily separated scales without a protective tunic.

Bulbs differ in their longevity. Tulip bulbs, for, example, are replaced annually. When the mother bulb dies after flowering, it is replaced by a number of daughter bulbs arising from buds between the scales of the parent bulb. In contrast, daffodil bulbs last for several seasons because they have a complex branching system which persists from year to year. A new bulb grows within the old one each year. After flowering, more than one new growing point may become active and two or more bulbs arise within the old scales.

Bulbs lend themselves to a number of special methods of propagation. The optimum time for carrying out these methods of propagation is when they are dormant. This is usually in late summer or early autumn for spring and summer flowering bulbs, and in spring for those that flower in autumn and winter. Hygiene is important because the fleshy scales are prone to rotting, especially when cut with a knife. They must, therefore, be treated with a fungicide before being placed into a suitable propagating compost. This should be free-draining, such as a mixture of perlite and peat. Once prepared and placed in the propagation medium, the propagules should be kept at a temperature of about 20°C until new bulblets form and roots emerge. In some cases, such as the Turk's cap lily, *Lilium martagon*, the newly formed bulbils are dormant and require a period of chilling before they begin to sprout.

Bulbs that have loosely packed scales, such as all lilies and some fritillaries, can be propagated by *scaling*. The outer scales are pulled off in succession and are inserted to half their depth in the propagating medium, which is covered to retain moisture. Alternatively, they can be placed in a suitable, moist medium in a sealed plastic bag, retaining as much air as possible to allow the scales to breathe. Usually, only a few of the outer scales are removed so that the parent bulb can be re-planted.

Bulbs with a tighter structure, such as daffodils (*Narcissus* spp. and cultivars) and hyacinths (*Hyacinthus* spp.), must be cut into pairs of scales, known as *twin scaling*. Any old, outer scales are removed and the bulb is cut vertically into several segments, the number depending on the size of the bulb. Beginning from the outside, each segment is then split into pairs of scales, using a sharp, sterile knife, making sure that each piece retains a portion of the base plate. They are then treated as for scales.

Small bulbs and non-scaly bulbs, for example *Hippeastrum* spp., can be propagated by *chipping*. The bulb is divided vertically with a sharp knife to produce up to sixteen chips, each retaining a portion of the base plate. They are then treated as for scales. As with twin-scales, hygiene is extremely important if chipping is to be successful.

Hyacinths and some other bulbs (e.g. Crown Imperial, *Fritillaria imperialis*) are frequently propagated by *scooping* or *scoring* the bulbs. Both involve removing or damaging the apex of the parent bulb and so removing its apical dominance. The production of small bulblets between the scales is, thereby, stimulated. In scooping, most of the basal plate and the main growing point of the bulb is scooped out from below, leaving the outer rim of the basal plate intact. Scoring involves making deep cuts in the basal plate and removing wedges of tissue from the basal plate and the base of the bulb. In both case, the wounded bulbs are placed upside down on trays of moist, coarse sand. The bulblets which form at the base of the parent bulb are detached when they are large enough to handle and treated as seeds.

Some bulb plants can be induced to form offsets, even where none occur naturally. Onion for example, can be induced to form bulbils by removing the flower buds from the seed head, leaving a 'ball of bristles'. Small bulbs then form between the bristles.

Pseudobulbs

Epiphyte (tree-growing) orchids have two methods of growth, namely *sympodial* (terminal growth ceases and growth is continued by a lateral bud) and *monopodial* (growth of the terminal bud continues). Sympodial orchids, such as *Cattleya* and *Odontoglossom* species and hybrids, have creeping rhizomes. Each season, growing points arise from new growths on the rhizomes and develop into new pseudobulbs. These can be divided to produce new plants. For most orchids, two or three new growths should be retained in each division. *Cattleya* and its relatives need at least one leafy pseudobulb on the back growth and a prominent bud at its base if they are to have a chance of success. *Lycaste* and *Odontoglossom* are usually separated into pieces with two or three pseudobulbs. *Cymbidium* species are easily propagated from old leafless pseudobulbs removed from the back of the plant. New plants propagated from pseudobulbs may reach flowering size in two to three years.

Corms

A corm, such as *Crocus* and *Gladiolus*, is a compressed stem covered with dry scales. The growing point of a dormant corm is in the depression at the centre of the upper, flattened surface and there are also axillary buds, both here and at the equator of the corm. Most corms produce several buds near the apex, each of which will form a new corm. Miniature corms, or *cormels*, may also develop during the growing season. These arise at the tip of stolon-like structures (see below) growing from the base of the mother corm. The stock can be increased by lifting and dividing the corms or cormels and re-planting in pots or in the garden.

Rhizomes

These are specialised underground stems, which also serve as storage organs. The main shoot dies at the end of the season and is replaced by one or more lateral buds in a typical sympodial branching pattern. These secondary rhizomes, in turn, produce an aerial shoot after a single node, as in Peruvian lily (*Alstroemeria*), or after many nodes, as in lily-of-the-valley (*Convallaria majalis*). Rhizomatous plants (e.g. rhizomatous irises) are propagated by cutting up the rhizome into portions, each with one or more growing buds. In the invasive weed, couch grass (*Elymus repens*), cultivation breaks up the rhizomes into fragments, which removes the dominance of the apical bud and allows each piece to regenerate. This is why couch grass is so difficult to eradicate by cultivation.

Stolons

These are also specialised stems and may arise either above or below ground. Unlike rhizomes, they do not have a storage function. Probably the best known garden plant that is propagated from stolons in the cultivated strawberry (*Fragaria* × *ananassa*). The above-ground stolons emerge from the mother plant and a new plant develops at every other node. Propagation usually involves pegging the stolon into the soil at the node until the young plants have rooted, when they can be cut from the parent and either planted or potted up. Strawberry stolons develop only in the long days of summer.

Tubers

There are of two types, both having a storage function. Root tubers, as typified by *Dahlia*, consist of swollen portions of roots near the base of the stem. The tuber itself is unable to develop shoots as it consists only of fibrous roots and is unable to produce buds. When propagating by division, therefore, it is important to see that each piece has at least one dormant bud and one tuber.

Stem tubers are pieces of swollen stem and can arise in different ways. Potato (*Solanum tuberosum*) tubers, for example, develop at the end of underground stolons, whereas the tubers of gloxinia (*Sinningia*) are modified, swollen hypocotyls which increase in size annually. The method of propagation from stem tubers depends on how the tuber is formed. Potato tubers, for example, are collected at the end of the season, stored and re-planted the following year. In *Cyclamen*, tuberous *Begonia* and gloxinia, which are all modified hypocotyls, the tuber can be divided when it is sufficiently large and more than one growing point has formed.

GRAFTING AND BUDDING

A graft is formed when a piece of shoot (the *scion*) is placed in contact with a root system from another plant (the *stock*) and the two pieces grow together as a single plant. Budding is a form of grafting in which the scion consists of a single bud. There are many reasons for grafting. As with cuttings, the use of adult scions bypasses the juvenile phase during which flowering and fruiting cannot occur, which means that trees will come into flowering and fruiting much more rapidly than those raised from seed. A second reason is that plants with complex genetic backgrounds, such as most fruit trees, do not breed true from seed, resulting in wide variation both in their fruit characteristics and growth habit. All tree fruit cultivars are clones derived from a single original parent and therefore all have the same genetic make-up. Most of them do not root readily from cuttings and grafting is a quicker and much more reliable method of obtaining new plants. Budding, as for example for rose cultivars on seedling rootstocks, is also a way of obtaining the rapid multiplication of new plants.

There are other special reasons for grafting. For ornamentals, it is possible to produce plants with enhanced decorative features. A pendulous form can be grafted to an upright rootstock at a suitable height to form an attractive weeping tree (for example, the Kilmarnock willow, *Salix caprea* 'Kilmarnock'). Or plants with different attractive features can be combined, such as *Prunus* × *subhirtella* 'Autumnalis' (winter flowers) grafted onto *Prunus serrula* (coloured bark). Several fruit cultivars can be grafted onto one plant to produce a 'family tree' to ensure pollination in small gardens where there is no room to plant several trees; in this case, the grafts are made once a framework of branches has developed.

One of the most important reasons for grafting is the effect of the rootstock on the growth of the scion. For some plants, rootstocks are available that can tolerate unfavourable soil conditions or soil-borne pests or diseases (for example, cucumbers (*Cucumis sativa*) can be grafted to the wilt-resistant *Cucurbita ficifolia*). For many purposes, both in ornamental gardens and in orchards, small trees are desirable and size-controlling rootstocks are available that cause the grafted tree to become partially or extremely dwarfed. Rootstocks for apple range from those that are very vigorous (e.g. MM109) to those that are extremely dwarfing (M9, M27 and EMLA 9), the resultant trees on the dwarfing stocks being only about one third the size of those on the vigorous stock. Dwarfing rootstocks also result in early fruiting.

The effect of the rootstock on the tree vigour can be explained by the fact that roots supply water, mineral nutrients, and hormones to the aerial part of the tree, whereas the resistance to soil-borne problems is a direct function of the root cells. Several experiments have indicated that the amounts of naturally occurring growth regulators supplied to the shoot may be implicated in the size-controlling effects of rootstocks. For example, for apple rootstocks, a reduction in auxins and cytokinins and an increase in inhibitors may be involved in the dwarfing effect.

Clonal rootstocks (such as the Malling (M) and East Malling, Long Ashton (EMLA) series) are frequently employed, especially for tree fruits. These have been selected for particular effects such as reduction in vigour and, because they are clones, they are highly uniform. They must, of course, be raised from some type of cutting, often hardwood cuttings taken from stool beds or hedges (see *juvenility* above). Seedling rootstocks, although not clonal, are used in some cases and have several advantages. In particular, the production of large numbers of seedlings is relatively simple compared with taking cuttings and, moreover, not all plants are easily propagated by cuttings. Although seedlings are likely to be more variable than clonal material, this can be reduced by careful selection of the seed source. Rose cultivars are usually budded onto seedling rootstocks.

Successful grafting depends on the development of a graft union that is physiologically functional and can transport substances between rootstock and scion. Many organic substances cannot cross by simple diffusion but must be transported by living cells, often in the phloem. Ultimately a vascular connection must develop between the two graft partners. This is well illustrated by the many studies on flowering that have been carried out using grafting techniques. Non-flowering plants can be caused to flower in many cases by grafting onto them a flowering shoot. However, this is only successful when a graft union has been established, demonstrating the need for the active transport of a flower-inducing substance across the union.

Incompatibility

Grafts are more likely to be successful when the plants are closely related botanically. For example, grafting between plants of the same clone is always possible whereas, very occasionally, even grafting between different clones of the same species fails. At the other end of the scale, successful grafting between plants of two different botanical families is usually considered to be impossible. Successful grafts between plants of different genera in the same family are rare. A good example of the complexity of the problem is that of quince and pear. Quince has long been used as a dwarfing rootstock for some pear varieties, but the reverse combination, quince grafted onto a pear rootstock, is never successful. When different species of the same genus are grafted together experimentally, the results are quite confusing. Some are successful, some are not, and some, like the quince/pear graft, are successful only in one direction.

When plants cannot produce a satisfactory and stable graft union, they are said to be *incompatible*. The graft union may fail completely or may initially appear to be successful but fail later, sometimes after several years, at the point of union. Two types of incompatibility have been recognised. In some cases, the problem can be overcome by inserting a piece of another plant between the two incompatible plants. In such cases, the incompatibility seems to depend on local factors at the site of the failed graft union. In other cases, incompatibily cannot be overcome in this way and it seems that some substance which can be transported across the inserted piece, or *interstock*, is the cause of the failure to establish a union. In at least one case, sweet orange on sour orange stock, this is known to be a virus present in the scion that is tolerated by only one partner.

Incompatibility is clearly related to genetic differences between the graft partners, but the underlying physiological mechanisms are not fully understood. Various types of structural abnormality can develop at the graft union and there must be some kind of adverse physiological response between the two graft partners. Hormones from the wounded regions are assumed to be involved in the regeneration process, leading to a connection between the between living cells of the graft partners, called a *symplastic* connection. It has been suggested that incompatibility may result when substances antagonistic to the action of hormones are produced. However, some kind of cell to cell recognition response involving 'signalling molecules' has also been proposed.

Grafting techniques

The origins of grafting can be traced back to ancient times and it was a well established and popular practice in the days of the Roman Empire. It is not surprising, therefore, that several different techniques have been developed over the years. The goal of grafting is to fit together two pieces of living tissue in such a way that they will unite. Irrespective of the technique used, there are a number of basic requirements for successful grafting. Clearly, the two plants must be related and genetically compatible. They must also be physically compatible in the sense that the cambial regions (see Chapter 1) of the stock and scion must be capable of being brought into direct contact, as it is the cambial cells that will divide to generate a successful union. This means that, usually, only gymnosperms and dicotyledons can be successfully grafted since they have a continuous vascular cambium between the xylem and phloem. Monocotyledons have scattered vascular bundles which are difficult to connect, resulting in a low percentage take. Obviously, the two cut surfaces must be held together by tying or taping.

The physiological status of the graft partners is also important; usually the best results are obtained when the scion buds are dormant and the rootstock is just coming into growth. Grafts with scions in leafy growth are rarely successful. It is essential to prevent drying out of the cut surfaces, either by placing the graft in a humid atmosphere, covering the union with a water-impermeable wax, or using special plastic ties. Subsequent management involves removing shoots from the rootstock and/or staking if the scion begins to grow vigorously.

A few popular types of graft are shown in Fig. 7.5 to illustrate the variety of techniques that can be used.

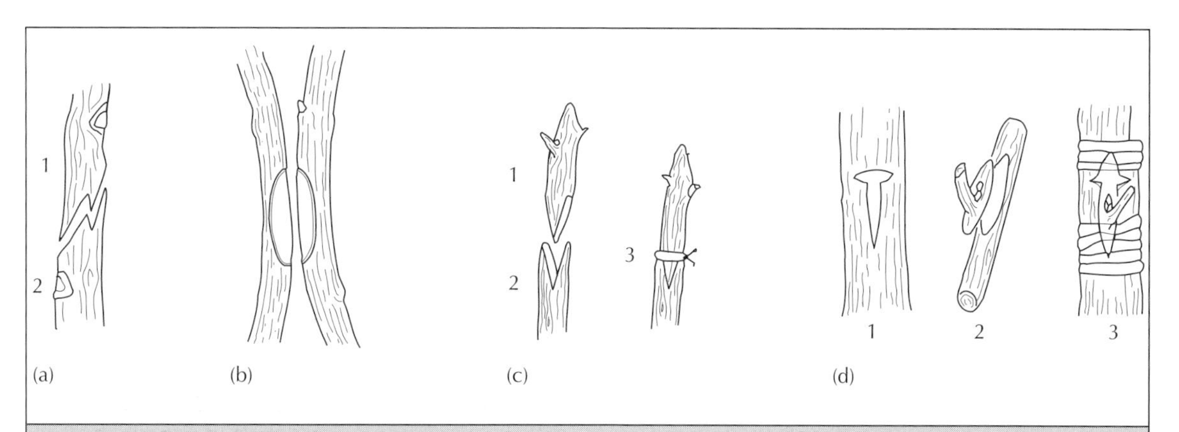

Fig. 7.5 Some common types of grafting.
(a) *Whip and tongue graft.* A short piece of stem is chosen as the scion and a tongue is cut into it as shown (1). The upper part of the rooted stock plant is removed with a slanting cut and a similar tongue is made in the cut surface at the top of the stock (2). The scion and stock are fitted together with the tongues interlocking and tied.
(b) *Approach graft.* This is usually carried out with two potted plants which can be closely approached together. Similar cuts are made in stock and scion and the cut surfaces are brought together and tied. Once union has taken place, the top of the rootstock *above* the union and the base of the scion *below* the union are removed.
(c) *Cleft graft.* The base of the scion shoot is cut into a blunt point (1) and the top of the rooted stock plant is removed, forming a cleft (2). The two are fitted together and tied (3).
(d) *Budding.* A T-shaped cut is made in the stem of the rootstock (1). The scion consists of a single bud removed with a small piece of bark – the *bud shield* (2). The bud is then inserted into the cut made in the stock and pushed under the two flaps of bark until the bud shield is completely covered (3) and the graft is tied.
In all cases, close contact between the cambium of stock and scion is essential for a successful graft union.

MICROPROPAGATION

This is a specialist technique in which small pieces of tissue, known as *explants*, are grown on an aseptic growth medium (i.e. uncontaminated by microorganisms). It is essentially a laboratory method and is practised by only a few dedicated units. Some plants propagated in this way are now appearing for retail sale still growing in small culture flasks on their nutrient medium (for example, some orchids). Such plants require careful handling in order to establish them as free-growing individuals. Because of the increasing importance of micropropagation methods, a short discussion is included here, even though it is not a technique that can be followed by most growers and gardeners.

Micropropagation is carried out for several reasons. One of the most important is the elimination of viruses. Another is the production of large numbers of new clonal individuals from small amounts of starting material, as, for example, the rapid multiplication of a new cultivar. It is also used for some difficult-to-root subjects and, in some instances, micropropagation can be successful for plants that it has not been possible to propagate by other means. It is also of interest because it is being used in the conservation of rare or uncommon plants and as a means of facilitating the exchange of plant material between countries. Immature embryos can also be grown in aseptic culture, a technique that has been used to 'rescue' embryos that would not continue to develop in the seed (see Chapter 3). Micropropagation is also essential for somatic hybridisation and for the insertion of new or modified genes into plants since, for both techniques (see Chapter 3), the starting points are protoplasts, which must subsequently undergo organogenesis in culture media in order to develop into new plants.

Micropropagation is applied extensively to fruit tree rootstocks and scions that are difficult to root and is an accepted method of rapid multiplication while maintaining the integrity of the clone. However, fruit trees propagated in this way show that physiological differences can develop in culture. In particular, it has been observed that there is a progressive increase in rooting potential when the material is repeatedly moved from one culture vessel to new culture medium in another (i.e., they are *sub-cultured*). For example, in the apple rootstock M9, which is difficult to propagate conventionally, rooting ability increased with repeated sub-cultures over a year. Based on the fact that some other plants in culture, such as the giant redwood, *Sequoiadendron giganteum*, have developed vegetative characteristics that are markers of juvenility, it is thought that the most likely reason for this is some degree of rejuvenation. It is an advantage for fruit trees as it produces plants that retain the property of easy rooting by conventional means.

Flowering is less affected than would be expected if the plants were truly juvenile and, moreover, flowering has actually been accelerated in some micropropagated plants. An interesting example of this effect has been observed in some species of bamboo (e.g. *Bambusa bambos*) where flowering occurred in culture within six months, whereas seedlings of the same species normally take some 30 years before they flower. The causes of these phenotypic changes are still debated but it is clear that some populations of plants produced from tissue culture show a range of genetic variations (see Chapter 3).

Micropropagation usually begins with a piece of tissue cut from the parent plant. One starting point is to remove a piece from the tip of a shoot (about 1 mm long) consisting of the apical dome plus two or three leaf primordia. This technique of *meristem culture* is used when virus elimination is required since it has long been known that very young, actively dividing cells are often free from many viruses and also certain bacterial diseases. It also results in fewer mutations than when other tissues form the starting point, since it does not require that non-meristem cells become organised into meristems. However, explants taken from any tissue can be used as a starting point for micropropagation, as can protoplasts and embryos.

Whatever its origin, the starting explant must be initially sterile (e.g. from inner tissues) or sterilised before placing it on some kind of a support, usually a solid gel made with agar, in a small glass flask. However, protoplasts are initially cultured in a liquid medium. Everything must, of course, be carried out under sterile conditions. Such small pieces of tissue do not photosynthesise so the basal medium contains sugar, as well as mineral nutrients and some vitamin supplements. Even though sugars are supplied and photosynthesis does not occur, light is still necessary for normal development of the emerging shoots, which would otherwise be etiolated (Fig. 5.5). Rather little light is needed since photosynthesis is not involved and fluorescent lamps ('warm white' or 'daylight') are usually employed.

One of the early findings from research was the fact that the way in which new plants developed could be altered by the addition of plant hormones to the culture medium. A high cytokinin/auxin ratio favours the initiation of shoots while a high auxin/cytokinin ratio favours the initiation of roots (Fig. 7.1). This knowledge from basic research still underpins the manipulation of root and shoot formation in micropropagation, whether plants are being produced for sale or for experimental purposes. Initially, a high level of cytokinin is used to regenerate multiple shoots, which are then divided and sub-cultured. The auxin level is increased later in order to allow root initiation to occur.

Once individual plants have been established in aseptic culture, they must be moved out of the culture flasks into soil. This has to be done gradually in order to minimise the shock of the transition from the high humidity of the culture flask to the open greenhouse. The roots that develop in agar gel are somewhat abnormal to begin with and, for this reason, shoot tissues are sometimes removed from the culture flask and rooted in a normal rooting medium as *micro-cuttings*.

FURTHER READING

Attridge, T.H. (1990) *Light and Plant Responses*. Edward Arnold, London.

Bleasdale, J.K.A. (1973) *Plant Physiology in Relation to Horticulture*. Macmillan, London.

Brickell, C. (ed) (1972) *The Royal Horticultural Society Encyclopedia of Gardening*, 540–41. Dorling Kindersley, London.

Gardiner, J. (1997) *Propagation from Cuttings*. Wisley Handbook, The Royal Horticultural Society, Wisley.

Goodwin, D. (1999) Bending the rules (layering). *The Garden*, **124**, 92–4.

Hartmann, H., Koster, D.E., Davies Jr., F.T. & Geneve, R.L. (eds) (1997) *Plant Propagation. Principles and Practices*. 6th edn. Prentice Hall International Inc.

Hide, D. (1997) Taking the straight and narrow (leafless hardwood cuttings). *The Garden*, **122**, 38–49.

Hide, D. (1997) A fresh look at propagation composts. *The Garden*, **122**, 268–70.

Hide, D. (1998) Simple but effective (leafy hardwood cuttings). *The Garden*, **123**, 731–3.

Hide, D. (1999) Splitting up. *The Garden*, **124**, 522–3.

Lumsden, P.J., Nicholas, J.R. & Davies, W.J. (eds). (1994) *Physiology and Development of Plants in Culture*. Kluwer Academic Publishers, Dordrecht.

McMillan Browse, P. (1997) *Plant Propagation*. Mitchell Beasley, London.

Rees, A.R. (1992) *Ornamental Bulbs, Corms and Tubers*. Commonwealth Agricultural Bureau International, Oxford.

Toogood, A. (ed) (1999) *Propagating Plants*. Dorling Kindersley, London.

Vince-Prue, D. (1975) *Photoperiodism in Plants*. McGraw Hill, Maidenhead.

8

Shape and Colour

- Colour: how colour is perceived by humans
- Plant pigments: carotenes, anthocyanins and other pigments
- Colour perception by pollinators
- Variegation: chimeras, transposons, pattern genes and viruses
- Environmental factors affecting colour
- Autumn colour
- Colour in the garden
- Shape: the effect of water, light, CO_2 and essential mineral nutrients on the speed of plant growth
- The genetic control of basic morphology and growth habit
- Modification of shape by the gardener
- Effects of environmental conditions, especially light
- Effects of temperature on plant height
- The use of growth modifying chemicals to reduce plant height
- Mechanical treatments practised by the gardener: pruning, root pruning, bending, nicking, notching and shaking and their effects

Having decided what kind of plant would be suitable for the site (see Chapter 5), the choice of plants for a border will largely be determined by their colour and/or shape. Shape and size matter whether growing plants in containers or in borders: short plants with a compact habit are increasingly popular, reflecting the trend towards smaller and smaller garden plots, especially in towns and cities. Plants which take up less room and do not require time-consuming staking are becoming more and more desirable. This chapter will look at some of the factors that determine the colours of plants, as well as their size and habit.

COLOUR

The perception of colour

To some extent, the sensation of colour varies with individuals. There is also a strong aesthetic component. For example, harmonious combinations such as blue and purple are attractive to some people, while others prefer contrasting colours, such as red and green.

The sensation of colour depends on the wavelength of light that reaches the eye. To human beings, light is that part of the electromagnetic radiation spectrum capable of stimulating the human eye and so enabling them to see. It is only a very small portion of the total electromagnetic spectrum and its limits are dependent on the absorption characteristics of *visual purple* (rhodopsin), the light-sensitive pigment in the rod cells of the eye that is responsible for vision. Pigments in the cone cells allow discrimination between different wavelengths of light and give rise to the sensation of colour. Because of differences in these pigments, not all people see colour in the same way and some individuals are colour 'blind' in that they are unable to distinguish between certain colours. Lack of discrimination between green and red is the most common form of colour blindness. The wavelength limits of human vision also differ slightly between individuals. Below about 380 nm, there is no sensation of vision. As the wavelength increases above this, a faint stimulus is produced which creates in the brain the sensation of the colour violet. The colour perception then changes with increasing wavelength through blue, green, yellow, orange and red until at between 700 and 750 nm the light is once more invisible (Plate 14). When all wavelengths are present in the light, the eye perceives the sensation 'white', while the absence of visible light is seen as 'black'. The sensitivity to light at any given wavelength also varies and, for human vision, maximum sensitivity occurs at about 550 nm in the green region of the spectrum (Fig. 8.1). The overall visual effect thus depends on both the sensitivity to light at the wavelengths reaching the eye and on the colour sensation that they evoke.

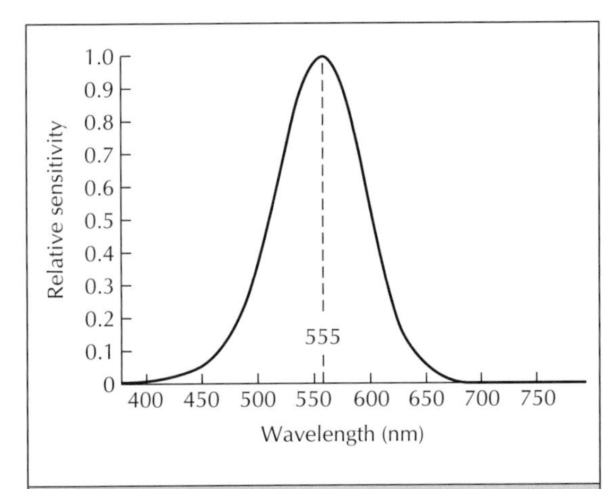

Fig. 8.1 The sensitivity curve of the human eye. This has its peak in the yellow-green part of the spectrum (555 nm) and decreases towards both the red and blue wavelength regions.

The limits of light in human terms lie between approximately 380 and 750 nm (it varies somewhat between individuals). The near infra-red region at wavelengths beyond 700 nm has a heating effect on biological tissue, while wavelengths shorter than 320 nm in the ultra-violet produce undesirable chemical reactions and are damaging to cells. The mutagenic effect of exposure to the sun's short-wave radiation (less than 290 nm) in causing skin cancer is now well recognised.

The vision of most animals is different from

that of humans and, from the point of view of the garden, the most important differences are found in animals that pollinate plants. Many, including bees, see further into the ultra-violet and it is evident that flowers frequently have markings that are visible to pollinators such as bees and butterflies but are invisible to humans. The way in which light is 'seen' by plants is also different. For example, the far-red region between 700 and 760 nm (which is scarcely visible to humans) is important for plants because it is absorbed by the pigment, phytochrome, and results in stem elongation in response to shading by a tree canopy (see Chapter 5).

Plant pigments

In order to produce a photochemical effect (whether vision or any other light-dependent process, such as photosynthesis), light must first be absorbed by a pigment. These are molecules that contain a *chromophoric group* which is responsible for their colour. They do not absorb light equally throughout the visible spectrum and consequently, when they are present in sufficient quantity, result in the colour of the perceived object. This is white when all visible wavelengths are reflected (i.e. none are absorbed). Three main groups of pigment are responsible for the colours of leaves, flowers and other organs of the plant. These are *chlorophylls*, *carotenoids*, and *anthocyanins*.

Chlorophylls

The most abundant pigments in plants are the chlorophylls, which absorb light mainly in the red and blue parts of the spectrum to give the green colour of leaves. Although chlorophylls are the major light-absorbing pigments for photosynthesis, the chloroplasts also contain accessory pigments which absorb light and transfer energy to the photosynthetic system. Most of these are carotenoids. The abundance of chlorophyll means that the red, yellow and purple colours associated with other pigments are usually not evident in leaves until the chlorophyll is broken down. The best example of this is autumn colour, which is discussed in more detail below.

Carotenoids

There are two groups of carotenoid pigments. The *carotenes* (first isolated from carrots, hence their name) consist of only carbon and hydrogen atoms and are mainly orange in colour. The *xanthophylls*, which have an oxygen atom in addition to carbon and hydrogen, give the yellow colours of petals and fruits. The carotenoid pigments may be located in the chloroplasts as accessory pigments for photosynthesis, or they may be found in special pigment cells called *chromatophores*.

Flavonoids

The *anthocyanins* belong to a group of chemicals called flavonoids, of which more than 2000 have been identified. Most flavonoids are soluble in water and accumulate in the vacuoles of the cells, although they are synthesised elsewhere. They are often confined to the epidermal or outer layer of cells and are particularly important in determining the colours of flowers, although they also occur in stems, leaves and fruits. The name anthocyanin is derived from the Greek words *anthos* (flower) and *kyanos* (dark blue) and these pigments are mainly red, blue, or purple in colour.

Anthocyanins consist of a coloured molecule, called an *anthocyanidin*, with a sugar group attached. The most common anthocyanidin, *cyanidin*, is purplish red and was first isolated from cornflowers (*Centaurea cyanus*). *Pelargonidin*, first isolated from a red *Pelargonium*, is orange-red, while *delphinidin* (from *Delphinium*) is bluish purple. However, the colour can vary depending on other factors. Anthocyanins are more red at low (more acid) cell pH, and more blue under alkaline conditions. An example of this is found in the flowers of Morning Glory (*Ipomoea tricolor*). In bud, the flowers are reddish purple but, as they open, the petals become bright blue. At the same time, the pH of the vacuole increases from 6.6 to 7.7. Confirmation that the increase in pH is the cause of the change in colour comes from a purple mutant that does not develop the blue colour of the petals and also fails to increase the pH of the vacuole.

Some anthocyanins can bond with metals to produce colours that are distinctly different from that of the free pigment. Probably the best known example of this is found in *Hydrangea*. The free pigment results in red or pink flowers but, when it is associated with aluminium, it becomes blue. The fact that hydrangeas are more blue on acid soils may seem to contradict the statement that a low cell pH produces a more red colour. The point is, however, that aluminium is only available to the plant in an acid soil (see Chapter 4) and it is the bonding to aluminium that changes the colour towards blueness. The pH of the cells is unlikely to be affected by the acidity of the soil.

Although similar in colour, the *betalains* are not chemically related to the anthocyanins, and the two groups do not occur together in the same species. The betalain, *betanin*, gives the colour to red beets.

Flavones and *flavonols* also belong to the flavonoid group. They range in colour from yellowish to almost white, but even the colourless ones absorb in the ultra-violet and affect the 'colour' that is visible to bees.

In addition to their function as attractants for pollinators, both flavonoids and carotenoids have a protective action against the tissue-damaging effects of ultra-violet wavelengths. With the loss of the ozone layer and the resultant increase in the ultra-violet content of sunlight reaching the earth, this function is likely to be of greater importance in the future.

Light-sensing pigments

In addition to photosynthesis, other important light-dependent reactions in plants are those which give information about their light environment; the two pigments involved are phytochrome, which absorbs mainly in the red and far-red regions of the spectrum, and BAP, which absorbs only blue light. Although in the red-light absorbing form, Pr (see Chapter 6), phytochrome is intensely blue in colour, it does not result in any colouring of the tissues because it is present at such low concentrations.

Pigments as attractants

In order to attract pollinators, plants need to stimulate the sense of vision and/or smell of the animal. The most common pollinators are insects, which have both a highly developed sense of smell and excellent vision. However, their sensitivity to colours differs from our own. Bees, for example, are much less sensitive to red light and are more sensitive to blue light than humans and they can see further into the ultra-violet. Undoubtedly bees see colours differently from humans and some wavelengths in the ultra-violet, which are invisible to humans, are visible to them. This becomes apparent when certain flowers are photographed using ultra-violet sensitive film, revealing nectar guides that are invisible when the flower is viewed by eye directly. Tests have revealed that some butterflies have distinct colour preferences; tortoiseshells, for example prefer yellow while swallowtails are attracted more to blue and purple. Flowers pollinated by animals that feed at dusk, such as moths and fruit bats, are normally pale in colour, or white, so that they are more easily visible in poor light. Some flowers are pollinated by birds, which have acute vision but no sense of smell. Their sensitivity to different wavelengths is more like ours than that of insects and they are attracted to bright colours, often in distinctive combinations such as the orange and purple of Bird-of-paradise (*Strelitzia reginae*) flowers.

Although this section is primarily concerned with 'colour' as an attractant for pollinators, we should also recognise the importance of other attractants such as scent and shape. The bee orchid strongly resembles the shape and form of a bee and orchids are among other plants that have scents which resemble those of *pheromones* (the sexual attractants for many insects). Plants pollinated by blow flies (such as the Dead horse *Arum*, which is native to parts of the Mediterranean) frequently resemble dead flesh in colour and smell of rotting meat, while flowers pollinated by night-flying creatures are often attractively and highly scented. Much is now known about the way in which flower shape is controlled by the plant's genes, especially with respect to flowers with *radial symmetry*, such as roses (*Rosa*), or *bilateral symmetry*, such as *Antirrhinum* (see Chapter 3).

VARIEGATION

Among the plants that are prized by many gardeners are those with leaves that are streaked, marbled or patterned in other colours. This is also true of some flowers, although to a lesser extent. The causes of these types of variegation include *chimeras, transposons, pattern genes* and, in a few cases, *virus infection*.

Chimeras

These are discussed in detail in Chapter 3 but a few points may be mentioned here. Chimeras are made up of genetically different cells. When the genes of these cells cause obvious differences such as, for example, a change in colour, then the different cell types become visible. Chimeras develop because flowering plants have a layered structure (see Chapters 1 and 3). Because of this arrangement, the outer layers of one partner in a graft can grow over the inner layers of the other, producing a plant made up from cell layers from two different genetic types (for example, +*Laburnocytisus adamii*). Chimeras can also occur when a mutation occurs in the growing point that causes part or all of a layer to be genetically different from other layers in the same plant. This type of chimera is usually stable and, when the mutations affect the production of chlorophyll, can result in good garden plants with attractively variegated leaves, such as *Hosta* and *Euonymus* (and see Chapter 3 for other examples). Although in evolutionary terms such a chimera does not confer any selective advantage and many may have arisen and died out, gardeners have selected them for their ornamental value, maintaining the variegation by taking shoot cuttings since they do not breed true from seed. However, they cannot be propagated by root or leaf cuttings, which disrupt the stable layered structure of the original growing point.

In some variegated chimeras, reversion to the all-green form may take place in one or more shoots. Because the variegated shoots have no chlorophyll in the white or yellow part of the leaf, they are less vigorous than the all-green shoots which, if left alone, will eventually swamp out the variegation. Thus it is important to prune out any all-green shoots as soon as they are seen.

Transposons

Another cause of variegation in flowers and leaves is the occurrence of *transposons* (see Chapter 3). Transposons can move around within the chromosome and occasionally 'jump' into a gene and inactivate it. If they affect a gene for flower or leaf colour, a plant with random spots, stripes or sectors may result. One of the best-known garden examples of transposons giving rise to an ornamental plant is that of Rosa mundi (*Rosa gallica* 'Versicolor') which has flowers with red streaks on a white background (see Chapter 3 for details of how the transposons cause the red streaks). Transposons also affect foliage patterns if they disturb the production of chlorophyll. For example, it is not uncommon for pale sectors to occur at random in leaves (e.g. in garden peas, *Pisum sativum*).

Pattern genes

The variegation patterns in some leaves are controlled by *pattern genes* and are a normal genetic characteristic of the species. In this case, the plants will breed true from seed. For example, in species of the tropical genus *Maranta* the green leaves are regularly patterned with dark blotches, while in *Actinidia kolomikta* the leaves are tipped with pink and white. Another form of seed-transmitted variegation is seen in Our Lady's milk thistle (*Silybum marianum*) where pockets of air along the veins reflect the light and give a silvery effect.

Viruses

Although not now of any great importance with respect to the ornamental value of plants, infection with viruses can result in some types of variegation. Two viruses are mainly responsible and both have a limited host range and so are not a threat to other plants. Infection with vein-clearing virus is the reason for pale leaf veins in *Vinca major* 'Reticulata' and similar forms of ground elder (*Aegopodium podagraria*) and sweet violet (*Viola odorata*). Abutilon mosaic virus only affects species of *Abutilon* producing a mosaic pattern of yellow/green on the leaves. The forms infected with this virus are *A. pictum* 'Thompsonii',

A. megapotamicum 'Variegatum', *A.* × *milleri* 'Variegatum' and it has also been introduced by grafting to the hybrids 'Ashford Red' and 'Kentish Belle'.

Tulipomania

Holland was the setting for one of the strangest episodes concerning a virus infection in plants. Although it happened to some extent in both England and France, 'Tulipomania' reached its peak in Holland between 1634 and 1637. The tulips that were introduced into Europe from Turkey were single colours and once the initial stocks had been bulked up, the bulbs were relatively cheap. However, such a plain and relatively valueless bulb could suddenly produce flowers that were marked in contrasting colours. For example, a plain red flower might emerge the following spring with petals that were marked with intricate patterns of white and deep red. It was this 'breaking' of the colour that produced the most sought after tulips of the time and resulted in bulbs that were sold for extraordinary prices. A single bulb of the variety 'Semper Augustus' (a white flower patterned with a red flame from the base to the tip of the petals) was already selling for 1000 florins in 1623 and by 1633 was valued at 5500 florins, finally reaching an estimated value of 10 000 florins at the height of Tulipomania (the average annual income at the time was about 150 florins). The highest price ever quoted for single bulb of 'Semper Augustus' was 13 000 florins, more than the cost of the most expensive houses in Amsterdam, and tulips of this kind became the ultimate status symbol of the time. Since the cause of breaking was a virus, it is not surprising that broken tulips tended to grow poorly and produced few offsets, thus increasing their value. Only offsets produced plants that were true to the original variety since the virus is not transmitted to the seeds.

At the time, the cause of breaking was a mystery. It occurred randomly and only in a small number of plants (perhaps one in a hundred) and it was probably this element of chance that drove the spread of Tulipomania in Holland. A plain, relatively valueless bulb might emerge one season miraculously feathered and flamed in contrasting colours and worth many hundreds of times its original cost. Since the cause was not known, the effects could not be controlled, although many attempts were made to induce breaking by treating the soil in various ways. It was not until the late 1920s that breaking was shown to be caused by a virus that works by partly suppressing the colour of the outer cell layers in the flower, allowing the underlying colour to show through. The latter is always white or yellow and the darker colour (red, purple or brown) looks as if it is painted on the petals from the base. This contrasting colour is always sharply defined, resulting in intricate flame- or feather-like patterns on the petals. Although they were the most sought-after plants of the seventeenth century, the fashion for broken tulips later declined and today growers try to eliminate the virus infection that causes 'breaking'.

ENVIRONMENTAL FACTORS INFLUENCING COLOUR

The intensity of colour developed by flowers and leaves is influenced to a considerable degree by environmental factors such as light and temperature. High light intensity intensifies colours, particularly where these depend on anthocyanins, which are produced as a by-product of photosynthesis. A good example of this is the reddest apples occurring on the outside of the tree. High temperatures have the opposite effect. The bronze colours of chrysanthemums tend to become yellow at high temperatures because the amount of red anthocyanin pigment is reduced, while the yellow carotenoids are unaffected. Similarly many pink flowers (where the colour is largely due to anthocyanins) can be almost white at high temperature. That this is exacerbated by low light is observed in some house plants, which may be deep pink when brought into the home but rapidly fade to almost white if kept in low light conditions at high temperatures.

Seasonal changes

One of the most significant seasonal changes in colour occurs during autumn, when leaves can range from the yellows and oranges of birch (*Betula* spp.) and Mountain ash (*Sorbus* spp.), through the vivid red of sugar maples (*Acer* spp.) and oaks (*Quercus* spp.), to the purples of some Japanese maples (*Acer palmatum*). The short days of autumn induce the cessation of growth followed by the onset of dormancy (see Chapter 9). In deciduous trees this is accompanied by leaf fall, which is preceded by changes in leaf colour as the chlorophylls begin to break down revealing the colours of the other pigments that are present in the leaf. The yellow/orange carotenoids also start to decompose rapidly in autumn but anthocyanins and other flavonoids accumulate leading to the development of yellow, bronze and red tones. This increase in anthocyanins occurs because, in autumn, the leaves contain more sugar than usual. Photosynthesis continues but the lower temperatures of the season reduce the movement of sugars out of the leaves, resulting in an excess of sugars in the leaf cells. These sugars are then converted to anthocyanins. Since photosynthesis is increased by high light intensity, bright autumn days result in a good display of autumn colour.

Although autumn colour is an annual occurrence, the intensity of the display varies widely from year to year and in different locations. Some of this variation is genetic since it has been found that the location of origin can have an effect on the intensity of autumn colour. For example, when plants of native north-eastern azaleas in the USA were raised from seed collected from different regions it was found that the intensity of leaf colour was related to the seed's origin. This emphasises the importance of selecting strains for good autumn colour. This is usually most successful when individual plants with good colouring are propagated vegetatively.

The environment also plays a significant part in the timing and intensity of autumn colour. Although the onset of dormancy is controlled by short days, the accompanying senescence and shedding of leaves is influenced to a greater or lesser degree by the lower autumnal temperature. In some species (e.g. in the Tulip tree, *Liriodendron tulipifera*, and the Tree of heaven, *Ailanthus altissima*) short days seem to be the most important environmental stimulus for leaf fall. In other cases (birch, *Betula pubescens*, and sycamore, *Acer pseudoplatanus*) low temperatures may affect the onset of leaf fall more than day-length.

Bright, cool days in autumn increase the intensity of leaf colour, while plants in the shade often colour poorly. For example, leaves of *Acer palmatum* 'Bloodgood' developed much brighter colours at a night temperature of 14–18°C than at 26°C. A good example of the effect of light intensity is displayed by *Fothergilla major*, where leaves in full sun turn a vibrant red but those in the shade only become pale yellow. Nutrition may be another factor. For example, high levels of nitrogen in the soil may decrease the amount of colour because the sugars, which would normally form pigments, combine with nitrogen to form proteins.

Autumn colour is unpredictable because it is influenced by many factors, several of which are beyond the control of the gardener. Those which are include the selection of the best cultivars, growing plants under the best light conditions and not over feeding with nitrogen. Other than this, the gardener can only hope for bright autumn days with cool nights which, together, will increase the free sugars in the leaves and so lead to a brilliant autumn display.

COLOUR IN THE GARDEN

Although the scientific basis of colour in plants is well understood, the use of colour in the garden ultimately depends on the likes and dislikes of the gardener. For many, it is harmonious combinations of colour that appeal and these are generally more restful to the eye than contrasting colours. Harmonious colour combinations are those which are next to each other on a *colour wheel* which begins with yellow and circles through orange, red, purple, blue, green and back to yellow. Contrasting colours are those that lie opposite to each other on the wheel, such as blue and orange, or green and red. There are sound biological reasons for the contrasting effect. If, for example, you stare at a red dot on a white background and then look away at the white area, you will see a faint green circle. The reason for this is that the cones in the retina of the eye become fatigued as they are repeatedly stimulated by the red light reaching them. As you look away from the red spot, the cones send a signal to the brain that the eye is looking at a red-deficient area, which the brain interprets as green because green and red are 'complementary' in terms of the signals that they send to the brain. When red and green plants are grown together in the garden, the messages from the eye continuously reinforce each other as the eye moves between red and green, with the result that both colours appear brighter.

The way colour is experienced depends on the attributes of the light falling on the plant. For example, the direction from which light falls on a border changes depending on the time of day. With the sun behind the viewer, the colour that is seen is reflected from the plant. When sunlight comes from the other direction, however, the colour that is seen has been transmitted through the plant. Viewing a plant with transmitted light increases the contrast between light and dark and results in a more dramatic effect. Reflected light can leave the border looking flat. The perceived colour also varies widely depending on whether the plant is in the sun or in the shade.

The colour composition of sunlight changes throughout the course of the day, especially early in the morning and late in the evening, and this has a strong influence on the way in which colours are seen. In the early morning and evening, the colours appear more intense while at midday the colour can appear washed out and leaves start to become more white and less intensely green.

Many plants change colour when they age, which can be a problem when choosing colours for the border. For example, *Allium* 'Beau Regard' is mauve when in flower, but the persistent seed heads are an intense straw colour.

Colour can be used to deceive the eye. Yellow has the effect of demanding attention and so draws the eye along a border or through a planting. The doyenne of colour-gardening, Gertrude Jekyll, planted bright yellow marsh marigolds (*Caltha palustris*) in the narrow canal at Hestercombe in order to move the eye along the line of the canal. Colour can also be used to distort the length and width of paths. Red tulips scattered in the borders on either side of a path makes it seem wider and shorter, whereas the pastel blue of catmint (*Nepeta racemosa*) makes a short path appear longer.

Ultimately the use of colour in the garden is a matter of taste but a knowledge of the factors that control the amount and composition of the pigments formed by the plant, as well as understanding how colour is actually seen by the eye, can help the gardener in the choice of plants to achieve the desired result.

SHAPE

Plants are characterised by an enormous variation in size and form. The basic elements for plant growth (water, light, CO_2 and minerals) determine how fast the plants can grow within their genetic constraints. Basic morphology, growth habit and factors such as flower and leaf shape are determined by their genetic code and understanding how genes control the pattern of growth may lead to the development of plants with shapes better tailored to the needs of the horticulturist. However, shape is also to some extent under the control of gardeners and growers who can bring about profound changes in plant development and morphology in a number of ways, including the application of chemicals, changes in the environment and mechanical manipulations such as pruning.

Environmental effects

Light

Among the most dramatic effects of environment of shape are those of light and day-length. The effects of light on the development of shape or *morphogenesis* have already been described earlier (Chapter 5) and only a brief resumé is needed here. In the absence of light plants are etiolated; following exposure to light, leaves expand, stem elongation is reduced and plants become green (Fig. 5.5). For these responses, light is a source of information, not of energy as in photosynthesis, so only small amounts of light are required. The main light-absorbing pigment controlling these responses is phytochrome (see Chapter 5) and so red light is the most effective. However, blue light absorbed by another pigment (the blue-light sensitive pigment or BAP) also modifies growth in a similar way.

Because of the control by phytochrome and BAP, the main influence of light on plant shape depends on the ratio of red (R) to far-red (FR) wavelengths in the light and the total amount of blue light. Many species growing in heavy shade from a plant canopy (i.e. in light with a low R:FR ratio) are tall and often have undesirably weak stems (Fig. 5.4). A further response to a low R:FR ratio is that axillary growth is suppressed so that the plants tend to grow without branching; this effect is seen in forestry practice where trees are deliberately planted close together so that the trunks develop without excessive branching. Increasing the spacing between plants can, therefore, be considered as a means of increasing branching and reducing overall height. However, there is a considerable variation in the degree of response.

Shade-avoidance plants (see Chapter 5) increase their stem length and reduce their branching when closely spaced because they detect and respond to the increased reflection of FR light from their near neighbours (Plate 13). However many shade-tolerant plants are largely insensitive to spacing. Other types of shading from walls or nearby structures reduce the overall content of blue light without affecting the R:FR ratio. This results in similar (but often less exaggerated) increases in height, with the development of thinner, larger leaves. In relation to the effects of the R:FR ratio of the light on stem length, there is considerable interest in the possibility of using coloured plastic filters to change the light quality received by plants. In order to reduce height and encourage branching, filters that remove the far-red wavelengths but transmit all other wavelengths in the visible spectrum are the most appropriate. The use of such filters to reduce stem elongation is likely to be of greatest importance for plants growing in the greenhouse rather than in the garden and experiments to test their effectiveness are being undertaken.

Other than the effects of shading, the effects of light on plant shape are relevant to horticulture in a number of ways. One of the most obvious is that etiolated, dark-grown plants have less fibre and so are less tough than those grown in the light. The exclusion of light (blanching) to achieve this is a long-established practice as, for example, in the production of chicory, sea kale and forced rhubarb. The latter is interesting because, although the crop is grown in dark sheds, some light is needed for harvesting. Based on an understanding of human vision and a knowledge of plant responses to light, it is clear that green light would be the best colour to use, since our eyes are most sensitive to green wavelengths (Fig. 8.1) and this is the region which has least effect on the responses controlled by phytochrome and BAP. In fact, green 'safe lights' have been used for many years in laboratories where research on the effect of light to prevent etiolation is carried out.

For plants growing entirely in artificial light (as in micropropagation) or where supplementary light is given during winter to increase photosynthesis, the light spectrum to which they are exposed may be very different from that of sunlight. Therefore, the effect of the light on plant shape must be taken into consideration when choosing lamps for these purposes. The use of artificial light is discussed further in Chapter 10.

Day-length

As discussed in more detail in Chapter 9, day-length can also profoundly modify the pattern of growth by inducing flowering (often accompanied by stem elongation in rosette plants), dormancy in woody plants (formation of bud scales instead of foliage leaves) and the development of storage organs (scales in onion bulbs instead of foliage leaves). Planting date can also affect plant height as it may determine the duration of growth before terminal flower buds form. In chrysanthemum, for example, later planting will mean shorter plants because autumnal short days induce flower bud formation earlier. Use is made of this in commercial 'All-year-round' production where the time at which plants are transferred to short-day conditions is determined by the length of stem desired (this is covered in more detail in Chapter 10).

Light direction

Finally, the direction from which light comes can also change plant habit. It is well known that plants grow towards the light (particularly young seedlings), a process known as *phototropism*. This can result in lopsided plants and it is important to ensure, as far as possible, that light is uniform from all directions. Placing a reflecting surface behind a pot (on a window sill for example) can help to reduce bending towards the outside light. A less well-known response is found in the seedlings of some tropical climbers, such as the Swiss cheese plant (*Monstera deliciosa*), which grow towards the relative darkness of the nearest tree.

Temperature

There are also some interesting responses to temperature. A higher temperature during the day has been found to be associated with longer stems, while a lower day temperature results in a more compact and sturdier growth habit. For some time it was widely assumed that the effect was due to the difference between day and night temperature (DIF). A day temperature lower than the night temperature (−DIF) would result in shorter plants, while a day temperature higher than the night temperature (+DIF) would result in taller plants. However, recent work at Horticulture Research International has shown that the effect on elongation is due to the absolute day and night temperatures, rather than to the difference between them. There was a strong positive effect on increasing height with an increase in day temperature in tomato (*Lycopersicon esculentum*) and chrysanthemum, while night temperature had little or no effect (Fig. 8.2). In many cases (e.g. chrysanthemum, poinsettia (*Euphorbia pulcherrima*), tomato and *Fuchsia* spp.) the effect of lowering the day temperature was greatest in the early part of the day. For example, in both tomato and poinsettia, stem length was decreased by dropping the day temperature from 20 to 12°C for four hours at the beginning of each day (this is now known commercially as DROP): lowering the temperature later in the day had much less effect. Unfortunately, this is not always the case and some plants (e.g. *Petunia* spp.) are more sensitive later in the day when it is more difficult to lower the temperature.

Because gibberellin-deficient mutants do not show a response to a drop in temperature, it is suggested that the dwarfing effect of lowering the temperature may be associated with a decrease in the production of gibberellin by the plant and, if so, it would be comparable with the effects of genetic dwarfing (Chapter 3) and growth retardant chemicals (see below). Although not practicable out of doors, lowering the temperature for part of each day, especially in the early part of the day, is feasible in the glasshouse and has been suggested as an alternative to costly chemical sprays.

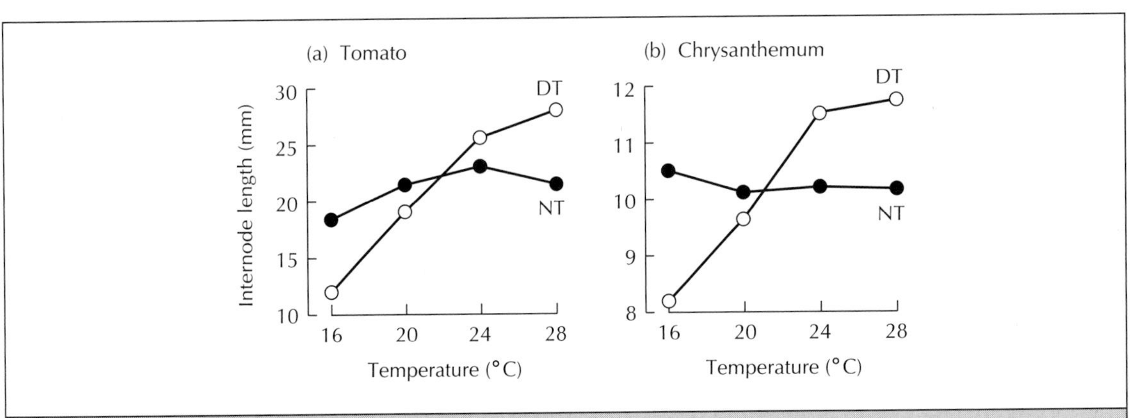

Fig. 8.2 The effects of day temperature (DT) and night temperature (NT) on stem elongation in tomato and chrysanthemum. Lowering the day temperature has a much greater effect to reduce height than lowering the night temperature. (From F.A. Langton and K.E. Cockshull, *Scientia Horticulturae*, **69**, 229–37, with permission from Elsevier Science, Amsterdam.)

Growth modifying chemicals

Gibberellins

The only plant growth hormones known to increase height when applied to intact plants are the gibberellins. This group of hormones was first discovered in Japan, where it was associated with a fungal disease of rice (called the 'foolish seedling' disease), in which the infected plants grew excessively tall. Subsequently, gibberellins were found to be widespread throughout the plant kingdom. Although they have many other effects on plants (for example in flowering (Chapter 9) and germination (Chapter 6)), the control of stem growth is a major function. Of the many known gibberellins, the most important endogenous regulator of stem elongation in higher plants appears to be GA_1. However, only GA_3 (gibberellic acid) and GA_{4+7} are readily available to growers (as Berelex and Regulex respectively). The majority of higher plants respond to the application of GA_3 but some conifers do not and, for these, GA_{4+7} is more effective. Gibberellins have been used commercially to promote stem growth in a few cases, such as producing standard *Fuchsia* plants and increasing flower stalk length in gerberas (*Gerbera jamesonii*), but are mostly used in other ways.

Growth retardants

Several groups of chemicals have been found to *reduce* stem length. Grouped together under the generic term of *growth retardants*, these are among the most widely used chemical growth regulators in horticulture, especially in the production of ornamental pot plants, such as poinsettia and chrysanthemums. In addition to reducing height, growth retardants have been found to accelerate flower bud production in some woody plants (especially in azaleas), to increase resistance to some stresses such as drought, and to increase the colour of leaves. Depending on which growth-retardant is used, it can be applied either as a soil drench to each pot, or as a foliar spray.

Several widely different groups of chemicals have been found to have growth retardant activity although no one category is effective on all plants. The choice, therefore, depends on the plant being treated. Chlormequat (Cycocel) is used commercially on poinsettias to reduce their height and on azaleas to increase the number of flower buds. Daminozide (B-nine) is used on chrysanthemums and azaleas (but is less effective than chlormequat on poinsettias), while ancymidol (A-rest) is effective on a wide range of plants, including chrysanthemums, poinsettias and many annuals. Paclobutrazol (Bonsai) is effective at low concentrations and is relatively persistent, as well as controlling elongation in a wide spectrum of plants.

Although growth retardants vary considerably in chemical structure, research indicates that they all act in a similar way and achieve their effects by reducing the amount of gibberellin in the plant. In many plants, the inhibition of growth by growth retardants can be completely overcome by the application of GA, suggesting that their major effect is to reduce gibberellin synthesis. However, individual growth retardants block different steps in the complex biosynthetic pathway leading to gibberellin production. The fact that growth retardants depress the biosynthesis of gibberellin means that treated plants are effectively *chemical dwarfs* and are comparable with those genetic dwarfs which under-produce or are insensitive to gibberellins (see Chapter 3). However, the essential difference is that the genetic dwarfs remain so throughout their life, while the dwarfing effects of a chemical treatment will ultimately be lost. This means that treated plants (bought from a commercial source) would not maintain their dwarf habit indefinitely if planted out in the garden, although the effect may be quite persistent in some cases.

Although they are widely used in commerce, growers are now trying to find alternatives to growth-retardant chemicals as a means of controlling height. Their application requires a lot of labour (for example, more than fifteen sprays of chlormequat are not unusual for poinsettias), they are expensive and it is easy to make a (possibly disastrous) mistake in the time of application or in the concentration of the chemical used. They are, moreover, regarded as undesirable, especially where the crop is edible. Alternatives that are being considered, or in some cases already used, are control of temperature (DROP discussed above), the restriction of water supply in vegetable seedlings, brushing plants (see below), and filters which reduce the amount of FR light received by the plant (see above).

Mechanical treatments

The modification of growth patterns by manipulating the plant has long been practised by gardeners who discovered by trial and error that certain treatments led to profound effects on the shape of the plant. With an increased understanding of the roles and movements of plant hormones and the operation of inter-organ signalling, we can now understand better why some of these treatments result in the observed effects. Among horticultural practices, pruning is one of the oldest and most widely used ways of achieving the desired shape, as well as being used to encourage maximum flowering and fruiting, and an understanding of the underlying basic physiological processes is essential if the correct results are to be obtained. Moreover, before undertaking any pruning treatment, it should be recognised that pruning always causes stress and is not necessarily the best way of achieving the desired results. It is better to choose the right plant for the right place and so avoid the need for remedial pruning. No amount of pruning will make a Leyland cypress (× *Cuprocyparis leylandii*) suitable for a tiny garden and the right solution is to grow plants that are genetically small, or have been grafted onto dwarfing rootstocks (see Chapter 7).

Pruning

The basis of all pruning treatments is apical dominance. The apical bud in many species, prevents or slows the growth of the lateral buds below it. If the apical bud is removed by pruning (or, under natural conditions, is damaged by wind or animals), the lateral buds are released from dominance and start to grow. The uppermost lateral bud grows more vertically and becomes the new leading shoot; in turn, this begins to exert dominance over the lower laterals. The presence of the shoot apex also makes the lower branches grow out more horizontally, while the apical bud itself grows vertically. We do not know if this effect results from a difference in response to gravity, or is dependent on some particular angular relation between the stem and the lateral branches. We do know, however, that auxins appear to be involved.

It has been known for some time that removing the youngest visible leaves is as effective as removing the whole of the apical bud and it is evident that one or more inhibitory factors are formed in them. However, the precise nature of the inhibitory factor(s) is still in some doubt. If the apical bud is removed and replaced with an auxin, the inhibition of lateral bud growth is re-imposed, indicating that apical dominance is due to auxin being transported downwards through the plant (see Chapter 7) and accumulating in the lateral buds. However, this does not seem to be the whole story since the amount of auxin required to re-impose dominance is some thousand times higher than the amount that is actually present in the apical bud. Moreover, when ^{14}C-labelled auxin is applied to a decapitated shoot, auxin does indeed travel down the plant as expected but it does not move into the inhibited lateral buds. When auxin is applied directly to these buds, they begin to grow out as if dominance has been released.

All these lines of evidence suggest that auxin moving from the apex into the lateral buds is not the immediate cause of dominance. One suggestion is that the dominant, actively-growing terminal shoot is a sink into which nutrients and hormones are diverted, resulting in the inhibition of lateral bud outgrowth. The involvement of auxins and also cytokinins in the control of apical dominance is supported by the results of genetic-modification experiments. When tobacco plants were genetically engineered to produce higher than normal amounts of cytokinins or lower than normal amounts of auxin, in each case they branched excessively when compared with unmodified plants. Thus the auxin/cytokinin ratio could be the factor that determines the degree of apical dominance and the extent of branching (see Chapter 7 for the effect of the auxin/cytokinin ratio on the development of roots and shoots).

One of the problems about pruning to restrict size is that plants maintain a constant root:shoot ratio and re-direct growth to maintain it. Consequently, hard pruning actually stimulates vegetative growth and increases the vigour of the pruned shoot. Thus hard pruning does not necessarily reduce size, although constant cutting back will ultimately weaken the plant. The general recommendation, therefore, is to prune vigorous shoots lightly and less vigorous shoots more drastically in order to achieve more balanced growth. It is better to thin out shoots by removing some of them entirely, rather than to cut them back. In order to achieve a shapely plant, cuts should be made to a bud that, as far as possible, faces in the desired direction.

Pruning is not undertaken only to control plant shape. In order to obtain maximum flowering, pruning must be considered in relation to the flowering habit of the plant in question. For plants that flower late in the year on the current season's wood, cutting back is normally carried out in the early spring in order to allow maximum growth before flowering, while pruning normally takes place after flowering for plants that flower earlier in the season on the previous season's growth. Detailed recommendations for individual plants are widely available in the horticultural literature.

Root pruning

Because of the relatively constancy of the root to shoot ratio, one method by which shoot vigour can be reliably controlled is by root pruning. Originally this was a common practice to reduce the vigour of orchard trees but has largely been rendered obsolescent by the introduction of dwarfing rootstocks. In the open ground, root pruning of over vigorous large trees should be carried out over two seasons, removing half of the root in each year. A 30–40 cm deep trench is dug out about 120 cm from (and half-way round) the trunk during the winter and large roots are cut through. The soil is then replaced and the tree should be mulched in the following spring. Root pruning is a necessary practice for container-grown plants that have become pot-bound. The outer roots should be pruned back by approximately two thirds and the plants re-potted. In order to maintain the balance of growth, it is also necessary to remove about one third of the top growth.

Nicking and notching

It is well known that vertical branches grow more vigorously than more horizontal ones and this fact has long been the basis of training fruit trees as espaliers and cordons. When branches are tied horizontally, apical dominance is weakened and the lateral buds begin to grow out, a necessary pre-requisite for the formation of fruiting spurs. The practices of nicking and notching (Fig. 8.3) during the training of fruit trees also has its origin in apical dominance. Removing a small wedge of tissue above a bud (notching) isolates the bud from the effect of the apex and increases the accumulation of materials into the bud from below; the result is to weaken the apical dominance and allow the bud to develop. It is particularly useful where it is desirable to encourage bud growth from low down on the plants. The opposite treatment is to remove a piece of tissue immediately below the bud (nicking), where the effect is to increase apical dominance and weaken bud growth. It is useful for fan-training, where the apical dominance effect results in the development of branches at a wider angle to the main trunk.

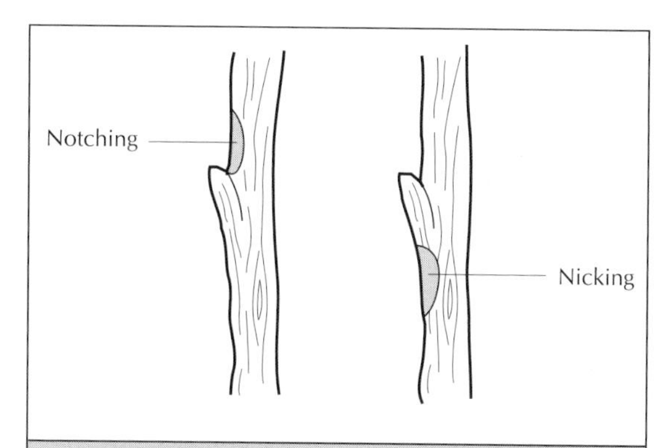

Fig. 8.3 Nicking and notching. Removing a wedge of bark below a bud (*nicking*) prevents that bud from developing into a shoot or weakens its growth. Removing a wedge of bark above a bud (*notching*) encourages that bud to grow out into a shoot.

Bonsai

Perhaps the ultimate restriction of tree size is found in *bonsai*, which depends for its success on the same kinds of manipulation, namely root pruning and/or restriction, the bending of shoots from the vertical (twisting round wire) and careful pruning to achieve the desired final form.

Shaking and stroking

Manipulation of size by pruning is an old established horticultural practice. More recently a very different approach has resulted from the observation that exposure to air movements can affect plant height. Although the effect is not very noticeable out of doors because of the prevalence of wind, shaking or stroking plants can reduce growth considerably in plants growing under glass. For example, shaking chrysanthemum plants for just 30 seconds twice a day was found to decrease stem length by as much as 30%. Since plants are continually being handled during pruning, staking, disbudding and other treatments, this shaking (*thigmo-seismic*) effect should not be ignored, although not all plants are equally sensitive. For example, marigold (*Tagetes* spp.) and sunflower (*Helianthus annuus*) are relatively sensitive to light brushing, while *Zinnia* and China aster (*Callistephus chinensis*) are not. Salvia (*Salvia splendens*) and Marvel of Peru (*Mirabilis jalapa*) have been found to be particularly sensitive with stems growing to only half the length of those in untreated plants. A relatively small amount of brushing can produce a noticeable reduction in size of young seedlings and further treatments increase the risk of damage.

In one recent trial with pansies, plastic netting was dragged back and forth across young seedlings, this being equal to one 'stroke'. A considerable reduction in height was obtained with only a few 'strokes' each day, while the response was saturated with about 10 strokes (Fig. 8.4). To obtain an effect it was necessary to flex the plants at each stroke and to begin the treatment at an early stage of growth. This has the added advantage that many plants can be treated with a single stroke. It was only necessary to treat the plants once each day (on at least five days in each week) and the time of day was not critical.

The effect of brushing is not new but has not, until recently, been tried commercially. It is an interesting technique because it is environmentally friendly (no chemicals), is effective to decrease height in most of the species tested so far and is within the reach of the amateur grower.

One important application of the effects of shaking on plants is the recommendation that stakes for newly planted trees should be relatively short, in order to allow for some flexing of the young trunks. This results in thicker trunks and sturdier growth.

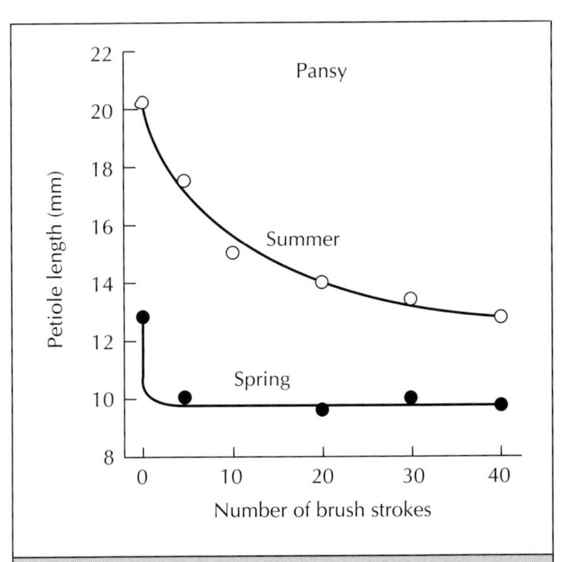

Fig. 8.4 The relationship between the length of petioles in pansy and the number of brush strokes given each day. The response was somewhat different in spring and summer but in both cases maximum effect was achieved by giving about 10 brush strokes each day. The strokes were given by brushing the young plants back and forth (one stroke) with nylon mesh, making sure that the plants were actually moved each time. (Data from A. Langton, Horticulture Research International, reproduced with permission.)

FURTHER READING

Brickell, C. (1979) 'Pruning'. *The RHS Encyclopedia of Practical Gardening*. Reed International Books, Ltd., London.

Brickell, C. & Joyce, D. (1996) *Pruning and Training*. (Royal Horticultural Society Handbook) Dorling Kindersley Ltd., London.

Burras, J.K. (1997) Streaks and Sectors. *The Garden*, **122**, 12–15.

Dourado, A. (1997) Unpredictable pigments, *The Garden*, **122**, 717–19.

Jekyll, G. (1988) *Colour Schemes for the Flower Garden*. Frances Lincoln, London.

Hanks, G. & Langton, A. (1998) Growing to length. *The Horticulturist*, **7**, 25–9.

Lawson, A. (1996) *The Gardeners Book of Colour*. Frances Lincoln, London.

Pavord, A. (1999) *The Tulip*. Bloomsbury, London.

Salisbury, F.B. & Ross, C.W. (1992) *Plant Physiology*, 4th edn. Wadsworth Publishing Company, Belmont, Ca.

9

Seasons and Weather

- Day-length: history; advantages and disadvantages of a response to day-length
- Flowering: long-day and short-day plants; the concept of critical day-length; perception of day-length by the leaves
- Chemicals and flowering: the flowering signal exported from the leaves; the role of plant hormones; gibberellins; ethylene
- The detection of day-length: measurement of the length of the night; circadian clock measurement of the critical duration of darkness
- Dark-dominant and light-dominant plants
- Possible effects of clouds, moonlight and street lights
- Storage organs: the control of tuber and bulb formation by day-length
- Leaf-fall and dormancy: the onset of dormancy in woody plants and its consequences for garden plants
- Changes with altitude and latitude
- Temperature and flowering: vernalisation and the requirement for chilling
- Planting times in relation to cold requirement and bolting
- De-vernalisation
- Interactions between vernalisation and day-length
- Breaking winter dormancy by chilling
- Direct effects of temperature on flowering
- Damage by below freezing temperatures: how frost damage occurs
- The acquisition of frost resistance

All gardens change with the changing seasons and, in this chapter, we will look at the ways in which some of these seasonal changes are brought about. Many environmental conditions alter during the course of the year but those which have the greatest effect on plants are temperature, rainfall, the intensity of light, and the daily duration of light. The way in which these vary depends on latitude and Fig. 9.1 gives examples of yearly changes in temperature, light and the length of day for two latitudes, one in the tropics at latitude 5°N and one in the northern hemisphere at 50°N, the latitude of London.

Most seasonal factors show some year to year variation depending on the weather patterns. In contrast, the length of the day is the one factor that changes with season but does not vary from year to year. It should come as no surprise, therefore, to find that many of the seasonal responses of plants (and animals) are controlled by the length of day (usually abbreviated to *day-length*). However, as we shall see later, they can also be modified by other environmental conditions leading to the differences, such as in the time of flowering, that occur from year to year.

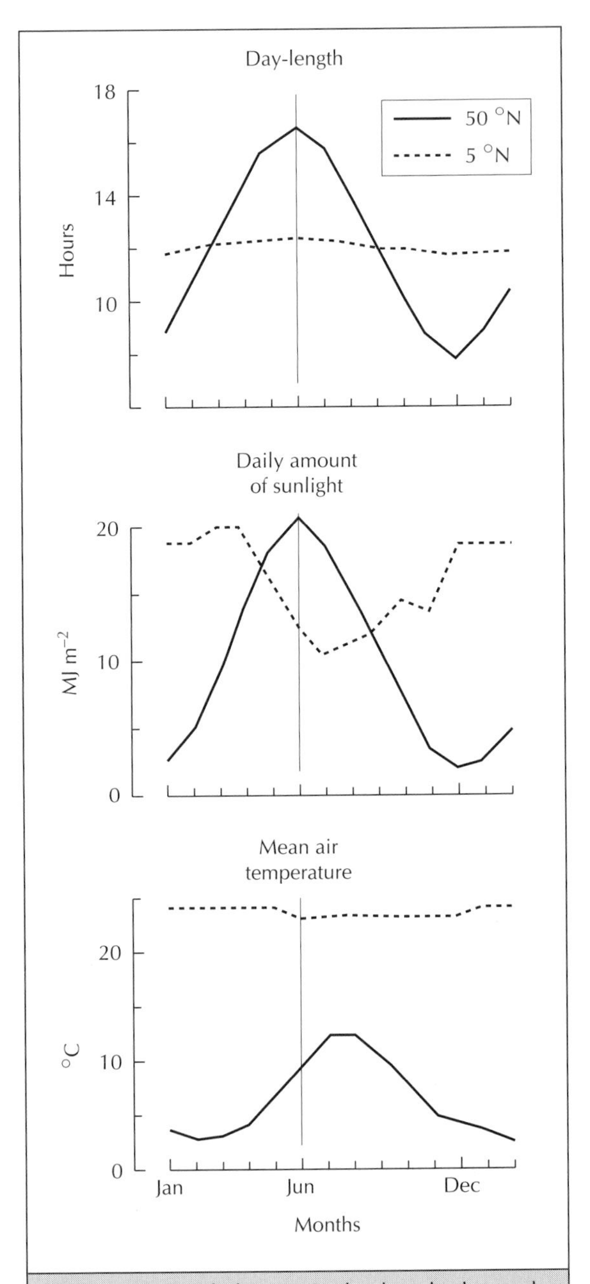

Fig. 9.1 Seasonal changes in day-length, the total amount of light each day and the daily mean air temperature at two latitudes.

DAY-LENGTH

Despite major changes in climate during the earth's history, day-length patterns have remained remarkably constant, so it is not surprising that plants should have evolved to respond to it in the regulation of their seasonal patterns of growth and development. Perhaps more surprising is that it took such a long time for biologists to realise this, for it was not until the beginning of the twentieth century that the biological effects of day-length began to be suspected. At about that time, Julien Tournois of Paris was puzzled as to why hop plants should flower so precociously when sown in greenhouses in winter. He later found that he could also achieve earlier flowering in spring by placing covers over the plants to reduce the hours of daylight. In a paper published in 1914, he concluded that the precocious flowering of hops in winter was caused by the short day-lengths or, more likely, by the longer nights that accompanied them. The latter was an inspired guess, since the importance of the duration of darkness was not confirmed until many years later. Tournois died in the war and it was not until the mid 1920s that the effects of day-length were clearly established.

Working at the US Department of Agriculture in Washington D.C., Wightman Garner and Henry Allard were trying to find out why a variety of tobacco, 'Maryland Mammoth', continued to grow throughout the summer without flowering, even though it reached a very large size. On the other hand, quite small plants flowered rapidly when they were kept in a greenhouse during the winter. They were also puzzled by the fact that, when the 'Biloxi' variety of soya bean was sown at intervals during spring and summer, all the plants tended to flower at the same time. After eliminating temperature and light intensity as possible causes of the differences, they concluded that day-length was the only other seasonal factor that might be having an effect. They tested their idea in a very simple way by putting plants of soya beans and tobacco into a light-tight shed for 17 hours each night during the summer. These plants flowered while their counterparts remaining in the open did not. The conclusion was that these varieties of tobacco and soya bean would flower only when the daily duration of light was sufficiently short.

One of the consequences of these simple experiments was the realisation that growth and flowering are to a large degree independent of each other. Under favourable day-lengths, plants may flower precociously without growing to any extent, as in Tournois' hop plants. In unfavourable day-lengths, they may grow to an enormous size without flowering, as in Mammoth tobacco.

Since these early experiments, the length of the day (or *photoperiod*) has been found to control an enormous number of responses in both plants and animals. Some examples for plants are given in Table 9.1. In addition to flowering, responses that are particularly important for gardeners include the onset of dormancy in many tree species and the formation of underground storage organs, such as bulbs and tubers.

The study of how day-length regulates flowering and other aspects of growth and development has acquired its own terminology. *Photoperiodism* may be defined descriptively as a response to the length of day that enables an organism (plant or animal) to adapt to seasonal changes in its environment. A more useful definition is that it is a response to the *timing* of light and darkness during the daily cycle. This definition is important because, as will be shown in Chapter 10, the time when a relatively short exposure to light is given to a plant during a dark period may have a profound effect. Indeed, such *night-break* treatments are used by commercial growers to regulate the flowering time of a number of crops.

With a mechanism sensitive to day-length, the seasonal timing of events can be precisely controlled. Indeed, a day-length difference of only twelve minutes has been found to switch between flowering and vegetative growth in some tropical plants which, contrary to what is sometimes suggested, are often highly sensitive to the relatively small differences in day-length that they experience in their native habitats. A point that has to be taken into account is that the day-lengths in spring are the same as those in autumn. Confusion between the two seasons can be avoided by an associated response to temperature or by having a different day-length requirement at different stages of development. For example, autumn-flowering chrysanthemums form terminal flower buds when the day-length is less than about 14.5 hours in either spring or autumn, but their further development into open flowers only occurs in shorter days of about 13 hours. These varieties are clearly well adapted to flowering in the autumn when the days are getting shorter. In contrast, in many varieties of strawberry, flowers are formed in the short days of autumn but their further development into open flowers is most rapid in long days, encouraging flowering in late spring. Some plants must experience a sequence of long and short days before they can flower. Where long days must be given before short days, plants will normally flower in the autumn, while the reverse sequence will result in late spring or early summer flowering. Another way is to couple a response to day-length with a response to low temperature, or *vernalisation* (see below). A requirement for exposure to cold for several weeks will prevent flowering until a winter has been experienced by the plant and so would ensure that flowering does not occur until spring or early summer. Biennial plants fall into this category.

Table 9.1 Some responses of plants to the length of day.

Response	Promoted by: Long days	Promoted by: Short days	Example
Flowering:			
initiation of flowers	+		*Fuchsia* 'Lord Byron'
		+	poinsettia (*Euphorbia pulcherrima*)
growth of flowers	+		strawberry (*Fragaria* × *ananassa*)
		+	*Caryopteris* × *clandonensis*
Sex of flowers:			
promotion of maleness		+	maize (*Zea mays*)
promotion of femaleness	+		*Begonia* × *cheimantha*
Dormancy in woody plants:			
formation of resting buds		+	birch (*Betula*); poplar (*Populus*)
shedding of apex		+	false acacia (*Robinia pseudoacacia*)
leaf fall		+	tulip tree (*Liriodendron tulipifera*)
increased frost resistance		+	dogwood (*Cornus*); spruce (*Picea*)
Formation of storage organs:			
underground stem tubers		+	Jerusalem artichoke (*Helianthus tuberosus*)
aerial stem tubers		+	*Begonia grandis* subsp. *evansiana*
root tubers		+	dahlia hybrids (*Dahlia*)
corms	+		*Triteleia laxa*
bulbs	+		onion (*Allium cepa*)
Seed germination	+		birch (*Betula* spp.)
		+	*Nemophila insignis*
Other changes:			
plantlets on leaves	+		*Kalanchoë daigremontiana*
runners	+		strawberry (*Fragaria* × *ananassa*)
rooting capacity	+		*Weigela florida*
		+	holly (*Ilex crenata*)
leaf size	+		China aster (*Callistephus chinensis*)
formation of coloured pigments (anthocyanins)		+	flaming Katy (*Kalanchoë blossfeldiana*)
increased stem elongation			
rosette plants	+		henbane (*Hyoscyamus niger*)
plants with stems	+		fuchsia; French bean (*Phaseolus vulgaris*)

What are the advantages to the plant of ensuring that a response occurs at a particular time of the year? One advantage is that synchronous flowering within a population of plants in the wild can ensure good cross fertilisation. Flowering can also be timed to occur when the environment is favourable as, for example, when water is plentiful during a rainy season, or in the summer when there is enough light for photosynthesis to meet the energy demands of seed production. Another important advantage is the ability to avoid the damaging effects of unfavourable conditions such as drought or low temperatures by shedding leaves and/or becoming dormant.

For gardeners, however, responses to day-length can be both advantageous and dis-advantageous. For example, many plants from lower latitudes require short days for flowering and so do not flower (or flower poorly) during a northern summer (e.g. poinsettia, Table 9.2). Similarly, plants from high latitudes are usually adapted to flowering in the long days of summer when temperature and light conditions are favourable; hence, they will not flower (or may flower poorly)

Table 9.2 Examples of the critical day-length for the formation of flowers. The critical day-length (in hours) is given for each plant. Short-day plants (SDP) flower in day-lengths *shorter* than the critical, while Long-day plants (LDP) flower in day-lengths *longer* than the critical. However, the critical day-length is not always defined in the same way. In **A**, the critical day-length is that above (LDP) or below (SDP) which some flower parts are formed, although open flowers may not always follow. This is the definition used by many plant physiologists. For the LDP in **B**, the critical day-length is that at and above which rapid and uniform flowering occurs. This definition is likely to be more useful to gardeners. The critical day-lengths in **B** were determined as part of a research programme at Michigan State University under the direction of A. Cameron, W. Carson and R. Heins.

It is important to emphasise that, irrespective of how it is defined, the critical day-length may vary in other cultivars or strains and under different environmental conditions. Moreover, some plants must be given a period at low temperature (see text, under *vernalisation*) before they become sensitive to day-length.

Short-day plants A	florists' chrysanthemum, 11–16 coleus (*Plectranthus fredericii*) 13–14 poinsettia (*Euphorbia pulcherrima*) $11-12\frac{1}{2}$ strawberry (*Fragaria* × *ananassa*) 11–16 soya bean (*Glycine max* 'Biloxi') 12 flaming Katy (*Kalanchoë blossfeldiana*) 12	tobacco (*Nicotiana tabacum* 'Maryland Mammoth') 14 *Perilla*, red-leaved, 14 *Perilla*, green-leaved, 16 Japanese morning glory (*Ipomoea nil* 'Violet') 15
Long-day plants A	dill (*Anethum graveolens*) 10–14 pot marigold (*Calendula officinalis*) $6\frac{1}{2}$ coneflower (*Rudbeckia bicolor*) 10 ice plant (*Sedum spectabile*) 13	campion (*Silene armeria*) $11-13\frac{1}{2}$ mustard (*Sinapis alba*) 14 spinach (*Spinacea oleracea*) 13
B	yarrow (*Achillea* 'Moonshine') 16 Japanese anemone (*Anemone hupehensis*) 16 thrift (*Armeria pseudarmeria*) 16 milkweed (*Asclepias tuberosa*) 16 astilbe (*Astilbe chinensis* var. *pumila*) 16 bellflower (*Campanula carpatica* 'Blue Clips') 16 tickweed (*Coreopsis verticillata* 'Moonbeam') 14 (*Coreopsis* 'Sunray') 13 blanket flower (*Gaillardia* 'Kobold') 13 *Gaura lindheimeri* 'Whirling Butterflies', 13 cranesbill (*Geranium* × *cantabrigiense*) 16 (*Geranium dalmaticum*) 16 *Gypsophila* 'Happy Festival', 16 *Helenium autumnale*, 16 *Hibiscus moscheutos* 'Disco Belle', 16	plantain lily (*Hosta* spp.) 14 *Leucanthemum* × *superbum* 'White Knight' & 'Snowcap', 16 *Lobelia* × *speciosa* 'Compliment Scarlet', 16 bergamot (*Monarda didyma* 'Gardenview Scarlet') 16 evening primrose (*Oenothera fruticosa*) 16 (*Oenothera missouriensis*) 16 *Phlox paniculata* 'Tenor', 'Eva Cullum', 14 *Physostegia virginiana* 'Alba', 16 coneflower (*Rudbeckia fulgida* var. *sullivantii* 'Goldstrum') 13 sage (*Salvia* × *sylvestris* 'Blaukönigin') 16 saxifrage (*Saxifraga* 'Triumph' (*Saxifraga* × *arendsii*) 14 ice plant (*Sedum* 'Herbstfreude') 16 Stokes' aster (*Stokesia laevis* 'Klaus Jellito') 13

in the tropics (e.g. *Salvia* × *sylvestris* 'Blaukönigin', Table 9.2). Fortunately much is known about the genetic control of photoperiodic responses and so breeding can often be used to eliminate such problems. A major advantage for commercial growers is the possibility of using artificial day-lengths in order to time plants for a particular market as, for example, poinsettias for Christmas, or to spread out the cropping period in order to reduce gluts and peak demands on labour.

FLOWERING

The time of flowering has been by far the most intensively studied of the many responses to day-length and most of our ideas about the underlying mechanisms come from these experiments. Plants in which flowering is responsive to day-length are usually classed in one of two groups (Fig. 9.2). These are *short-day plants* (SDP) which only flower, or flower earlier, when days are shorter than a particular duration (known as the *critical day-length*) and *long-day plants* (LDP) which only flower, or flower sooner, when the days exceed the critical day-length. Those plants in which the time of flowering is independent of day-length are called *day-neutral*. A few plants are known to flower only (or more rapidly) when the days are neither too long nor too short. These are called *intermediate-day plants*.

It is important to understand that the difference between short- and long-day plants does not depend on the actual value of the critical day-length. For example, the critical day-length for some varieties of marigold (*Calendula officinalis*) is only 6.5 hours; nevertheless they are LDP and will flower only when the day-length *exceeds* this value. In contrast, the critical day-length for some varieties of the Japanese Morning Glory (*Ipomoea nil*) lies between 15 and 16 hours, but these are SDP and will flower only when the day-length is *shorter* than this value. Thus, in any one day-length, both long- and short-day plants may be able to flower and their classification can only be determined by growing them experimentally in a range of different day-lengths (Plate 15). Under natural conditions, flowering in LDP will be

Fig. 9.2 Flowering responses to day-length. Plants were grown either in short days (8 hours, plants on left) or long days (16 hours, right). The short-day plant (*Kalanchoë blossfeldiana* (a), flowered rapidly in short days but was markedly delayed in long days. The long-day plant *Antirrhinum majus* (b) flowered much more rapidly in long days.

delayed until the critical day-length is exceeded during the lengthening days of spring and early summer. Many SDP will flower in autumn, when the day-length becomes shorter than the critical duration. The critical day-lengths of some common horticultural plants are given in Table 9.2.

Although the changes that lead to flowering take place at the shoot tips (either the main or lateral ones), it is the leaves that actually 'see' the day-length signal. This is also true for other responses to day-length such as the onset of winter dormancy in woody plants and the formation of underground storage organs. The special role of the leaves was demonstrated as early as 1934 when it was shown that exposing only the leaves of the LDP spinach to long days resulted in the formation of flowers at the shoot tip. Perhaps the most spectacular result is with the SDP *Perilla* (red-leaved perilla) where leaves were exposed to short days *after* they had been cut off from the plant (Fig. 9.3). These leaves were able to cause flowering when they were subsequently grafted on to vegetative plants growing in long days. It follows from these experiments that a leaf exposed to appropriate day-lengths is able to transmit some kind of signal, presumed to be a chemical, that is able to travel to the shoot apex and trigger a flowering response there. We can, therefore, separate the process of the day-length control of flowering into three components:

(1) the detection of day-length and subsequent changes in the leaf (known as *photoperiodic induction*)
(2) the transmission of some kind of signal from the leaf and
(3) the changes at the shoot tip that result in flowering (known as *evocation*).

It is highly likely that the events at the shoot tip are not unique to photoperiodism but also occur where the onset of flowering is independent of day-length, as in day-neutral plants.

From the point of view of horticulture, there are two major questions to be addressed. First, what is the identity of the signal that is exported from the leaves? This obviously has implications for the possible control of flowering by the use of appropriate chemical treatments. The second question concerns the way in which day-length is detected by the leaves, since this becomes important when growers use artificially imposed day-lengths to control flowering and other responses to day-length.

Chemicals and flowering

Several approaches have indicated that the signal exported from leaves following photoperiodic induction is a chemical that appears to move mainly in the phloem (the sugar transporting

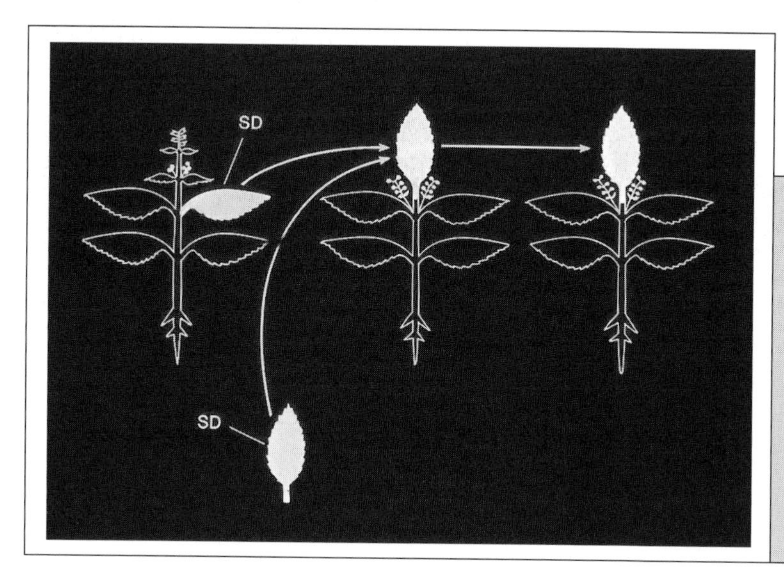

Fig. 9.3 The detection of day-length takes place in the leaf. A leaf of the short-day plant, *Perilla*, was exposed to short days either while still attached to the plant (top left) or after removal (bottom). When grafted, this *induced* leaf (see text) was able to cause flowering in a receptor plant maintained in long days (top middle) and retained the ability to cause flowering when detached and re-grafted to another plant (top right).

system, see Chapter 1). Unfortunately, the identity and nature of this chemical signal remains a mystery. Since it moves from a site of production (the leaves) to a site of action (the shoot tip), the floral stimulus has the characteristics of a hormone. The term *florigen* is sometimes used for this hypothetical hormone but *floral stimulus* is probably the better term. *Florigen* implies that the action of the hormone is uniquely concerned with the formation of flowers, whereas the existence of such a flower-specific hormone has not yet been demonstrated.

One interesting line of experiments into the nature of the floral stimulus has been to use the horticultural technique of grafting. The stimulus will move from a LDP grafted to a SDP and, conversely from a SDP grafted to a LDP. To take a typical example, the LDP henbane (*Hyoscyamus niger*) will flower in short days when grafted to the SDP tobacco (*Nicotiana tabacum*); this will only occur, however, when the donor plant, tobacco, retains its leaves and is kept in short-day conditions. Thus only when they are themselves induced by exposure to the appropriate day-length can the leaves of the tobacco plants bring about flowering in a plant of a different day-length response group.

The fact that the floral stimulus is equivalent in plants of different day-length response groups and different genera is one of the pieces of evidence that led to the suggestion that there may be a chemical that is effective in a wide range of plants and perhaps in all, irrespective of their day-length response. There are, however, two problems with this proposal. One is that grafting is only possible between closely related plants where any floral stimulus might be expected to be similar. The second is that, so far, all attempts to isolate such a universal flowering hormone have been unsuccessful.

There could be many reasons for this failure. *Florigen*, if it exists, could be unstable or even a mixture of compounds. Or inhibitors of flowering (*anti-florigens*) could be present in plant extracts and interfere with attempts to test them for their possible effects on flowering. Indeed, there is good evidence from similar grafting experiments that anti-florigens do exist. For example, when a day-neutral variety of tobacco was grafted onto the LDP henbane (*Hyoscyamus niger*), flowering in the tobacco shoot was prevented when the grafts were kept in short days and the henbane failed to flower. This implies that an inhibitor of flowering moved from the henbane to the tobacco, since this variety of tobacco normally flowers in short days. As the inhibitory effect increased with the number of leaves remaining on the henbane plant, it is clear that, like the floral stimulus, the inhibitor originates in the leaves.

At present we have to accept that the existence of a unique and specific floral stimulus is unproven and is beginning to appear unlikely. Moreover, the signals for the control of flowering have been shown to include both promoters and inhibitors. This has led many research workers to examine the effects on flowering of the known plant hormones. This approach has been used by horticulturists for many years because of the enormous practical implications that would arise from the discovery of chemical regulators of flowering that are relatively inexpensive and easy to apply.

Although hormones belonging to the *auxin* and *cytokinin* groups may be involved in the changes that are undergone at the shoot tip during the transition to flowering, they have little or no effect when applied to intact plants. In contrast, hormones in the *gibberellin* group, have been found to bring about flowering in several species, while the gaseous hormone, *ethylene,* is highly effective in one particular family, the Bromeliaceae.

Gibberellins

It has been known for a long time that the application of a solution of gibberellin (GA) can substitute for a particular day-length in many plants (Table 9.3). In particular, GAs cause flowering when applied to many long-day plants that grow as rosettes in short days. The flowering response to either GA or long days is often accompanied by elongation of the flowering stem. The problem of interpretation here is that the normal function of GA is to promote stem elongation, so that the flowering response may simply be a consequence of this. There are, however, a few LDP where GA application can bring about flowering without stem elongation, which suggests that the two responses are separate.

GA application can also substitute for a low

Table 9.3 Gibberellin affects flowering in many different ways. The application of gibberellin can promote, inhibit, or have no effect on flowering, even with plants that have the same environmental requirements for flowering.

	Response to gibberellin application	Plant
Long-day plants	stimulates flowering in short days	henbane (*Hyoscyamus niger*)
	inhibits flowering in long days	*Fuchsia* × *hybrida*
	no effect	white mustard (*Sinapis alba*)
Short-day plants	stimulates flowering in long days	*Zinnia elegans*
	inhibits flowering in short days	strawberry (*Fragaria* × *ananassa*)
	no effect	soya bean (*Glycine max*)
Day-neutral plants	stimulates flowering	cypress (*Cupressus* spp.)
	inhibits flowering	apple (*Malus domestica*)
	no effect	evening primrose (*Oenothera* spp.)
Plants requiring vernalisation		
long-day plants	stimulates flowering in long days	oat, winter strains (*Avena sativa*)
short-day plants	no effect	florists' chrysanthemum
day-neutral plants	stimulates flowering	cauliflower (*Brassica oleracea*)
	no effect	*Saxifraga rotundifolia*

temperature signal in the *vernalisation* (see below) of several biennials and in other cold-requiring plants such as tulip bulbs. In addition, striking effects to accelerate flowering in juvenile plants have been seen in several conifers. Thus, in an impressive list of plants, the application of GA has been found to substitute for the seasonal signals of day-length and low temperature, as well as for the internal trigger of 'age'. It is also well established that day-length affects certain steps in the biochemical pathways leading to the synthesis of GA and so alters the content and composition of GAs within the plant. At least 84 different GAs have been detected and several can occur together within the same plant; is the floral stimulus, therefore, one or more of the many known GAs?

Unfortunately, any simple interpretation of the role of GA in flowering is ruled out because not all plants respond in the same way and, in many cases, the application of GA has no effect or may even be inhibitory. Even in the same plant, GA can have opposite effects. Flowering in many temperate grasses requires exposure to short days, followed by long days. Here GA is inhibitory to flowering when applied during the short-day treatment, but promotes flowering when applied after the short-day requirement has been completed. Some examples of the different flowering responses to the application of GA (usually GA_3, known as gibberellic acid, which is one of the few gibberellins that are generally available) are given in Table 9.3 but few of these have yet been found to be of any practical use in horticulture.

Ethylene

This gas has many different effects on plants but it has also been found to modify flowering in ways that have found practical uses. Although the application of ethylene gas to plants is difficult, chemicals such as 2-chloroethyl phosphonic acid (CEPA) that release ethylene within the plant are available and can be applied as sprays. Ethylene and ethylene-releasing compounds *inhibit* flowering in several SDP, including chrysanthemums and Japanese Morning Glory (*Ipomoea nil*). Ethylene also inhibits flowering in sugar cane and CEPA has been used commercially to prevent flowering (which leads to cane senescence) and thus increase the yield of sugar.

The most striking effect of ethylene to *promote* flowering occurs in pineapple and other members of the family, Bromeliaceae. This was discovered when it was found that turning pineapple plants on their side resulted in rapid flowering. This is now known to be associated with the accumulation of auxin at the under side of the shoot apex under the influence of gravity, because it has been established that a high concentration of auxin results in the production of ethylene by cells. As far as is known, this effect of ethylene applies to all bromeliads and it has been exploited commercially by applying ethylene-releasing compounds to bring about flowering in pineapple and ornamental members of the Bromeliaceae such as *Aechmea* and *Billbergia*. An intriguing variant is the use of smoke from burning green wood to cause flowering in pineapples grown commercially under glass in the Azores. Since ethylene is known to be produced during fires, it seems likely that the flowering response is due to ethylene and related gases in the smoke. Ethylene is also the only hormone with a significant effect on flowering in bulbs, a response that has been used commercially in forcing Tazetta narcissus and bulbous iris.

The control of flowering in the bromeliads also involves prevention when not desired. A single treatment with an inhibitor of ethylene synthesis within the plant can prevent flowering for several months. Treatment with a chemical that leads to the production of ethylene by the plant will then induce flowering.

Despite its quite dramatic effect on flowering in some species, particularly in the bromeliads, the mechanism through which ethylene causes flowering is unknown. It does not appear to substitute for, nor to interact with, day-length; because it is a gas, it is most unlikely that it is part of any floral stimulus. This does not, however, exclude chemicals that are required for the synthesis of ethylene by the plant. One or more of these could be exported from the leaves and converted to ethylene gas at the shoot apex where flowering takes place.

HOW IS DAY-LENGTH DETECTED BY THE LEAF?

Under natural conditions, the nights get longer as the days get shorter and so it is not possible to say whether plants are responding to the duration of darkness or to the duration of light in each daily cycle, or perhaps to both. To answer this question, it was necessary to carry out experiments in which the durations of darkness and light were varied independently. When, in 1934, Karl Hamner and James Bonner carried out such experiments with the SDP cocklebur (*Xanthium strumarium*), they found that it was the duration of darkness that was decisive, as had already been suspected by Julien Tournois in 1914. Plants grown in long days would flower, provided that the duration of darkness exceeded a critical value. Conversely, they failed to flower in short days when these were coupled with nights that were shorter than the critical value. Even a single long night was enough to cause flowering in cocklebur, irrespective of the length of the associated light periods. A similar pattern of response was later seen in other SDP and in some LDP (Table 9.4).

We know now that the measurement of the duration of darkness is controlled by a biological clock of the same kind and with many of the same properties as that which controls other time-dependent processes in plants and animals, such as the daily rhythms of leaf movement in plants and the 'jet-lag' effect in people. This internal clock keeps time by going through a cycle that returns to the same point at approximately 24-hour intervals. The cycles do not repeat in exactly 24-hours and so the clock is usually called *circadian*, which is the Latin for 'about a day'. Because the repeating cycles are longer, or shorter, than 24 hours, time as measured by the endogenous clock would drift in relation to the real time by the sun unless the clock is re-set by some external signal. In plants, the times of sunrise and sunset are the daily signals that re-set the clock to local time.

The way in which the circadian clock measures the critical duration of darkness is thought to be as follows. In plants that are sensitive to day-length, each cycle goes through a period when exposure to light triggers a switch between 'flowering' and 'not-flowering'. This occurs at a certain number of hours from the beginning of darkness, although the exact number can vary depending on the species. When the dawn (daylight) arrives before this light-sensitive period occurs, the switch is triggered by the light and the critical night-length is not reached. However, when the light-sensitive period occurs before dawn (i.e. during darkness), the switch is not

Table 9.4 Long nights rather than short days are important for the control of flowering in plants that are sensitive to day-length. Short-day plants flower with *long* dark periods, whereas long-day plants flower with *short* dark periods.

Light treatment		**Flowering response**	
Day-length	Night-length	Short-day plants	Long-day plants
8 hours	16 hours	Flowering	Not flowering
16 hours	16 hours	Flowering	Not flowering
8 hours	8 hours	Not flowering	Flowering
16 hours	8 hours	Not flowering	Flowering

triggered and the critical night-length is exceeded. As we shall see in Chapter 10, this is also the basis of the *night-break* response in which a relatively short exposure to light given at a particular time during the night triggers flowering in LDP but inhibits flowering in SDP (Fig. 10.9). The switch between flowering and not-flowering occurs only when light is given during the light-sensitive period of the circadian clock.

So far, the discussion about the way in which day-length is detected has concentrated on the responses of those plants in which the duration of darkness is decisive. Most SDP and a few LDP behave in this way, which is sometimes called *dark-dominant*. However, when people began to look at LDP in more detail, it was found that many of them have certain characteristic features which differ from those of SDP. In particular, it seems that the spectral quality of light during the day is important in many long-day plants, even when the duration of darkness is favourable for flowering. For this reason, they are sometimes termed *light-dominant*. The differences between light-dominant and dark-dominant responses are not important under natural conditions in sunlight but assume considerable significance when artificial light is used to manipulate day-length. Since artificial light is mainly used for plants growing under glass, a discussion of the differences between light- and dark-dominant plants and the relevance to lighting practice in horticulture is deferred to Chapter 10.

Irrespective of the particular mechanism, plants that are sensitive to day-length flower seasonally in response to changes in the natural photoperiod. For plants growing in the open, the transitions between light and dark are not abrupt but occur through a gradually changing intensity of twilight. At what point, then, does a plant begin to respond to darkness in the evening, or to light at dawn? Is the effective length of day influenced by morning or evening clouds, or moonlight, or even street lamps? The answers have been shown to vary with species and also with the kind of light given. This makes it very difficult to make any kind of generalisation but a few points are worth noting.

The presence of clouds during twilight and dawn might influence the time at which a threshold intensity between light and darkness is reached, allowing the night to begin earlier or end later. This could lead to small changes in the time of flowering from year to year. In some varieties of Japanese Morning Glory (*Ipomoea nil*), for example, the effective day-length on clear days was found to be longer by 20–30 minutes than on cloudy days. For more sensitive species, however, the presence of clouds has little effect and errors in timing would be very small in such plants. The longer twilights of high latitudes and the short ones in the tropics must, of course, also influence how long it takes to go from above to below a threshold value of light. Because of the twilight effect, critical day-lengths determined under artificial conditions when light is switched on and off instantaneously may differ from those actually experienced by plants growing out of doors.

As far as the moon is concerned, present evidence indicates that even light from the full moon is below the threshold necessary to influence flowering. Street lights are, however, a different story. Although there are considerable differences between species, for many plants, the threshold light level above which flowering was promoted (LDP) or inhibited (SDP) was less than 100 mW m^{-2} (about 50 lux) in experiments where light from tungsten-filament (incandescent) lamps was given continuously throughout the night (Table 9.5). The possible effect of street lamps depends on the light level reaching the plant, the sensitivity of the species to light, and the spectral composition of the lamp. Sodium-vapour lamps emit light containing a high proportion of active wavelengths and so are more likely to have an effect than mercury-vapour lamps, which emit less light in the active part of the spectrum. The present trend of keeping street lights on during the entire night makes it more likely that they could influence flowering or other responses to day-length. For example, street lights may be sufficiently bright to delay the onset of dormancy in trees and so increase their susceptibility to freezing injury (see later), especially in northern latitudes where autumn frosts begin early.

STORAGE ORGANS

One of the seasonal features of some plants is the formation of resting structures in which food is stored. Such storage organs (Table 9.6) arise by lateral swellings of a number of different tissues including stems (tubers and corms), roots (tuberous roots) and leaf bases (bulbs). Their formation is usually accompanied by the cessation of active growth followed by death of the rest of the plant. Once formed, storage organs often become dormant, during which time they are more resistant to unfavourable environments such as water-stress, or extremes of temperature. In many cases, the formation of storage organs depends on or is accelerated by exposing the plants to particular day-lengths and, with the exception of bulbs in the genus *Allium*, most are favoured by short days (Table 9.6).

As with other responses to day-length, the formation of storage organs is completely

Table 9.5 The minimum level of light necessary to influence flowering is not the same for all plants. Most long-day plants require higher light intensities to stimulate flowering than is required by most short-day plants to inhibit flowering. There is, however, a lot of difference between plants and the values only apply for these particular conditions, i.e. when light from tungsten-filament lamps was given continuously throughout the night. They would be different with other light sources or when light is given at different times and/or durations.

Short-day plants	Minimum light level needed to prevent flowering in mW m^{-2}
florists' chrysanthemum	90
flaming Katy (*Kalanchoë blossfeldiana*)	90
poinsettia (*Euphorbia pulcherrima*)	21
Japanese morning glory (*Ipomoea nil*)	4–40
soya bean (*Glycine max*)	0.4
Long-day plants	Minimum light level needed to promote flowering in mW m^{-2}
turnip rape (*Brassica campestris*)	4515
campion (*Silene armeria*)	31–90
barley (*Hordeum vulgare*)	10–20
China aster (*Callistephus chinensis*)	4–12

mW m^{-2} = milliwatts of light energy reaching a square metre of the surface.

Table 9.6 The formation of storage organs is often influenced by day-length.

Favoured by short days	
peanut (*Apios tuberosa*)	root tubers
Begonia socotrana	aerial stem tubers
tuberous begonia (*Begonia tuberhybrida*	underground stem tubers
dahlia (*Dahlia* hybrids)	root tubers
artichoke (*Helianthus tuberosus*)	underground stem tubers
potato (*Solanum tuberosum*)	underground stem tubers
Favoured by long days	
shallot (*Allium cepa*)	bulbs
onion (*Allium cepa*)	bulbs
garlic (*Allium sativum*)	underground and aerial bulbs
Triteleia laxa	corms

dependent on exposure to appropriate day-lengths in some plants, but only accelerated in others. In potato, for example, varieties differ considerably in the extent to which they respond to day-length. In many European and North American varieties, tuber formation occurs in both long and short days but occurs earlier in the latter. The early dieback of the haulms associated with the formation of tubers often results in a lower yield in short days, despite the fact that the onset of tuber formation is accelerated. This is a good example of the need to match the response of the plant to the day-length conditions in any particular geographical location.

The regulation of tuber formation by short days appears to be very similar to the induction of flowering in many SDP, although detailed studies have been carried out with relatively few species. The duration of darkness is the controlling factor and the formation of tubers will only occur when this exceeds a critical value. Thus, in most cases, tubers develop during the lengthening nights of late summer or early autumn. In contrast, bulb formation in onions is dependent on exposure to long days (Fig. 9.4). In lower latitudes, such as in Egypt, many varieties commonly grown in northern Europe do not form bulbs and the so-called 'short-day bulbing' types must be grown. However, these are also LDP, the difference being that they have a shorter critical day-length (Table

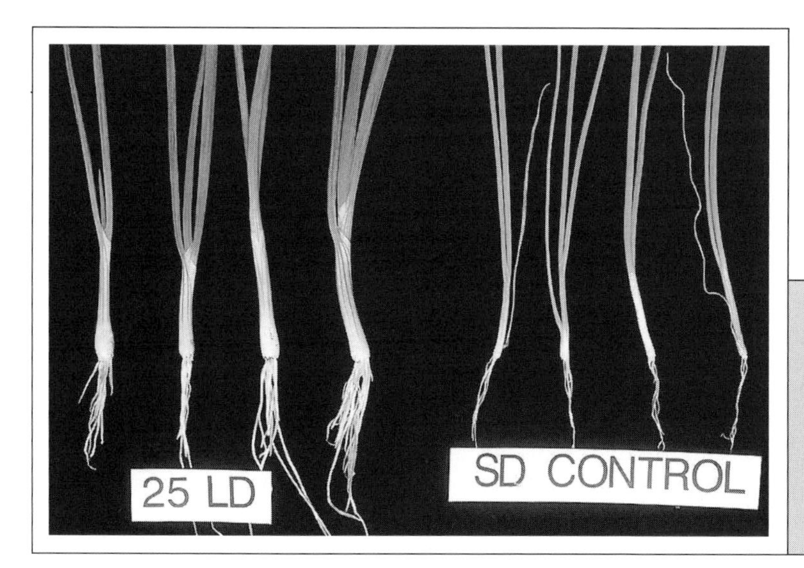

Fig. 9.4 The formation of bulbs in onions require exposure to long days. Bulb formation had already begun in plants given 25 long days (left), while those in short days (right) showed no signs of bulbing. (Photo, courtesy of Brian Thomas, Horticulture Research International.)

Table 9.7 Different varieties of onion have different day-length requirements for bulbing. The figures refer to the percentage of plants with bulbs. Although the variety 'Sweet Spanish' forms bulbs on shorter day-lengths than 'Yellow Zittau', both only form bulbs when the day-length is longer than the critical value. This is about 14 hours for 'Yellow Zittau' but only 12 hours for 'Sweet Spanish' which can, therefore, be successfully grown at lower latitudes.

Variety	Day-length in hours				
	10	12	13	14	16
'Sweet Spanish'	0	29	77	93	100
'Yellow Flat Dutch'	0	0	15	92	100
'Yellow Rijnsburg'	0	0	9	46	100
'Yellow Zittau'	0	0	0	40	100

9.7) and so will form bulbs in the shorter days of lower latitudes. When these varieties are grown at higher latitudes, the critical day-length may be attained early in the summer leading to rapid bulb formation in young plants, with a consequent reduction in bulb size. As with tuber formation in potato, therefore, the yield is reduced when day-lengths appropriate for bulbing occur too early in the life of the plant.

The formation of storage organs depends, in the same way that flowering does, on exposing the leaves to the appropriate day-lengths. Again, some kind of chemical stimulus must be exported from the leaves to the region where the storage organ develops. One intriguing observation arising from grafting experiments is that the formation of a tuber-forming stimulus is not confined to the leaves of plants which themselves are able to develop tubers. In one experiment, shoots of sunflower (*Helianthus annuus*) were grafted on to Jerusalem artichoke plants (*Helianthus tuberosus*). Artichokes normally develop tubers only when their leaves are exposed to short days. Although sunflowers do not form tubers themselves, they were able to cause tubers to develop on the artichoke plants to which they were grafted when the leaves of the sunflower donors were maintained in short days. Moreover, there is some evidence that the signal for flowering and tuber formation may be the same, or similar, the nature of the response being dependent on the plant itself. For example, when shoots of tobacco varieties with different day-length requirements for flowering were grafted on to potato plants, tubers formed in the latter only when the leaves of the donor tobacco plants were in day-lengths that induced flowering in them (Table 9.8).

What is the nature of this tuber-forming stimulus? The fact that it is also produced by plants which are themselves incapable of forming tubers suggests that we are not dealing with a unique tuber-forming substance but rather with a non-specific compound. Consequently many substances have been tested for their possible effects to bring about the formation of storage organs, especially when these are economically important crop plants.

Although many chemicals, including sugars and known plant hormones, have been found to have some effect on the formation of storage organs in particular species, the results are often conflicting. It is, however, well established that gibberellins play a significant role as a chemical signal for the control by day-length of tuber formation in the potato. For example, a dwarf mutant with a low content of gibberellins formed tubers in both long and short days, as did plants treated with a chemical that prevents the production of gibberellins. The likely explanation is that the formation of tubers is prevented in long days by an inhibitor coming from the leaves, which is presumed to be one of the many gibberellins. So far, there appears to be no role for any of the other known plant hormones in potato

Table 9.8 The formation of tubers on potato plants is affected by grafting them to flowering or non-flowering plants of tobacco. Tubers formed on potato plants only when the tobacco donors were flowering ('Mammoth' in short days, *N. sylvestris* in long days), indicating that the signal which causes flowering in tobacco may also cause tuber formation in potato.

Tobacco donor	**Day-length conditions in which grafted plants were maintained**	
	Short days	Long days
Nicotiana tabacum 'Mammoth' (Short-day plant)	Tubers formed	No tubers formed
Nicotiana sylvestris (Long-day plant)	No tubers formed	Tubers formed

and discovering the identity of the tuber-forming signal that is indicated from the grafting experiments remains an elusive goal.

LEAF FALL AND DORMANCY

Perhaps the most majestic of all seasonal displays is that of autumn colour (see Chapter 8), which precedes leaf fall and entry into a rest period (*dormancy*) in deciduous trees of high latitudes. In most cases this is a response to the longer nights and shorter days of autumn and, under their influence, stem and leaf growth stop and winter resting buds develop. The value to the plant is that the seasonal signal of short days gives an advance warning of low temperatures to come and enables plants to increase their resistance to freezing temperatures. Thus the time at which they become dormant and develop cold hardiness may be crucial for survival in a particular location.

In some plants, including many evergreens, dormancy involves only metabolic changes without the development of any specialised structures. In others, protective bud scales are formed from overlapping leaf bases or by the development of modified leaves. Perhaps the most obvious morphological change is the shedding of leaves in deciduous trees. However, although leaf fall is directly influenced by day-length in some plants, temperature is also important. In the tulip tree (*Liriodendron tulipifera*) leaves are always shed in short days but, in the False acacia (*Robinia pseudoacacia*), leaves are retained when the temperature remains high.

The response to day-length is of great importance for the adaptation of trees to the latitude in which they originate. That this is controlled by the plant's genes can be seen when seeds are collected from several different latitudes and grown together in one place. In one experiment, as many species as possible were collected from three different latitudes in Norway and Denmark and grown together in day-lengths ranging from 12 to 24 hours. It was found that plants originating from the same geographical area had approximately the same critical day-length for the maintenance of stem growth and the prevention of dormancy, irrespective of species (Table 9.9).

The shorter critical day-lengths of more southerly populations of the same species means that they continue growing for longer if they are moved further north and may be damaged by autumn low temperatures. For example, the onset of dormancy and the development of cold resistance in a species of willow (*Salix pentandra*) from 60°N (with a critical day-length of about 14 hours) was delayed when plants were grown outdoors in the longer days of Tromsø in Norway (at about 70°N) and they were severely damaged by frost. The critical day-length tends to be longer for trees growing at higher altitudes where winter comes early. Consequently, they stop growing earlier in the season before the temperature drops (Plate 16).

Table 9.9 The critical day-length for the onset of winter dormancy varies with the latitude at which the plants originate. When seeds were collected from trees growing at different latitudes in Scandinavia, seedlings from the far north ceased growing and became dormant when grown in day-lengths shorter than 22 hours, while seedlings from further south at latitude 55°N only ceased growing when the day-length was shorter than 15 hours.

Latitude of origin	Approximate critical day-length
55°N	15 hours
60°N	17 hours
65°N	$19\frac{1}{2}$ hours
70°N	22 hours

Stunting of growth is often seen when plants adapted to grow at high latitudes are grown in shorter day-lengths further south. For example, plants from the far north at latitude 70° have a critical day-length of about 22 hours, which may never be achieved when they are grown further south. Consequently, the shoots cease to elongate and become dormant after making only a small amount of growth in the spring, following the breaking of dormancy by winter cold (see under temperature below). This dwarfing effect is very clearly seen in birch trees from seeds collected at 68°N and grown in Uppsala, Sweden, at latitude 60°N (Plate 16).

Winter cold is not the only seasonal stress that can be avoided by a response to day-length. In the Mediterranean regions of hot, dry summers and cool wet winters, the shedding of leaves in the summer allows plants to survive a period when water is in short supply. In some cases, the shedding of leaves has been shown to be speeded up in long days. Summer dormancy is also associated with regions that have hot, dry summers and may be triggered or accelerated by long days as, for example, in the Crown anemone (*Anemone coronaria*).

Bud dormancy is a complex process which, depending on the species, may include the cessation of shoot growth, the development of resting buds with various types of morphological modification, entry into dormancy, increase in frost hardiness, increase in drought resistance, and leaf fall. Each of these processes has been shown to be influenced to some extent by day-length but there are considerable differences between species in the degree of control that day-length can exert. Although low temperatures are important for the later stages in the development of deep dormancy and greater frost tolerance, short days are the main signal for the first stage, i.e. the cessation of shoot growth and some frost tolerance. This day-length dependent stage goes faster at mild temperatures and can actually be inhibited if the autumn temperature is too low.

As with flowering and the formation of storage organs, it is the leaves that detect the day-length signal and it is presumed that some kind of chemical stimulus is produced in them and exported to the buds where the changes leading to dormancy take place. Given that the overall process is so complex, it is perhaps not surprising that several substances appear to be involved. The known plant growth hormones, particularly the gibberellins, have been found to influence many aspects of dormancy and are generally thought to be the main controlling factors.

TEMPERATURE

Changes in the average temperature are also seasonal factors, although from year to year the actual temperature is strongly influenced by the immediate weather conditions. Temperature has a major effect on plants since all plant processes are influenced by it to some extent. The growth rate is speeded up at higher temperatures, provided that other factors, such as the amount of light, are not limiting and that the temperature is not so high as to be damaging. Temperature may also alter the plant's response to day-length. Temperature as well as day-length changes with latitude (Fig. 9.1) and plants are often sensitive to both factors. Thus, although day-length may be the overriding factor controlling whether or not a particular response can take place, the magnitude and even the timing of the response can be altered not only by temperature, but also by the amount of light and the supply of water.

In the temperate and cold regions of the world, temperatures below freezing during the winter are damaging, except for plants with adaptation or avoidance mechanisms. Some of these mechanisms (e.g. winter rest and the formation of storage organs) have already been discussed and later in this chapter we will look at the physiology of freezing damage. For some plants, however, exposure to low temperatures during winter is an essential step in their development. Two of these processes, *vernalisation* and the breaking of winter dormancy, have important implications for gardeners.

Vernalisation

Not everyone agrees on the precise definition of *vernalisation* but, from the point of view of understanding the underlying mechanism, it is probably best restricted to the increase in flowering that occurs in response to a cold treatment given to seeds or young plants. It is sometimes used to refer to the cold requirement of some seeds for germination (see Chapter 6) but this is a different process and should not be called vernalisation (although it is often erroneously referred to as *seed vernalisation*). Vernalisation is usually an 'inductive' process in the sense that, after the cold treatment is complete, no physical changes can be detected in the plant and the formation of flowers begins only after plants have been returned to higher temperatures and, in many cases, to particular day-lengths. Exceptions are Brussels sprouts and some bulbs, where the flower initials are actually formed during the cold or cool treatment.

The requirement for vernalisation is most commonly, but not only, found in long-day plants. These may be annuals (e.g. winter cereals), biennials (e.g. carrot, beet) or perennials (e.g. perennial ryegrass). Without exposure to an adequate period of cold, these plants show delayed flowering or flowering may fail completely, and they often grow as rosettes (Fig. 9.5). The optimum temperature for vernalisation lies between about 1 and 7°C, with the effective temperature ranging from just below freezing to about 10°C. When different parts of the plant were given localised cooling treatments, it was found that cold is sensed by the shoot tip and its effect seems to be largely independent of the temperature experienced by the rest of the plant.

Most studies indicate that, unlike day-length, the effect of vernalisation is restricted to the cells that directly experience cold and no transmissible substances are involved Once vernalisation is complete, it is perpetuated through subsequent cell divisions and all of the cells arising from those that were originally exposed to cold are also vernalised. In annual and biennial plants, the requirement for vernalisation is established again during the process of reproduction. Perennial plants, in contrast, have to be vernalised again each winter if they are to flower. In chrysanthe-

Fig. 9.5 Vernalisation in the long-day plant *Coreopsis* 'Sunray'. In the absence of a cold treatment (vernalisation) plants failed to flower and continued to grow as rosettes. Stem elongation and flowering occurred rapidly after exposure to 5°C, but only in long days. The cold treatment began when the seedlings had developed 12 leaves. (Photo, courtesy of Royal Heins, Michigan State University.)

mums, plants become 'de-vernalised' during the preceding summer while, in some perennial grasses, the vernalisation effect is not perpetuated indefinitely through cell divisions and the tillers that are formed late in the summer are not vernalised.

The nature of the cellular changes that lead to the semi-permanent vernalised state are unknown. Because vernalisation leads to stem elongation (bolting) in plants such as carrot (*Daucus carota*), which grow as rosettes in the absence of a cold treatment, it is thought that gibberellins may be involved. Indeed, the application of gibberellin to plants often substitutes for low temperature and causes bolting and flowering (e.g. foxglove, *Digitalis purpurea*). However, it seems unlikely that gibberellins are involved in all vernalisation responses since, in many plants, applying GA does not cause flowering, even though the stems may elongate (e.g. Canterbury bell, *Campanula medium*).

Seed and plant vernalisation

Many annual plants, such as the winter forms of rye and oat, can be vernalised by treating imbibed or germinating seeds. Although some biennial plants can also be vernalised as seeds, many biennial root crops only become sensitive to cold after they have reached a certain size. The duration of this *juvenile phase* during which they cannot be vernalised can vary considerably between varieties and has important implications for horticulture. A long juvenile phase can allow earlier sowing and so enable gardeners to take advantage of a longer growing season without the problem of bolting in the first year, with consequent failure to produce a harvestable storage root. This is particularly important where the spring temperatures are low and within the range that can cause vernalisation. For example, when sugar beet (*Beta vulgaris*) was sown in Scotland in mid-March, some 40% of the plants bolted but, when sown in mid-April, most did not. In contrast, hardly any plants bolted from the earlier sowing in a non-bolting type.

The duration over which a cold treatment is required is also important and non-bolting types may require a longer exposure to low temperature in order to vernalise them, as well as, or perhaps instead of, a long juvenile phase. If flowering is required for the production of seeds or in plants grown for their ornamental value as cut flowers, pot plants, or garden plants, the need for a long exposure to low temperature can prevent successful cultivation in regions with mild winters. Prior vernalisation is also essential if cold-requiring plants are to be grown for flowering under heated glass (Fig. 9.7).

De-vernalisation

A further complication is that periods of higher temperatures (above about 15°C) may reverse the vernalising effect of a previous exposure to cold. This *de-vernalization* by higher temperatures, notably high daytime temperatures, progressively decreases as vernalisation proceeds and is finally lost once the vernalisation process is complete, usually after several weeks of cold. After two weeks of cold, seedlings of winter rye were completely de-vernalised by exposing them to 35°C for three days, whereas the same treatment had no effect after eight weeks of cold.

Vernalisation and day-length

It is perhaps not remarkable that vernalisation is linked with photoperiodism since this enables plants to respond more precisely to seasonal changes. Only a few plants are known to require exposure to short days after vernalisation (e.g. some chrysanthemum varieties). The majority of plants requiring vernalisation also require subsequent exposure to long days in order to induce flowering (Fig. 9.7). This combination of responses ensures that flowering occurs during the longer days of summer when conditions are more favourable for the energy demands of seed production. It prevents the flowering of small seedlings in the previous summer and allows the plant to build up food reserves, often in the form of storage organs, for seed production in the following year (e.g. in biennial crop plants such as carrot).

Other interactions between day-length and vernalisation are known and, in particular, exposure to short days can often substitute either partly or entirely for cold (winter strains of wheat and rye; Canterbury bell, *Campanula medium*). All plants where short days can substitute for low temperature are LDP and the short-day or low-temperature treatment must precede the exposure to long days. The ecological significance of the substitution of short days for low temperature may be to allow some flowering to occur after mild winters when vernalisation is incomplete. One interesting horticultural example of the interaction between day-length and temperature is seen in celery, where exposure to long days during vernalisation delays flowering although, after vernalisation is complete, long days accelerate flowering. As a result, early celery grown under unheated glass in the UK is sometimes given artificial light to lengthen the day during the period of winter cold, thereby preventing bolting.

Breaking winter dormancy

Once they have become fully dormant, most tree species from the temperate zone will not resume growth until dormancy has been broken by exposure to low temperature. In most cases, day-length has no effect once dormancy has been completely broken by cold, although long days may accelerate bud-break in trees that have received some low temperature but which are still in a state of partial rest. This is seen in beech (*Fagus sylvatica*), where exposing dormant cuttings to artificial long days in November and December had no effect on bud burst, whereas such treatments became increasingly effective during February and March. This effect of day-length to influence the time of bud-burst could be an effective strategy for breeding plants that would be more tolerant of early spring frosts. Plants with a strong requirement for long days to accelerate bud-burst would leaf out later and so be less likely to be damaged.

The most effective temperatures for the breaking of winter dormancy are similar to those for vernalisation and range from just above freezing to about 10°C. The cold requirement varies considerably between species and varieties and the failure to break dormancy can be a considerable problem in regions with mild winters. This is particularly true for temperate-zone fruit trees, such as apple, where the duration of winter cold may be insufficient to break dormancy completely, resulting in erratic or delayed bud growth. This affects the timing of crop spraying programmes and, in the worst cases, reduces the leaf expansion to such a degree that growth is affected. Consequently, it is commercial practice in some areas with mild winters to promote dormancy-breaking by the use of various chemical sprays, including mineral oils, often together with dinitroorthocresol. Their action in dormancy-breaking is not fully understood, although the effect of mineral oils has been associated with the imposition of anaerobic conditions within the dormant buds. In some cases, cultivars with shorter chilling requirements are available and are more suitable for warmer climates. These could become increasingly important if winters become warmer due to climate change.

Direct effects of temperature on flowering

As we have seen, there are numerous interactions between day-length and temperature in their effects on flowering. However, there are also many plants in which flowering is directly influenced by temperature with often quite specific requirements. Some cultivars of chrysanthemum, for example, require a minimum temperature to ensure uniform and regular bud formation. Other plants (for example calceolaria, *Calceolaria* Herbeohybrida Group and cineraria, *Pericallis* × *hybrida*) require temperatures below a certain critical value in order to flower.

Probably the most detailed information regarding specific temperature requirements for flowering relates to bulbs and is widely used to prepare them for forcing. In the major bulb crops, notably cultivars of hyacinth (*Hyacinthus* spp.), tulip (*Tulipa* spp.) and daffodil (*Narcissus* spp.), the production of normal flowers requires first a period of high temperature for the initiation of flower buds, followed by a period of cool temperatures (8–12°C) in order to trigger stem elongation. However, other bulbs (e.g. Dutch irises) initiate flowers only at relatively cool temperatures. Several different mechanisms are likely to be involved in these diverse responses to temperature and most of them are not well understood.

DAMAGE BY BELOW-FREEZING TEMPERATURES

A major problem for plants of the temperate zones is exposure to temperatures below freezing and many plants have evolved mechanisms that enable them to survive these adverse conditions. This may simply be avoidance by, for example, shedding of leaves or die back and over-wintering by underground organs, or it may involve changes in the cell that increase the resistance to damage.

Ice crystals first begin to form in the spaces between cells and in the cell walls, where the concentration of dissolved substances is lower than inside the cells themselves and the freezing point is, therefore, higher. Water from within the cell then diffuses out and condenses on the growing ice-masses. As water diffuses out, the increasing concentration within the cell further lowers the freezing point and so prevents the formation of ice crystals in the living protoplasm. The loss of water increasingly dehydrates the cell and this, in itself, can be damaging to the living tissues.

When thawing occurs, particularly if this is gradual, the ice crystals outside the cells melt and water goes back into the cell. However, in non-hardy plants damage to membranes and other cellular components may already have occurred so that water does not re-enter the cells completely and normal metabolism cannot be resumed. Damage is particularly likely to occur with fast thawing, with ice crystals melting before the cells have regained their ability to absorb water sufficiently rapidly. For this reason, direct exposure to early sunlight, such as against an east wall, often leads to frost damage, while those plants where thawing occurs more slowly can escape.

The ability of plants to develop frost resistance is quite remarkable. Some woody plants have been found to be able to withstand temperatures as low as that of liquid nitrogen (−196°C) when in the dormant, winter-hardy state and yet, when actively growing, the same plants can be killed at temperatures of −3°C, only just below freezing. We can readily see this in the damage to young expanding leaves, which is often observed when spring frosts occur soon after bud-break. In woody plants, winter hardiness is, in part, induced by the short days of autumn, but it is also dependent on subsequent exposure to low temperatures. In herbaceous plants, frost-hardiness typically develops during exposure to relatively low temperatures (usually about 5°C) and the gradual 'hardening-off' of plants raised under glass is an important practice, especially if they are likely to be exposed to temperatures below freezing. Hardening-off is, however, also important for plants that have no tolerance to frost but can be damaged by chilling temperatures.

In some cases, frost resistance is based on tolerance to severe dehydration following the formation of ice outside the cells. This has many features in common with the ability to tolerate water and salt stress and all three processes appear to involve the plant hormone, abscisic acid (see Chapter 5 for the role of abscisic acid in water stress). The experimental application of abscisic acid to plants leads to the accumulation of proteins, called *dehydrins*. These also accumulate in plants in response to any environmental influence that has a dehydration component, such as freezing, salinity and drought. However, the ice crystals themselves can damage the cells, especially when they are large. This potential source of damage is reduced in frost-hardy plants by the presence of other proteins that bind to the surface of the ice-crystals and prevent them from growing larger. Finally, the properties of the cell membranes may be altered in a way that enables them to function at low temperatures.

WATER AND LIGHT

The daily amount of light (as well as its duration) and the availability of water also vary seasonally and, like temperature, are modified by the immediate weather conditions. Both have significant effects on plant growth and are considered elsewhere. Light is discussed in relation to the limitations to growth imposed by low winter light on plants growing under glass in Chapter 10, and in relation to the effect of high light intensities to increase water loss from leaves in Chapter 7. Adaptations to water stress are considered in relation to the selection of plants for growing in dry conditions (Chapter 5), and to the avoidance or lessening of water stress by shedding leaves and the development of summer dormancy (elsewhere in this chapter).

FURTHER READING

Atherton, J.G. (ed.) (1987) *Manipulation of Flowering*. Butterworths, London.

Burroughs, W. (1998) Degrees of damage. *The Garden*, **123**, 249–51.

Burroughs, W. (1999) Winter frieze. *The Garden*, **124**, 22–5.

Gates, P. (1999) Burgeoning beginnings. *The Garden*, **124**, 28–33.

Halevy, A.H. (ed.) (1985 & 1989) *Handbook of Flowering, Vols. I–VI*. CRC Press, Boca Raton, Fl.

Jackson, S. & Thomas, B. (1997) Photoreceptors and signals in the photoperiodic control of development. *Plant Cell and Environment*, **20**, 790–95.

Salisbury, F.B. & Ross, C.W. (1991) *Plant Physiology*, 4th edn, Wadsworth Publishing Company, Belmont, Ca.

Vince-Prue, D. (1990) The control of flowering by day-length. *The Plantsman*, **11**, 209–24.

Vince-Prue, D. & Cockshull, K.E. (1981) Photoperiodism and crop production. In: *Physiological Processes Limiting Plant Productivity*. C.B. Johnson (ed.), pp. 175–7. Butterworths, Oxford.

Vince-Prue, D. & Thomas, B. (1997) *Photoperiodism in Plants*. Academic Press Ltd, London.

10

Gardening in the Greenhouse

- The greenhouse environment: light, air circulation, carbon dioxide concentration, humidity
- The consequences for photosynthesis in C-3 and C-4 plants
- Carbon dioxide depletion and enrichment in the greenhouse
- Air movement and its effects
- Greenhouse management: optimum temperatures; water management; ventilation; growing media; light
- Supplementary artificial lighting during winter: how to measure light; the choice of lamp; how much extra light?
- Day-length: the use of day-length controls to regulate flowering time
- The requirement for far-red light: far-red absorbing filters and their effect on growth
- Lighting levels needed for day-length control

THE GREENHOUSE ENVIRONMENT

Plants in a greenhouse are growing under conditions that are quite different from those in the open air. There are two main reasons for this: the properties of glass itself and the fact that the plants are growing in an enclosed space.

When sunlight passes through glass, the amount of ultra-violet (UV) light at wavelengths shorter than about 380 nmis progressively decreased until, below about 310 nm, there is no UV light. Consequently, the glasshouse is a low UV environment. This affects the way plants grow, especially their height. Conversely, the high UV levels in mountainous regions are thought to be one of the factors that contribute to the dwarf habit of many alpine plants.

Sunlight transmitted through glass is absorbed by the internal surfaces of the greenhouse and re-radiated at longer wavelengths as radiant heat. Since glass does not allow these wavelengths to pass through it, the re-radiated heat is trapped within the greenhouse and warms the internal atmosphere significantly, particularly on sunny days. Plastics do not have the same transmission characteristic as glass; some are more transparent to radiant heat and so may not trap the re-radiated heat as effectively.

Many commercial growers and some gardeners utilise flexible film plastics, attached to a supporting frame (for example, *polytunnels*). In the past, film plastics have had a poor reputation. Heat retention was poor and it was possible to have a *temperature inversion*, with frost inside a polytunnel, yet none outside. Modern film plastics have, however, been substantially modified, with better heat retention and less condensation, and they often out-perform glass in the amount of light that they transmit. Although film plastics have to be replaced at intervals, modern UV-inhibited plastics should last for at least four seasons.

Both plastics and glass reduce the amount of light reaching the plants compared to those growing in the open air. Clean glass transmits about 86% of the incident light, while dirty glass significantly reduces this and can cause more than 20% loss of available light. The importance of this is emphasised by the fact that, for many crop plants such as tomato, the results of experiments indicate that a 1% loss of light results in a 1% loss in yield. The amount of light transmitted through glass also depends on the solar angle. At low solar angles some 30% of the incoming light is reflected and only about 58% passes into the greenhouse. Glazing bars and any nearby shading structures also contribute to the overall reduction in the available light. Because of the strict relationship between light and photosynthesis, it is essential to maximise the available light by paying attention to factors such as the siting of the greenhouse, clean glass, and minimal structural interference with light transmission.

Siting the greenhouse

A major consideration is the nearness of potential shading objects such as trees and buildings. To minimise light loss, a rule of thumb is that the greenhouse should be no closer than four times the height of a nearby shading structure. It is also important to minimise heat loss, which is increased by air movement over the glass. It is best to site a greenhouse where it is protected from cold northerly and easterly winds, remembering that wind-breaks also cast shade. Lean-to structures are the least demanding of heat input, because the house wall provides complete insulation on one side. Twin-walled polycarbonate (a type of rigid plastic) provides better heat retention than single glass, but has reduced light transmission. Another way to save heat is to line the greenhouse with bubble polythene, although this also results in the loss of available light.

A final point concerns orientation. In small gardens there is usually little choice, but it has been demonstrated that, for a single free-standing greenhouse, an east-west orientation transmits more light in the winter months (when light is the main limiting factor for growth) than does a north-south orientation.

Ventilation

The fact that plants are growing in an enclosed environment is also important. When there is no ventilation there is little exchange of carbon dioxide (CO_2) between the external atmosphere and that of the greenhouse; this results in a reduction in atmospheric CO_2 levels inside the greenhouse due to uptake into the plants by photosynthesis, particularly in bright light. In a closed greenhouse the CO_2 level at midday can drop to a level well below that necessary to maintain maximum growth rates (Fig. 10.1). An enclosed environment also reduces air movement and plants within a greenhouse are growing under much less turbulent conditions than those growing outside.

EFFECTS OF THE GREENHOUSE ENVIRONMENT

The transmission properties of the cladding, and the fact that the greenhouse is an enclosed space, mean that the internal conditions are quite different from those outside. We can now consider the effect of these differences on plant growth. Light, temperature and CO_2 concentration are the major environmental factors that affect the rate of photosynthesis and this, in turn, is a major factor limiting the rate of growth (see Chapter 1). Light also modifies the general appearance of plants. Under low light conditions, stems tend to be longer and their strength is reduced, while leaves are thinner and often larger than in plants growing under better light conditions.

Light

Photosynthesis is a complex process, which involves reactions that are directly dependent on light, as well as dark reactions that are dependent on CO_2 and temperature (see Chapter 1). Consequently, the responses to these environmental factors are also complex. When a single leaf is exposed to different amounts of light at normal atmospheric levels of CO_2 (about 350 parts per million), the rate of photosynthesis increases with increasing light levels until, above a certain value, there is no further increase. This value is known as the *light-saturation point* and varies with the species and the previous light history of the plant. Plants with C-4 photosynthesis (see Chapters 1 and 5) usually have much higher light saturation values than plants with C-3 photosynthesis (Fig. 10.2).

Species native to shady habitats have low rates of photosynthesis in bright light and the light-saturation value is also lower than in other plants. Many do not tolerate full sunlight and their leaves can be scorched under these conditions. However, under very low light conditions, they usually photosynthesise at higher rates than do other

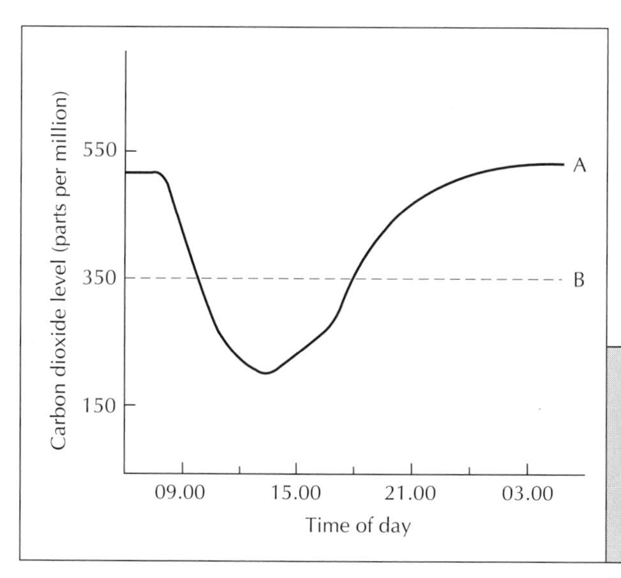

Fig. 10.1 Changes in the concentration of carbon dioxide within a closed greenhouse during the course of the day. Rapid photosynthesis during the period of high light intensity between noon and 15.00 hours leads to a drop in the concentration of carbon dioxide in a closed greenhouse (A) to well below that in the outside air (B).

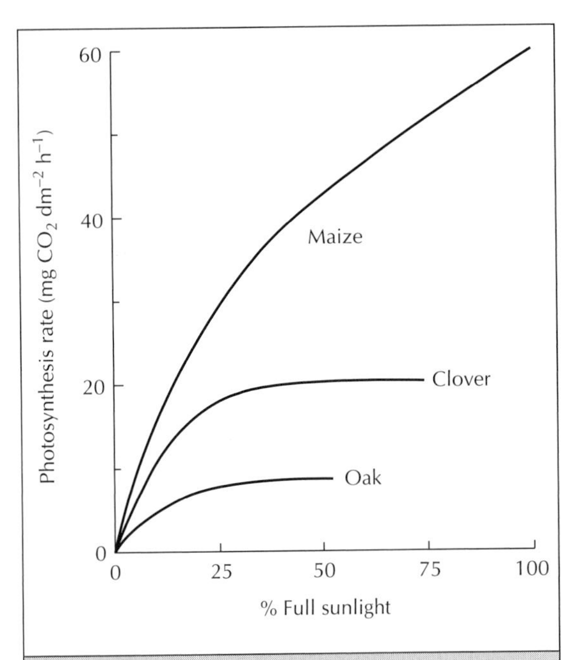

Fig. 10.2 The relationship between the amount of light and the rate of photosynthesis. Photosynthesis is saturated at much lower light levels in a typical C-3 plant such as clover (*Trifolium* spp.) than in a C-4 plant such as maize (*Zea mays*).

species. Many ornamental house plants (for example, African violet, *Saintpaulia ionantha*) fall into this category and are the plants of choice when light is limiting.

At the other extreme are 'sun' plants, which can only grow satisfactorily in bright light. However, there are many sun plants that, when grown in the shade, can *acclimate* and develop so-called shade-leaves; these are thinner and more expanded than leaves of the same species developing in full sunlight and they have more light-harvesting chlorophyll (see Chapter 1) per unit weight of leaf. These characteristics increase the ability of the leaf to capture light, enabling the plant to grow in shady conditions. However, the growth rate will normally be slower and undesirable morphological changes (such as long, weak stems) may still occur. This ability to acclimate to low light may be particularly important in large plants and in closely planted crops, where new leaves on the lower branches develop under shadier conditions than those on the upper ones.

Plants can also acclimate to an *increase* in available light, developing smaller, thicker leaves with a higher light-saturation value. At high light intensities, such leaves are able to photosynthesise at a faster rate than those developing in shade and so enable the plants to utilise the better light more efficiently. It has been shown that the plants detect in some way the total amount of light received each day (the *light integral*) and not the *irradiance* (brightness) at any one time. This means that 50% of the light for twice as long results in the same degree of acclimation as 100% for half the time. These changes can occur within a few days when plants are transferred from low to high light levels, although leaves previously grown in shady conditions can be severely damaged and bleached if they are suddenly exposed to bright sunlight.

Carbon dioxide

Provided that light is not limiting, the rate of photosynthesis increases with an increase in the amount of CO_2 in the atmosphere until a certain level is reached. The actual value at which saturation occurs depends on the amount of light; when there is more light, saturation occurs at a higher concentration of CO_2. Plants with C-3 photosynthesis show a strong response to CO_2 and, provided that light is not limiting, net photosynthesis in most C-3 plants continues to increase until quite high levels of CO_2 are reached (Fig. 10.3). These can be considerably above normal atmospheric concentrations. As greenhouse crops frequently lack enough CO_2 for maximal growth, some commercial growers add CO_2 to the greenhouse atmosphere. Levels are not usually allowed to exceed about 1000–1200 parts per million (i.e., more than three times the normal concentration in the air), as higher concentrations cause stomata to close and can be toxic. They are also uneconomic, especially when the ventilators are open. However, different plants respond in unpredictable ways to enrichment with CO_2. Tomatoes, for example. show a high response, while cherry seedlings do not.

We should also note that C-3 and C-4 plants respond to increased CO_2 concentration in different ways. Photosynthesis in C-4 plants is generally saturated at about 400 parts per million

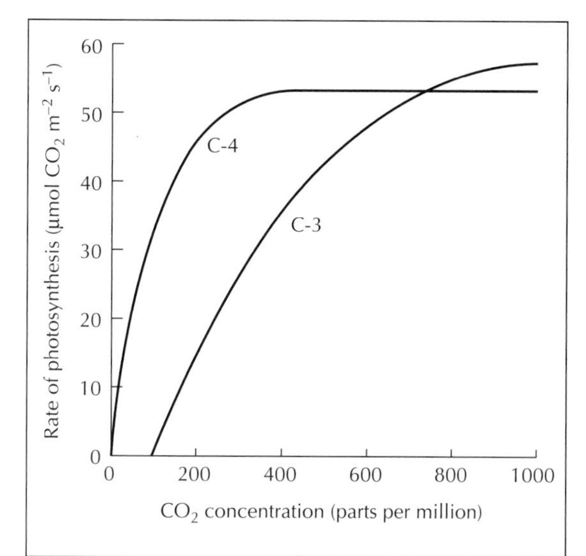

Fig. 10.3 The rate of photosynthesis depends on the concentration of carbon dioxide. Plants with C-4 photosynthesis are saturated at a much lower concentration of carbon dioxide than plants with C-4 photosynthesis. They can also carry out photosynthesis as lower levels of carbon dioxide than C-3 plants.

(i.e., just above the normal content in the air), even at high light levels when the demand for carbon dioxide is greatest (Fig. 10.3). At the other end of the scale, C-4 plants are able to continue to carry out photosynthesis at very much lower levels of CO_2 than is possible for C-3 plants.

Temperature

The third factor that influences the rate of photosynthesis is temperature and, as one might expect, the effect of temperature depends on the conditions under which the plant is growing as well as on the species. In general, the optimum temperatures for photosynthesis are similar to the daytime temperatures at which the plants usually grow in their natural habitats. Although there are exceptions, optimum temperatures are generally higher for C-4 plants than for C-3 plants.

Temperature influences the rate of carbon fixation in the 'dark' reactions of photosynthesis and, consequently, the true rate of photosynthesis usually increases with increase in temperature until the enzymes begin to become denatured. However, the loss of CO_2 through respiration also increases with temperature. Because of this, the *net* amount of photosynthesis is not promoted by increased temperature nearly as much as one might expect, especially in C-3 plants. In contrast, C-4 plants exhibit a much stronger response to temperature and have a higher optimum value. This is because, in addition to normal dark respiration, C-3 plants carry out a process known as *photorespiration* in the light (see Chapter 1). Like dark respiration, the rate of photorespiration (i.e., loss of carbon) increases with increasing temperature, so that the *net* rate of photosynthesis (i.e., carbon fixed by photosynthesis minus carbon lost by respiration) is decreased. C-4 plants do not carry out photorespiration, so that this source of carbon loss is absent and net photosynthesis increases with increasing temperature to a much higher value than in C-3 plants.

Air movement

The lack of air movement in the enclosed space of the greenhouse affects plants in a number of ways. In still air, the atmosphere immediately around the leaf may be depleted of CO_2, especially in bright sunlight. Even a slight air movement can enhance photosynthesis by displacing this CO_2-depleted air (Table 10.1). Because of the high moisture content of the still air around the leaf, cooling by transpiration is reduced and leaf temperatures can increase substantially. This increase

Table 10.1 The wind velocity affects the rate of photosynthesis.

Wind velocity (cm per second)	Rate of photosynthesis (cc carbon dioxide fixed per square cm of leaf, per hour)
10	79
16	88
42	101
100	109
300	114
1000	118

in leaf temperature in turn will increase water loss because of the temperature differential between leaf and air (see Chapters 5 and 7). Although this will increase transpiration and so cool the leaf, the overall effect of high temperatures in still air is to increase the rate of water loss from the leaf. This often leads to temporary wilting, even when the water supply is plentiful. Still air, especially at high humidities, may also increase the incidence of some fungal diseases.

MANAGING THE GREENHOUSE ENVIRONMENT

Temperature

The first two sections of this chapter have considered how the environment within a greenhouse differs from that outside and the effects of these differences on plants. Managing the greenhouse is largely concerned with minimising the undesirable aspects of the enclosed environment, as well as maximising the desirable ones. To take the latter first, the most obvious advantage of growing under a glass or plastic structure is the ability to maintain a minimum temperature. In order to conserve energy, this should be as low as possible with regard to the stage of growth of the plant and the season. High temperatures are undesirable in winter because the loss of carbon by respiration is increased, whereas the rate of photosynthesis is limited by light. No more than protection from freezing temperatures is often sufficient during periods of low light, or when plants are in a dormant or semi-dormant state. However, there are several plants that can be damaged by exposure to low temperature, even when this is above freezing and it is important to know which plants fall into this category. Common examples include tomato, cucumber and pepper as well as ornamental plants such as African violet (*Saintpaulia ionantha*).

The optimum temperature for any plant depends not only on the species, but also on the stage of growth and the amount of light and CO_2. There is also the problem that in most greenhouses there is a marked variation across the greenhouse, both in light and temperature. Moving plants around within the greenhouse can help to some extent, and the variation can also be used to some advantage by growing plants with different requirements in different parts of the greenhouse. To do this effectively the gardener needs to be able to make reasonably accurate measurements of both light (see below) and temperature.

Water

A second, and perhaps less obvious, advantage is the fact that the water supply is under the gardener's control. Correct water management is one of the most important and sometimes most difficult aspects of greenhouse culture. This is less tricky for plants growing in soil, since they have a larger reservoir of water to access and there is less danger of over-watering. However, for plants growing in containers, both over-watering and under-watering can be a problem.

A major difference between the soil and containers is that the soil has a continuous system of pores, allowing excess water to drain out, whereas a pot has a base where the pore system is discontinuous. The effect of this is that the growing medium in the pot retains water and does not drain as freely. In winter, temperatures and light are low, so that the rates of water loss by transpiration are also very low. Under these conditions it is very easy to over-water pots and allow the soil to become water-logged. Because the overall growth rate is slow (due to low rates of photosynthesis) root growth is restricted; this also limits water uptake and accentuates the problem. So, in the winter, care must be taken to restrict the supply of water.

The summer situation is obviously quite different; high temperatures lead to heating of the leaf and increase the rate of water loss, especially in modules and small pots, both of which can dry out extremely quickly. Plants are also growing much more rapidly under the better light conditions and so the demand for water is much greater. Because of the high temperatures of the interior, the greenhouse environment in summer is a dry one, with a low moisture content. The gradient of water vapour between the inside of the leaf and the surrounding air is increased by both the high leaf temperature and the fact that the content of water vapour in the surrounding air is low. This results in rapid water loss by transpiration.

Lowering the leaf temperature is a major consideration, and there are several ways in which the gardener can achieve this. Adequate ventilation is probably the most important factor in reducing the greenhouse temperature. Not only does ventilation allow exchange between the warmer air inside and cooler air outside, but the increased air movement lowers leaf temperature and, in this way, reduces the rate of water loss. Cooling the interior surfaces by damping down also helps, especially where the floor consists of concrete or paving. Some commercial growers use evaporative cooling systems (known as *fan and pad cooling*) in which the outside air is drawn across wet pads into the greenhouse. This is a very effective way of lowering the temperature but is outside the range of most gardeners who must rely on simple ventilation (which can be increased by fans).

Another way of reducing the heat load on the leaf is to shade the greenhouse, either by painting the glass with a shading material or by the use of blinds. However, shading has the disadvantage of decreasing the amount of light reaching the leaf. Even in summer, the amount of light reaching the lower, more shaded leaves on the plant is less than that needed for maximum rates of photosynthesis, so that shading necessarily results in some reduction in growth. Permanent shading, which remains in place even on cloudy days, is less satisfactory than blinds which can be lowered in bright light to reduce the heat load and raised to allow maximum light when the sun goes in. Where permanent shading is used, it is essential to ensure that it is removed as soon as possible at the end of the summer.

Ventilation

Ventilation not only lowers the air and leaf temperature, but also increases the air movement inside the greenhouse. This enhances photosynthesis by replacing CO_2-depleted air in the atmosphere around the leaf (Table 10.1). Ventilation also increases the concentration of CO_2 within the greenhouse by exchange with the incoming fresh air. In a closed greenhouse, the CO_2 concentration is reduced by photosynthesis, especially in bright light, and, as discussed above, this often limits the rate of photosynthesis in the summer. When ventilation is delayed in cold weather, CO_2 can fall below the normal atmospheric level even in winter. Air movement can also reduce plant height in the same way as shaking (see Chapter 8) and the effect is more marked under glass where air movement is normally restricted. Finally, ventilation can play an important part in preventing some fungal diseases, which can become a major problem in stagnant air with a high relative humidity.

Growing media

Many soils possess positive features that are helpful in supporting plant growth. They contain

- a large reserve of nutrients
- living organisms that cycle nutrients and assist in creating soil structure
- a range of pore sizes that allow excess water to drain, gases to flow and water to be stored
- colloids that allow pH to be 'buffered' and change only slowly.

An ideal potting compost would contain all of these features, but achieving this in a controlled and cost-effective way is very difficult. When plants are grown in containers, their roots are confined in a smaller volume than when grown in the garden, and this means that the demands made on the potting medium for water and nutrients are more intense. Using garden soil in containers gives poor results unless the physical and chemical properties are substantially enhanced.

The development of the John Innes composts was a substantial step forward because it produced a medium that had good chemical and physical properties for plant growth. A key feature of the compost was that it contained 'loam' (i.e. soil with properties that ensure a good supply of water and sufficient clay content to provide a buffered source of nutrients and regulation of pH). The advantages of a loam-based compost are that the nutrient supply is good, and minor element deficiencies are not common. The disadvantages are that good quality loam is in short supply, the material must be steam sterilised, and the composts are heavy and difficult to handle. These substantial disadvantages have led to the development of many loamless composts, usually based on organic materials.

Peat is the most commonly used organic material in loamless composts, but the growing concern about peat supplies is leading to trials with other materials, such as coconut fibre. Loamless composts are generally less variable in composition, do not require sterilisation and are cheaper and lighter to handle. Their disadvantage is that nutrient supplies are often lower and less well buffered, resulting in a greater dependence on supplementary feeding. Some commercial producers grow their crops entirely on inorganic substrates, such as rock wool. In this case, all the essential nutrients are supplied by liquid feeding and complex computer control is usually employed, together with frequent nutritional analyses, in order to ensure that the correct nutrition is maintained at all stages of the plant's growth.

Light

Light is often the factor that limits the growth of plants growing under glass. At higher latitudes in winter, there is much less available light for photosynthesis because of shorter days, low solar angle and many more overcast days (cloud cover can reduce light to less than 20% of bright sunlight). In the British Isles, the total light received on a winter day may be only 10% of that received on a summer one (Fig. 10.4). Even if the upper leaves are at light saturation in the summer, photosynthesis in the shaded lower leaves will still be limited by the available light. Although a single leaf may be light-saturated at about 90 watts per square metre ($W m^{-2}$), up to twice that amount can be required to achieve light saturation in a crop. Obviously, the closer together the plants are, the greater is the shading effect, so that one way of increasing the available light for individual leaves is to increase the spacing between the plants. This is particularly important in the winter when light is severely limiting. Some management techniques can also help, such as cleaning the glass and lowering the temperature.

SUPPLEMENTARY ARTIFICIAL LIGHTING

The problem of low winter light during the winter months (usually from November to March in

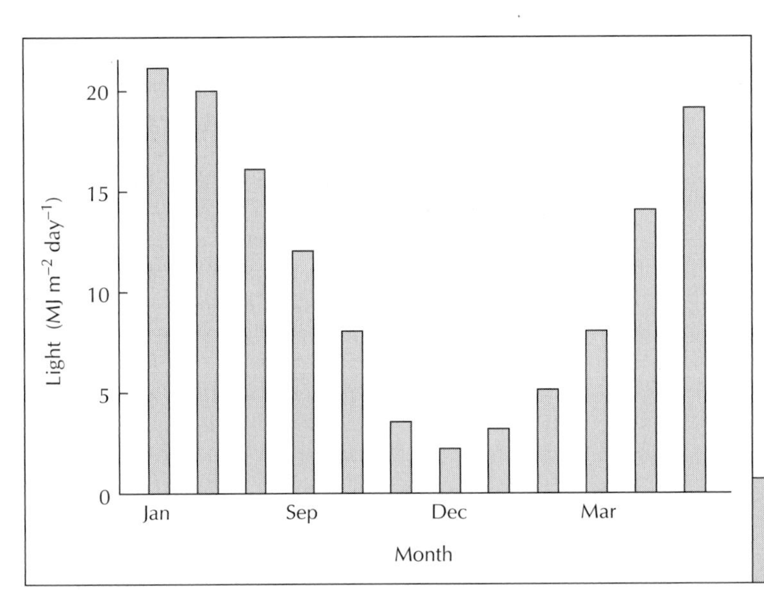

Fig. 10.4 The average total daily light received throughout the year at a site on the south coast of the UK.

Northern Europe) can be overcome to some extent by the use of artificial light to supplement natural daylight. In some facilities (e.g. in some specialist propagation units), plants are grown in insulated windowless structures and natural daylight is replaced entirely. However, artificial light is more commonly used to 'top-up' winter daylight in a greenhouse, by adding to the natural light during the day, by giving extra light during the night, or by both. This is especially useful for seedlings in the winter and early spring, when large numbers of plants can be illuminated in a small area. Extra light at this stage gets the seedlings off to a good start and the light levels are generally improving by the time the plants get bigger and need to be spaced out. The additional light results in a faster growth rate and better quality, sturdier plants.

The primary effect of the additional light is to increase the rate of photosynthesis and so the major requirement is to give sufficient extra light to make a difference. This can be expensive and, in order to make the best and most economic use of artificial light, several points need to be taken into account.

- How should light be measured in relation to plant responses?
- What type of lamp gives the best results?
- How much light is necessary for a satisfactory response and how is this best achieved?

In order to answer these questions, it is necessary to have some understanding of the basic properties of light and the way that it interacts with plants.

Light measurement

Measuring the amount of light that is available for plant growth is not straightforward. Many measuring devices are calibrated in *photometric* units and measure light in terms of the sensitivity of the human eye to different wavelengths (see Fig. 8.1). The light is expressed as *illuminance* and the units are *lux* (an earlier unit that is sometimes still used is the foot candle; 1 ft candle = 10.76 lux). The human eye is most sensitive to green light and is relatively insensitive to red and blue, which are the most important wavelengths of light for plants. Consequently, light measurements expressed in lux can be very misleading where plants are concerned. Light can also be measured (using a radiometric detector such as a thermopile) in terms of radiant energy per unit area (usually as watts per square metre, $W\,m^{-2}$). Once again this can be misleading because most detectors of radiant energy measure radiation not only in the visible wavelengths that are active in photosynthesis, but also in the ultra-violet and infra-red regions beyond the limits of plant responses. One advantage of detectors that measure in lux is that they measure only visible light, which is in the range that is also detected by plants. Illuminance values (lux) can be converted to irradiance ($W\,m^{-2}$) by using the appropriate conversion factors given in Table 10.2.

To be effective in photosynthesis, or any other light-dependent process in plants, light must first be absorbed by a pigment. For photosynthesis (see Chapter 1), the chlorophylls and some associated carotenoids are the light-absorbing pigments and they absorb at wavelengths throughout the visible spectrum, although not equally at all wavelengths. Absorption is least in the green part of the visible spectrum, hence the green colour of leaves. Moreover, it is important to remember that light is absorbed by the pigments in the form of discreet units of energy, called *quanta* (or *photons* for energy within the visible spectrum) and that the amount of energy in each quantum is inversely proportional to the wavelength of the light. This means that, for the same amount of measured energy, there will be fewer quanta available for photosynthesis in blue light at an average wavelength of 450 nm than in red light at 650 nm (Plate 17).

The most useful measurement of the amount of light available for photosynthesis is the number of quanta (photons) per unit area (the *photon flux density*) in the visible spectrum between 400 and 700 nm. This is sometimes called *Photosynthetically Active Radiation* (PAR). However, when only sunlight is being considered, the irradiance, expressed as $W\ m^{-2}$, between 400 and 700 nm is also a useful way of comparing the amounts of light available for photosynthesis at different times and in different places; for example, to look at the shading effects in different parts

Table 10.2 Factors used for converting illuminance (lux) to irradiance (milliwatts per square metre, mW m^{-2}).

Type of light	Conversion factor (multiply by this factor to convert measurements in lux into mW m^{-2})
Natural sunlight	4.0
Tungsten-filament lamp (100W)	4.2
High-pressure Sodium lamp (SON/T)	2.4
Tubular fluorescent lamps	
warm-white*	2.8
de-luxe warm white*	3.6
daylight*	3.7

* The precise conversion factors for the various colours of tubular fluorescent lamps may vary with the manufacturer, but the values given here are adequate for use in practice.

of a greenhouse, or to determine how much light is being transmitted through the glass.

The choice of lamp

Because the effect on photosynthesis and other photochemical reactions is dependent on the number of absorbed photons, photosynthesis will be stimulated more for the same amount of measured energy by red wavelengths (more quanta per unit of energy) than by blue or green ones with fewer quanta per unit of energy. So, other things being equal, lamps emitting red light would seem to be the best ones for supplementary lighting. However, other things are not equal. The horticultural market is small and lamp manufacturers are mainly concerned with lighting for human vision, which is most sensitive to green wavelengths and poorly sensitive to red. A few lamps designed specifically for plants are available but they are considerably more expensive and hardly offer sufficient advantage to justify their extra cost. There is also the question of the efficiency of the lamps; for the same input of electrical energy, some have a higher output of light than others.

In addition to photosynthesis, it is important to take into consideration the many other effects which light has on plant growth. These are considered in more detail elsewhere (see Chapters 5 and 8) and here it is only necessary to re-state the points that are relevant to the use of artificial light. Two regions of the spectrum have major influences on the growth patterns of plants, namely the red/far red (R/FR) ratio of the light and the absolute amount of light in the blue and near-ultraviolet part of the spectrum. Stem growth in many species, especially 'sun' plants, responds to the ratio of R/FR in the light (the phytochrome response) (see Chapter 5); the greater the *relative* amount of FR light, the more stems elongate.

Not all species are sensitive to the R/FR ratio and in some, stem growth appears to be mainly controlled by the amount of blue light. This is well illustrated by the response of lettuce seedlings (Fig. 10.5). When grown under low-pressure sodium lamps (which do not emit any blue light) some cultivars develop an extremely elongated and unsatisfactory habit of growth. Only when blue light is added do the seedlings grow in a normal manner. Other cultivars, however, are clearly dependent on the R/FR ratio of the light and do not require blue light. These cultivars grow normally under low-pressure sodium lamps, which establish the same amount of the biologically active form of phytochrome as red light. It is clear that plant responses have to be considered individually when choosing a suitable lamp type.

Of the lamps currently available, probably the most satisfactory, from the point of view of the wavelengths that they emit, are the various types

Plate 11 (*left*) Drought avoidance. Annual composites of the South African Cape survive the summer drought conditions in the form of seeds, which germinate only after a sufficient winter rainfall. This results in synchronous flowering in the following spring as shown here.

Plate 12 (*left*) Drought tolerance. Small leaves (left foreground) reduce water loss and are characteristic of the shrubs of the Mediterranean Maquis shown here. Reflective or grey leaves (middle foreground) reduce the heat load on the leaf and so reduce water loss.

Plate 13 (*above*) The transmission, reflection and absorption of red and far-red light by leaves.

Light from the sun contains both red (R) and far-red (FR) wavelengths. Red light is absorbed by the leaf but FR light is transmitted through the leaf and is also reflected to neighbouring plants.

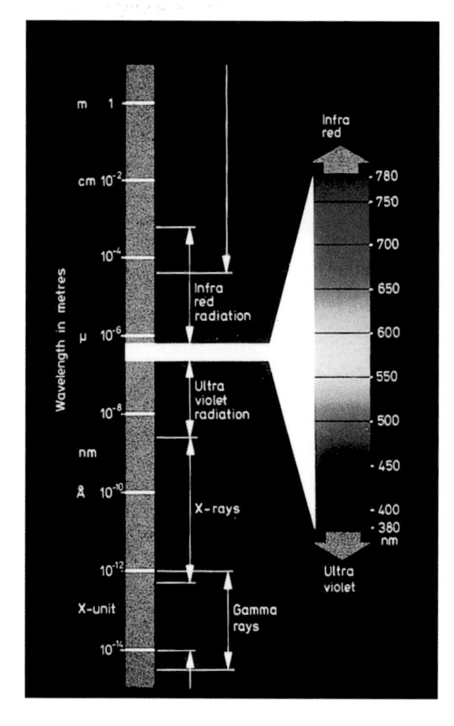

Plate 14 (*above*) The electromagnetic spectrum showing the region visible to the human eye.

Plate 15 Critical day-lengths for flowering. *Rudbeckia fulgida* var. *sullivantii* 'Goldsturm' (*right*) is classified as a long-day plant with a critical day-length between 13 and 14 hours. It flowered only when the days were longer than this.

Leucanthemum x *superbum* 'Snowcap' (*below*) is also a long-day plant but, in this case, the critical day-length lies between 14 and 16 hours.

Echinacea purpurea 'Magnus' (*below right*) belongs to the relatively rare group of intermediate-day plants and flowered rapidly in day-lengths between 13 and 15 hours, but not in longer or shorter ones.

All 3 plants also flowered rapidly with a short exposure to light (NI) given during a long dark period.

Photograph courtesy of Royal Heins, Michigan State University.

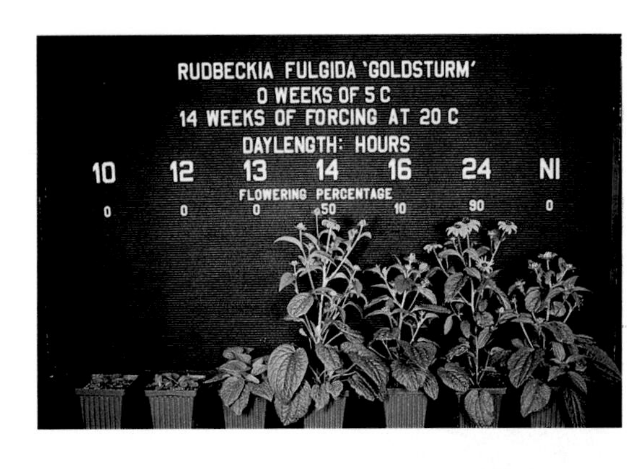

Plate 16 (*right*) The latitude of origin affects growth, leaf fall and dormancy in birch trees. Seeds were collected from different locations in Sweden (from left to right, latitude 68°, 65°, 63°, 63° (mountain) and 60° N) and grown together in Uppsala (latitude 60° N).

Trees on the far left (origin 68° N) are stunted and had already shed their leaves on 25th September when the photograph was taken. In contrast, those on the far right, which originated locally in Uppsala (60° N) are much taller and the leaves were still green. Trees from a high altitude (2nd from right) have a longer critical day-length and behave like trees originating in lowland sites further north (2nd from left).

Photograph courtesy of David Clapham, Uppsala Genetic Centre.

Plate 17 (*right*) The relationship between the wavelength of light, the energy per quantum and the number of quanta per unit energy. Each quantum of blue light (at a wavelength of 450 nm) has a shorter wavelength (*left*) and more energy (*centre*) than a quantum of red light (at 660 nm). Consequently, for the same amount of energy, there are fewer quanta in blue than in red light.

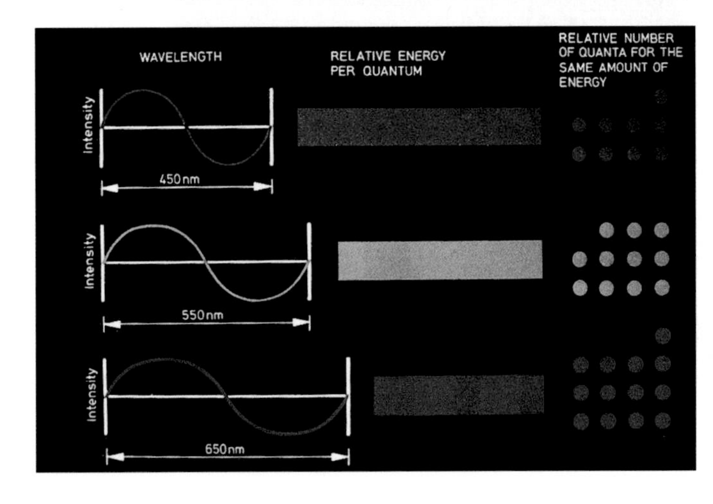

Plate 18 (*left*) 'All-year-round' production of chrysanthemums. During the winter months, plants are given artificial long days (night-breaks with tungsten-filament lamps) to prevent premature budding. When the stems are sufficiently long, the lights are switched off and the plants returned to short days to allow bud formation. The short days are either given naturally in winter or obtained by covering the plants with a black cloth each day during summer.

Plate 19 (*below*) Lesser celandine (*Ranunculus ficaria*), a native woodland perennial that is also a weed, spreading by bulbils and root tubers.

Photograph courtesy of Royal Horticultural Society.

Plate 20 Mind-your-own-business or baby's tears (*Soleirolia soleirolii*), an example of a plant introduced to the UK as an ornamental which has escaped and become a weed. Photograph courtesy of Royal Horticultural Society.

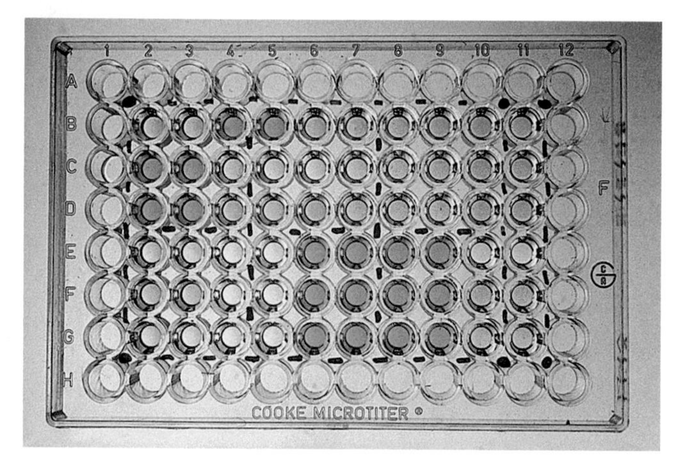

Plate 21 (*above*) A DAS-ELISA diagnostic test plate showing some positive (dark yellow) reactions. Photograph courtesy of N. Spence, Horticulture Research International.

Plate 22 (*above*) A codling moth pheromone trap. Photograph courtesy of Royal Horticultural Society.

Plate 23 (*above*) A scanning electron micrograph of a pustule of rose rust (*Phragmidium* sp.) on the underside of a rose leaf, showing a mass of large resting spores (*teliospores*) which carry the fungus through the winter and germinate in the spring to reinfect roses. Photograph courtesy of Georgina Godwin-Keene (CABI Bioscience); Royal Horticultural Society.

Plate 24 (*above right*) Horticultural fleece covering narcissus at RHS Wisley to prevent attack by large narcissus bulb fly (*Merodon equestris*). Photograph courtesy of Royal Horticultural Society.

Plate 25 (*above*) The predatory mite. *Phytoseiulius similis* feeding on the red spider mite. Photograph courtesy of Royal Horticultural Society.

Plate 26 (*left*) Cultivars of *Iris germanica* hybrids in a demonstration at RHS Wisley, showing large differences in resistance to Iris rust (*Puccinia iridis*). The one on the left was severely affected and would require a level of fungicide protection that most gardeners would find unacceptable, the other two were hardly affected. Photograph courtesy of Royal Horticultural Society.

Fig. 10.5 The responses of different lettuce cultivars to light from low-pressure sodium lamps. Some cultivars (B and C) grown in light from low-pressure sodium lamps with added red light (SOX + R) show elongated and abnormal growth, while others (A) grow normally. When blue light is added (SOX + B), all of the cultivars are normal. (Photo courtesy of Brian Thomas, Horticulture Research International.)

of 'white' fluorescent lamps. Most of these have a very low FR emission and emit light throughout the visible spectrum. They satisfy both the blue light and the phytochrome responses, producing a sturdy habit of growth. Indeed, because the R/FR ratio of most of these lamps is considerably higher than in sunlight, plants growing under them often develop a shorter habit of growth, with dark green leaves (Fig. 5.4). These lamps are the ones most often used where daylight is replaced completely, in micropropagation and some other specialist uses, where spectral quality is a more critical issue than when some daylight is present to supplement the spectrum.

In other respects, fluorescent lamps have both advantages and disadvantages. A major advantage, especially in small glasshouses where there is little headroom, is that fluorescent lamps emit little heat and have a low surface temperature. This means that they can be suspended quite close to the plants without damaging them. Because of their linear shape, they also have a uniform light distribution. Their disadvantage is that, although they are efficient in converting electrical energy into light, their overall light output is low (lamps are no more than 125 watts). This means that several lamps are usually needed to obtain light levels that are high enough to stimulate photosynthesis. In turn, this means that they can cast a considerable amount of shade and thus reduce the amount of natural light reaching the plants. This shading effect is substantially increased when the lamps are used with reflectors to increase the amount of light that goes downwards.

For most commercial growers, the high-pressure sodium lamp has become the preferred choice, because such lamps have a high light output, a long useful life, and a spectrum that produces a satisfactory habit of growth when used to supplement daylight. The lamp and its associated control gear are housed in specialised 'luminaires' that incorporate a reflector to throw the light downwards onto the plants and are designed to minimise the obstruction of natural light. In order to produce a uniform light distribution and prevent overheating, it is usually necessary to suspend the lamps at a distance of 2–3 metres above the plants, which is not normally possible in a small greenhouse. However, specially designed luminaires are available that throw light over a wider angle and these can be suspended from 0.75 to 1.5 metres above the plants.

Tungsten-filament (TF) lamps such as ordinary domestic electric light bulbs, or incandescent lamps are not satisfactory for supplementing daylight. They have a R/FR ratio considerably lower than that of daylight (Table 10.3) and result in an excessively 'leggy' type of growth in many plants (Fig. 5.4). Additionally, their high output

Table 10.3 The ratio of red to far-red light in lamps that are commonly used in horticulture.

Type of light	Ratio of Red to Far-red light
Tungsten-filament (incandescent)	0.7
Tubular fluorescent	
Warm white (Colour 29)	22.7
De-luxe warm white	3.6
High-pressure sodium (SON/T)	3.7
Sunlight*	1.15

* Sunlight is given for comparison.
The red:far-red ratio of tungsten-filament lamps is less than sunlight, resulting in excessive stem elongation in some plants.
The red:far-red ratio of most fluorescent lamps is much higher than sunlight, resulting in a compact habit of growth.

in the infra-red region of the spectrum results in a considerable heat load on the leaf.

The spectra of some common lamp types are shown in Fig. 10.6 and their R/FR ratios in Table 10.3.

How much 'extra' light?

The amount of supplementary light needed for satisfactory winter growth in high latitudes varies with the plant. In practice, light levels between 5 and 19 W m^{-2} produce satisfactory results when used for between 8 and 24 hours each day. It is generally true that plants respond to the total amount of light received each day, so that lower light levels from the lamps can be compensated for by increasing the daily duration of extra light. However, the daily duration of light can have a marked influence on plants, so this must be taken into account when deciding on the precise

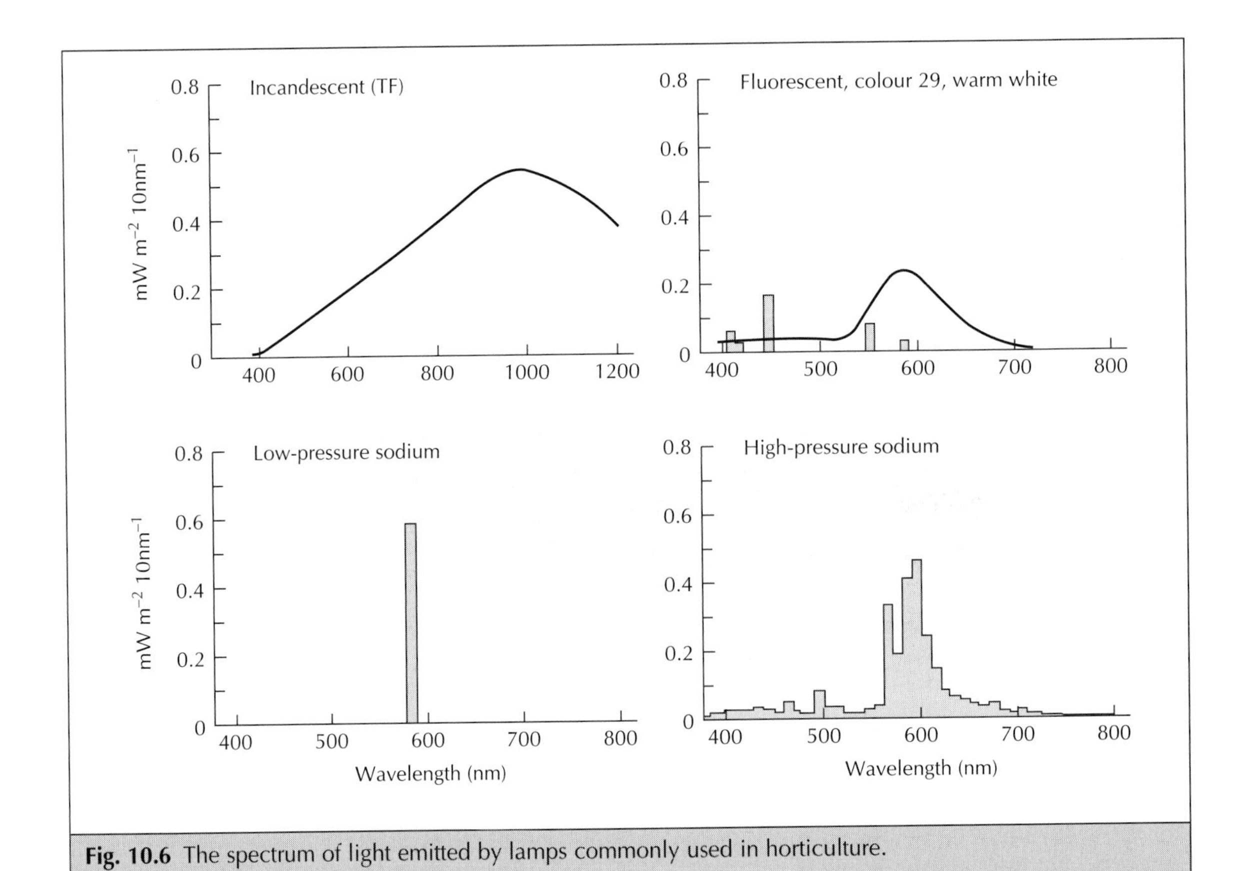

Fig. 10.6 The spectrum of light emitted by lamps commonly used in horticulture.

schedule for supplementary lighting. For example, tomatoes should not be lit continuously for 24 hours each day as this can result in damage to the leaves. For economy lamps can be switched off during bright periods in late autumn and early spring.

The precise design of the lighting installation depends on the type of lamp and reflector, its wattage, and other factors such as headroom and the plants being lit. In order to achieve an irradiance of 12 W m^{-2}, for example, a 400 W high-pressure sodium lamp will illuminate an area of about 3.5–4 m^2 when suspended about 2 metres above the bench. Detailed recommendations for particular installations and geographical locations are outside the scope of this book but are given in several manuals (such as *Lighting for Horticultural Production*, published for the UK by the Electricity Council).

DAY-LENGTH LIGHTING

The physiology of photoperiodism is discussed in Chapter 9 in relation to the effects of changes in the natural day-length. However, the use of artificial day-lengths to manipulate flowering is largely confined to the greenhouse (at least in the UK) and so needs to be considered here. The short days of winter can be extended by using artificial light, while the long days of summer can be decreased by covering the plants each day with a light-excluding material. Both are practised commercially.

Summer-flowering plants (such as *Antirrhinum*, *Calceolaria*, *Alstroemeria*, and even perpetual-flowering carnations which flower much more profusely in summer) can be brought into bloom in the winter. Chrysanthemums that flower in the short days of autumn can be marketed at any time of year by a combination of covering them in the summer (to allow flowering) and giving long days with artificial light in the winter to delay flowering and obtain a satisfactory length of stem for a cut flower (Plate 18). 'All-year-round' flowering (AYR) of chrysanthemums is now a major industry but, before about 1950, growers in the UK had great difficulty in holding back flowering in order to obtain the higher prices of the Christmas period. Other short-day plants for which artificial long days are used commercially to time the crop or to achieve a desirable size include poinsettia (*Euphorbia pulcherrima*), flaming Katy (*Kalanchoë blossfeldiana*, Fig. 10.7) and winter-flowering *Begonia* spp.

Short-day plants

Covering plants in summer in order to obtain summer-flowering of short-day plants (SDP) poses no major difficulties, although the actual

Fig. 10.7 Delaying the time of transfer to short days increases plant size in the short-day plant Flaming Katy (*Kalanchoë blossfeldiana* Tetra Vulcan). Plants were grown in long days for (from left to right) 0, 6, 8, 10, 12, 14 or 16 weeks before transfer to short days for flowering. When short days began early, flowering occurred early on small plants. Transfer to short days at a later stage of growth delayed flowering and resulted in much larger plants. (Phot. courtesy of K.E. Cockshull, Horticulture Research International.)

duration of covering depends on the plants. SDP will flower when the days are shorter than a certain critical length but, unfortunately, this varies with species and even cultivar (some examples are given in Table 9.2). However, a rule of thumb is that the majority will flower on a 10-hour day (or shorter). Species differ considerably in their sensitivity to light, so it is important to exclude daylight as much as possible. Domestic lighting is bright enough to delay or prevent flowering in some SDP, such as poinsettia, especially when the lights are on for several hours in each evening. Flowering can be achieved by placing the plants in a dark cupboard for about 14 hours each night, but they must be removed into the light for several hours each day in order for the photoperiodic mechanism to function.

Night-break lighting

Lighting to provide long days in winter is much more complicated and, to be successful, it is necessary to have some understanding of the underlying physiological mechanism. In the so-called *dark-dominant* plants (which includes the majority of SDP), flowering depends on whether or not a critical duration of darkness (the critical night-length) is exceeded (see Chapter 9). Flowering in these plants can be controlled by giving a relatively short exposure to light at a particular time during the night. Such a *night-break* produces the same effect as a long day. Night-break lighting in winter is used commercially to regulate the flowering of a number of SDP, the major use of such lighting techniques being to delay flowering until the plants have achieved the desired size. Lamps are then switched off to allow flowering.

If responses are to be manipulated by giving a night-break, it is clearly important to know when is the optimum time, how long the exposure should be, and what kind of light is needed.

Night-break timing

The time at which a plant is most sensitive to a night-break (NBk) varies with individual plants. Many practical manuals recommend that the NBk should be given near midnight in order to 'break the night up into two periods, each shorter than the critical night length'. However, this is not how the NBk mechanism works. Time-measurement during darkness begins at dusk, and plants become sensitive to NBk light after a number of hours (determined by the biological clock) have elapsed. If the duration of darkness is increased, the time at which a NBk is most effective still occurs at the same number of hours after dusk, even though the subsequent period of darkness far exceeds the critical night-length (Fig. 10.8). In other words, the plant 'counts' time from dusk until it reaches the time when it becomes sensitive to a NBk. The subsequent duration of darkness is unimportant. However, if the NBk is very long (from 6–8 hours, for example), the plant recognises this as a 'new day' and counting begins again from the beginning of darkness.

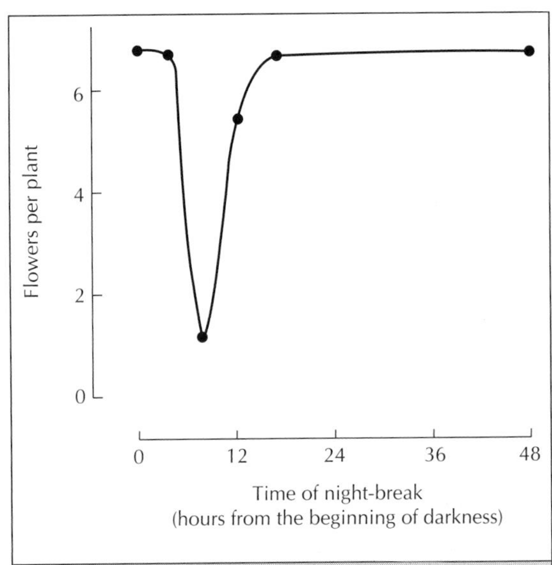

Fig. 10.8 The time of maximum sensitivity to a night-break in Japanese Morning Glory (*Ipomoea nil*) when the duration of the dark period is extended to 48 hours. A brief exposure to light (night-break) strongly inhibits flowering when given after 8–9 hours of darkness, even though the subsequent period of darkness (39–40 hours) far exceeds the critical night-length (8–9 hours in this plant).

The greatest effect of a NBk occurs at around the middle of a 16-hour long night in many plants (Fig. 10.9), but this is not always the case and it can be earlier, or later. For example, in sugar cane, a NBk given near the beginning or near the

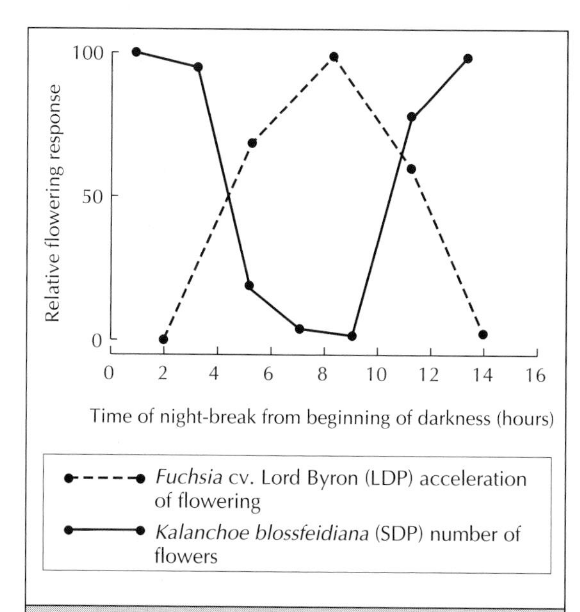

Fig. 10.9 The time of maximum sensitivity to a night-break to promote flowering in Fuchsia (*Fuchsia* × *hybrida* 'Lord Byron', a long-day plant) or inhibit flowering in Flaming Katy (*Kalanchoë blossfeldiana*, a short-day plant) when the duration of darkness is 16 hours. In both cases the greatest effect of giving a short exposure to light occurred near the middle of the night, i.e. after about eight hours of darkness as in Japanese Morning Glory (Fig. 10.8).

end of the night delays flowering more than when light is given near the middle. These variations make it unwise to assume that the 'best' time to give a NBk is always around midnight. Over the normal growing range, temperature has rather a small effect on the time of greatest sensitivity to a NBk, although it may have a considerable effect on other aspects of flowering.

What kind of light?

The night-break effect on flowering was one of the first responses shown to be under the control of phytochrome, with red light being the most effective. Giving far-red light immediately after the red night-break prevented its effect (Table 10.4), demonstrating that, as well as controlling germination and many growth responses to light (Chapters 6 and 8), the pigment phytochrome also controls the NBk response. Consequently,

Table 10.4 Giving a night-break with red light inihibits flowering in the short-day plant, *Chrysanthemum* 'Honeysweet', while exposure to far-red light reverses the effect of red light and promotes flowering. When the sequence of light treatments ended in **red**, flowering was prevented; when the sequence of light treatments ended in **Far-red**, flowering occurred.

Night-break Treatment	Response
Darkness for 16 hours (no night-break)	Flowering
3 minutes of **red light**	Vegetative
3 minutes red, followed by 3 minutes of **far-red light**	Flowering
3 minutes red, 3 minutes far-red, followed by 3 minutes **red light**	Vegetative
3 minutes red, 3 minutes far-red, 3 minutes red followed by 3 minutes **far-red light**	Flowering

lamps with a high R/FR ratio (such as 'white' fluorescent lamps) are the most efficient light sources for night-break lighting, although tungsten-filament (TF) lamps are also quite effective and are widely used because they require no control gear and are cheap to install.

Night-break duration

Although very short exposures to light can be effective (Table 10.4), longer exposures are usually given in practice, especially when TF lamps are used. The recommendation for chrysanthemums in the UK is to use TF lamps for 2 hours during September and March, increasing this to 5 hours in December.

Cyclic lighting

In large commercial installations the cost of electricity becomes important and this can be reduced by the practice of *cyclic lighting*. Experiments have shown that giving light intermittently can be just as effective as giving light continuously throughout the night-break, provided that the intervals over which light is given are sufficiently short. Only TF lamps can be used in this way, as other lamp types cannot be switched on and off so often. The underlying physiology is that the active form of phytochrome (which is formed in the light) continues to act during the following dark interval, provided that this is not too long. Recommendations for chrysanthemums are that the cycle should be at least 10 minutes but not more than 30 minutes long, with the light given for 20 to 50% of the time. For example, 20% light in a 30-minute cycle is 6 minutes light followed by 24 minutes dark; the cycle is then repeated for the duration of the night-break. Unfortunately, as with many other aspects of biology, not all plants behave in the same way. Carnations show a very weak response to intermittent lighting even when the dark intervals are extremely short and cyclic lighting offers no obvious advantage for this crop.

Long-day plants

In contrast to most SDP, flowering in many long-day plants (LDP) depends not only on the duration of the dark period but also on the kind of light received during the day. These so-called *light-dominant* plants are less responsive to a night-break and require quite long exposures to light of several hours duration in order to achieve a long-day response. Having said this, the same quantity and duration of light given during the middle of the night is usually more effective than the same amount given at the end of the day. So, in practical terms, to promote winter flowering in these plants the most efficient schedule is to give light for several hours (usually 4–8 hours) each day, in the middle of the dark period. However, in some cases, lighting throughout the night from dusk to dawn gives the most satisfactory results. In carnations, for example, 'all night' lighting for 2–4 weeks results in a flush of bloom some 3–5 months later.

Light-dominant plants have another important characteristic. Even when the duration of darkness is favourable, the kind of light during the daytime influences the response and, in order to be effective, a long day must contain some FR light. The reason for this is not fully understood, although it is thought that one particular type of phytochrome may be involved. This phytochrome has been shown to be particularly associated with sensing far-red light. (Another type of phytochrome may be involved in the perception of day-length in SDP, where FR is not required.) Sunlight is rich in FR wavelengths (Table 10.3) and so, under natural conditions, only the duration of darkness is important for LDP. When using artificial light to increase the day-length, TF lamps, which have a high content of FR, are the lamps of choice. Unfortunately, TF lamps have a lower R/FR ratio than sunlight (Table 10.3) and, when used over several hours each day, can lead to the development of 'leggy' plants, with long, weak stems.

Because of the reduction in light under glass and the higher temperatures, greenhouse-grown plants are often undesirably tall even in natural daylight. Recently, filters have become available that absorb FR light and so considerably increase the R/FR ratio of the light that is transmitted through them. This has the desirable effect of reducing internode elongation and producing a more compact habit of growth without the use of costly growth-inhibiting chemicals (see Chapter 8). Unfortunately, the removal of FR light means that flowering in 'light-dominant' long-day plants can be considerably delayed and may even be prevented entirely. So far, this problem has not been resolved. The amateur gardener needs only to be aware that the use of such filters to improve the habit of growth may pose problems for flowering in some long-day plants.

Day-length lighting in practice

Only relatively low irradiances are necessary for day-length lighting because light is only acting as a signal for the pigment phytochrome and is not supplying an energy source for photosynthesis. Although species do differ somewhat in their sensitivity to light, photoperiodic lighting is usually installed at between 0.25 and 0.5 $W\,m^{-2}$, which is adequate for most species. Supplementary lighting for photosynthesis can also be used to obtain long days and, in this case, a much higher irradiance is needed. It is important to take care to avoid overspill to other nearby plants if they are responsive to day-length.

From the above discussion, it is evident that much remains to be discovered about the mechanism(s) through which plants 'see' day-length and it is still necessary to experiment with individual species and cultivars, using different durations, timings, and lamp types, in order to obtain the best results.

Conclusions

This chapter has considered ways in which a greenhouse modifies the plant's environment and described the main physiological effects of these changes. The general guidelines suggested for managing the greenhouse are based on an understanding of how plants respond to the major factors that can be controlled in this environment, particularly those of temperature, light, carbon dioxide, water, air movement, humidity, nutrition and day-length.

FURTHER READING

Anon. (1964) *Commercial Glasshouses. Siting, Types, Construction and Heating*. Bulletin No. 115, Ministry of Agriculture. HMSO, London.

Anon. (1987) *Lighting for Horticultural Production: Grow Electric Handbook*. Electricity Council, Stoneleigh.

Canham, A.E. (1966) *Artificial Light in Horticulture*. Centrex Publishing Company, Eindhaven.

Hanan, J.L. Holley, W.D. & Goldsberry, K.L. (1978) *Greenhouse Management*. Springer Verlag, New York.

Revell, R. & Henbest, R. (1999) A bright future for plastics. *The Horticulturist*, **8**, 29–30.

Payne, C. (1997) Horticulture in the next millennium – the contribution of R & D. *The Horticulturist*, **6**, 25–28.

Salisbury, F.B. & Ross, C.W. (1991) *Plant Physiology*, 4th edn. Wadsworth Publishing Company, Belmont, Ca.

Thomas, B. & Vince-Prue, D. (1997) *Photoperiodism in Plants*. Academic Press, London.

Vince-Prue, D. (1990) The control of flowering by day-length. *The Plantsman*, **11**, 209–24.

Vince-Prue, D. & Canham, A.E. (1983) Horticultural significance of photomorphogenesis. In: *Photomorphogenesis. Encyclopedia of Plant Physiology, New Series*, **16B**, pp 518–44, Springer Verlag, New York.

Vince-Prue, D. & Cockshull, K.E. (1981) Photoperiodism and crop production. In: *Physiological Processes Limiting Plant Productivity* C.B. Johnson (Ed.), pp. 175–7, Butterworths, Oxford.

11

Controlling the Undesirables

- Pests: nematodes or eelworms, molluscs, mites, insects, birds and mammals
- Diseases (bacteria, fungi and viruses): opportunists, destructive parasites and biotrophic parasites
- Weeds: annuals, perennials, weed ecology, weed origins, useful weeds
- Deciding whether an organism is a pest
- Deciding when a pest is a pest
- Deciding when disease is a problem
- Weeds as pests
- When to take control measures
- How to control pests, diseases and weeds: cultural controls; resistance; chemical controls; biological control; integrated pest, disease and weed management
- The future

INTRODUCTION

In the imagined arcadian wilderness before gardening was invented there were no undesirables, only a rich biodiversity. Today's gardeners find this richness excessive and relabel some of it pests, diseases and weeds. Remembering, therefore, that a pest, disease or weed is simply biodiversity being over-assertive, it may be necessary to take some corrective action.

Pests

Throughout the year plants in gardens, greenhouses or on windowsills are at risk from pests. Some, like woodlice (Isopoda) and millipedes (Diplopoda), feed mainly on decaying plant material and only occasionally become pests of seedlings and other soft plant growth. Other pests cause more serious damage to plants of all ages, resulting in distorted growth, loss of vigour and a reduction in the yield of flowers and crops. The principal types of pest encountered by gardeners are listed below.

Nematodes or eelworms

These are mostly microscopic, worm-like animals that live in the soil or within plants. Plant-feeding nematodes have piercing mouthparts that are used to suck sap from plant cells. Some soil-dwelling nematodes cause stunted root growth and may transmit virus diseases. Leaf and bud nematodes live mainly inside the foliage and cause a distinctive pattern of brown wedges or islands where the infested tissues are separated from the rest of the leaf by the larger leaf veins. Narcissus and phlox nematodes (races of *Ditylenchus dipsaci*) cause distorted and stunted growth which is often followed by the plant's death. Cyst and root knot nematodes (*Globodera* and *Heterodera* spp., and *Meloidogyne* spp.) develop inside plant roots, causing poor growth as a result of severe disruption in the uptake of water and nutrients.

Molluscs

Slugs and snails are mainly active after dark. They eat holes in foliage, flowers and stems; some slugs are a problem underground where they feed on bulbs, potato tubers and other root vegetables.

Mites

Plant-feeding mites suck sap and mainly feed on the foliage. The most frequently encountered is the two-spotted mite (*Tetranychus urticae*) which causes a pale mottled discoloration of upper leaf surfaces. Tarsonemid mites attack the shoot tips of glasshouse and some garden plants. This leads to a progressive stunting and distortion of growth. Gall mites or eriophyids are microscopic animals that secrete chemicals which induce their host plants to produce abnormal growths in which the mites develop.

Insects

Insects are by far the most numerous and diverse type of plant pest. The more important insect orders containing plant pests are the Dermaptera (earwigs), Hemiptera (aphids, whiteflies, scale insects, mealybugs, leafhoppers, froghoppers, capsid bugs), Thysanoptera (thrips), Lepidoptera (butterflies and moths), Diptera (flies), Coleoptera (beetles) and Hymenoptera (sawflies, ants, gall wasps). Between them they exploit every conceivable ecological niche that can be provided by plants. Apart from the wholesale consumption of foliage and other plant parts, some insects feed as leaf miners or stem borers, or they may specialise in attacking seeds or fruits. Plant-feeding insects in the Hemiptera and Thysanoptera are all sap-feeders. Some excrete a sugary substance, honeydew, which soils the foliage and allows the growth of black sooty moulds. Sap-feeding insects can also spread plant virus diseases on their mouthparts when they move to a new host plant. Insects of many orders, but especially in the Hemiptera, Diptera and Hymenoptera, can induce abnormal growths, known as galls, on leaves, stems, buds and other plant parts. In the Lepidoptera and Diptera it is always the larval stage that causes the damage; in the other orders damage is often caused by both adult and immature stages.

Birds and mammals

Most birds are welcome in gardens, but wood pigeons (*Columba palumbus*), house sparrows (*Passer domesticus*) and bullfinches (*Pyrrhula

pyrrhula) sometimes damage plants. Bullfinches take flower buds from fruit trees and bushes and some ornamentals during the winter. Pigeons can defoliate brassicas and peas, while sparrows shred the petals of crocus and primulas. Other birds may damage ripening fruits. Mice (*Apodemus* and *Mus* spp.), voles (*Clethriomys*, *Microtus* and *Arvicola* spp.), grey squirrels (*Sciurus carolensis*), rabbits (*Oryctolagus cuniculus*), foxes (*Vulpes vulpes*), deer (especially roe deer *Capreolus capreolus*) and moles (*Talpa europaea*) are the most important mammalian garden pests. Rabbits and deer will eat most plants; these pests and squirrels and voles can also kill woody plants by removing bark from the stems. Mice, voles and squirrels also feed on seeds, fruits and bulbs. Moles can create havoc in lawns and seed beds with their tunnels and production of molehills.

Diseases

Disease is a harmful disturbance of normal function and in plants may be caused by infectious microorganisms, non-infectious agencies such as nutrient imbalances, adverse physical conditions such as waterlogging, or by pollution such as ozone, sulphur dioxide and 'acid rain'. The main groups of infectious agents will be familiar to gardeners because they also contain the causal agents of animal and human disease; but those organisms attacking animals, such as the influenza and pox viruses, do not attack plants, and vice versa. The relative importance of the groups is not the same for animals and plants. Viruses and bacteria are both important, but there are no protozoan parasites of plants to rival the mammalian malarial parasites in severity, and fungal *pathogens* like *Phytophthora* and *Armillaria* are generally more numerous and damaging than those attacking animals. Plants are also attacked by phytoplasmas, which have some properties of viruses, and by parasitic plants.

The microorganisms that cause disease are simple in structure and have only a few measurable characteristics, in comparison with the higher organisms for which conventional taxonomic methods were first worked out. This makes identifying and classifying them difficult. This is not merely an academic matter, because the name is the key to the scientific literature on the organism, and it has often been pointed out that this literature is now so vast that it sometimes seems easier to make a scientific discovery than to discover whether it has already been made.

The largest and most diverse group of infectious microorganisms attacking plants are the fungi, which are now classified into three groups: the true Fungi; the Chromista which contain the very important downy mildews, *Phytophthora* and white blister fungi; and the Protozoa which contain the slime moulds including *Plasmodiophora*, the cause of club root in the Brassicaceae (cabbage family). Within the true Fungi the classification is based primarily on the sexual structures, as in higher plants, but the fungi pose difficulties because some rarely form these structures, and some never do. Where sexual structures are not present various classifications have been based on asexual structures, mostly the dispersal spores (*conidia*) and the structures that produce them.

The problem of accurate naming becomes more acute with the smallest organisms, the viruses and bacteria. At present, much reliance is placed on biochemical tests that distinguish species by the enzymes they contain (for bacteria) and by physical and serological characteristics (for viruses). Morphological and biochemical characteristics are ultimately determined by the unique base-pair sequence of each organism's DNA and RNA, and the sequences of selected parts of the DNA or RNA will probably soon provide the definitive taxonomic criteria.

Microorganisms exploit different strategies to live and reproduce at the expense of plants. Some examples are as follows.

Opportunists

Botrytis cinerea causes grey mould of soft fruit. The spores are always present, because this fungus also lives and grows on dead plant remains. It can infect living tissues when they are wounded, weakened by stresses or lacking some of the normal defences. Thus spores infect the styles of fruits like strawberries at flowering, but the fungus is suppressed by the chemical defences in the unripe fruit and remains latent until ripening. Then, as the fruit's sugar content rises and its resistance to infection falls, the fungus becomes active, breaks out and causes mould.

Destructive parasites

These invade living tissue, kill it rapidly and then disperse, leaving the dead remains to others – *Phytophthora* potato and tomato blight does this. These parasites usually have no active life outside their hosts because the available food there is dead plant and animal material and they are unable to compete with other microorganisms, called saprophytes, which are adapted to using this material. They therefore pass the time between attacks on susceptible host plants as dormant resting spores.

Biotrophic organisms

Examples are rust fungi (Uredinales) and viruses. They extract nutrients from living cells without killing them and may subtly alter the metabolism of the host to their own advantage. Viruses are very specialised in this respect; they have no ability to reproduce outside the cells of the plant host and rely on *vectors* ('carriers'), often insects, to move them from one host to another. These *biotrophs* cause enormous economic and aesthetic damage to plants, but seldom kill them.

Symptoms caused by the different types of pathogenic microorganisms

The disease which all these organisms cause, the 'harmful disturbance of normal function', leads to abnormal growth patterns in plants which are recognised as symptoms. These may appear suddenly: *Phytophthora* leaf blight of tomatoes, for example, frequently devastates outdoor tomatoes in the UK while the unfortunate owners are away for the late summer bank holiday. Others develop slowly as a gradual spread of disease weakens the plant; root infections in trees may cause a slow decline in vigour and dieback of branches over several years.

Weeds

In spite of gardeners' best efforts, weeds very successfully interfere with the cultivation and enjoyment of gardens. The astonishing resilience of weeds is no accident; their life strategies are finely tuned to exploit the opportunities gardeners inadvertently provide. Realistically, weeds cannot be eliminated, but they can be managed. The key to this is to understand their strengths and exploit their weak points.

Weeds are very diverse and examples can be found among the flowering and non-flowering plants, mosses and liverworts. Perennial and annual life cycles are also to be found amongst weeds, sometimes in the same species, as for example annual meadow grass (*Poa annua*). Woody and herbaceous plants can be weeds. Sometimes weeds are wild plants: tree seedlings, for example sycamore (*Acer pseudoplatanus*), tuberous perennials such as lesser celandine (*Ranunculus ficaria*) or bulbs like bluebells (*Hyacinthoides non-scripta*) fit this category. Sometimes domesticated ones such as potatoes (*Solanum tuberosum*) and horseradish (*Armoracia rusticana*) run amok. As well as dry land, ponds and other aquatic environments are home to weeds, such as duckweed (*Lemna* spp.) or algae.

Annual weeds

These are the great survivors of cultivated ground. Fat hen (*Chenopodium album*), groundsel (*Senecio vulgaris*), shepherd's purse (*Capsella bursa-pastoris*) and annual nettle (*Urtica urens*) are common garden examples. Although not deliberately domesticated, many annual weeds almost appear to be, as they are highly dependent on human activities for their success.

Massive fecundity characterises annual weeds: they produce huge numbers of seeds which are also very mobile, constituting a ubiquitous 'seed rain'. Groundsel for instance has hairy parachutes that carry their seeds some distance. Other short range spreading techniques include explosive seed capsules. The Himalayan balsam (*Impatiens glandulifera*), and hairy bittercress (*Cardamine hirsuta*) are examples of this. Persistence of seeds in the soil, as a 'seed bank' of viable seeds, is another feature of weeds.

Perennial weeds

These colonise cultivated ground to some extent, but in gardens they are a particular menace in plantings of perennial plants. Examples of these are couch grass (*Elymus* (formerly *Agropyron*) *repens*), ground elder (*Aegopodium podagraria*), horsetail (*Equisetum* spp.), brambles (*Rubus* spp.) and creeping buttercup (*Ranunculus repens*).

Perennial weeds produce seeds but also supplement them with successful vegetative propagation: for example, some *Oxalis* spp. and *Ranunculus* spp., (Plate 19) with numerous bulbils or tubers, and bindweeds (*Convolvulus* spp.) with their highly viable fragments of rhizome. Tree seedlings are a particular menace in gardens: birch (*Betula pendula*) and willow (*Salix* spp.) seeds are long range drifters, while ash (*Fraxinus excelsior*) and sycamore are a nuisance over shorter distances.

Weed ecology

Most plants show no weedy tendency and weedy species are a distinct minority in the plant world. Weeds have lifestyles astonishingly suited to existing in cultivated areas despite gardeners' best efforts at removing them. Some survive intense soil disturbance by cultivation, others exist easily in mown herbage or shelter in the competitive environment of borders and around the bases of woody plants. Their adaptability in exploiting unpromising environments is impressive. Some weeds, like groundsel and annual nettle, have short life cycles ideally suited to exploiting bare soil.

Garden plants are selected for food value, flavour, attractive flowers, structure and foliage rather than their ability to survive. Their survival potential has been further eroded by breeding for features opposed to survival: dwarf, leafless and variegated forms, improved palatability and removal of seed dormancy for example. These characters leave them ill-suited to compete with weeds. Interestingly there is now some enthusiasm among gardeners for creating self-sustaining communities of cultivated plants that leave no room for weeds.

Using biotechnology, it is possible to engineer crops genetically so that they possess features that weeds do not have – for example, resistance to broad-spectrum herbicides. In this situation weeds' fine tuning to adapt to the crop environment is no help. A herbicide-resistant crop can be treated at almost any time, leaving weeds heavily exposed to damage. There may be some drawbacks to this approach. Herbicide-tolerant weeds may be selected by overuse of a particular herbicide. Alternatively, the genes for herbicide resistance might perhaps cross into related weed species or the crop itself may become a weed in subsequent crops. However, proponents of genetically modified crops suggest means of ensuring that this does not happen, by limiting how frequently modified crops are grown. Also no instances so far have been reported of problems in weed control, despite the widespread use of genetically modified crops, particularly in North America. In spite of the current reluctance to accept such modified crops by the European public, gardeners are likely to benefit eventually from this technology.

Weed origins

Despite the ancient origins of gardening, and agriculture, weeds probably have not adapted to gardening, but actually possessed their weedy invasive characteristics before cultivation began. When human activities created the opportunities weeds were ready to move in. Thus, large numbers of easily dispersed seeds and rapid germination and growth are useful characteristics for plants adapted to colonising transient habitats such as fresh soil exposed by natural erosion or periodic changes in river levels. These same characters are equally useful when bare soil is created by farmers, and gardeners.

Many annual weeds are believed to have inhabited the harsh, brief periods when glaciers left soil uncovered. Cold weather does not greatly inhibit them and they grow in winter. Others have sneaked into cultivated areas as passengers in food crops or by other man-made agencies. Fat hen is believed to have arrived as a weed in crop seeds. Other weeds were originally introduced for their medicinal or food qualities or even for their ornamental value. Ground elder (*Aegopodium podagraria*) was thought to alleviate gout. Fennel (*Foeniculum vulgare*), horseradish and leaf beets (*Beta vulgaris*) are vegetables gone native and are now persistent weeds. Unwise garden introductions have given gardeners some invasive and persistent weeds, for example Japanese knotweed (*Fallopia japonica*), which was esteemed and widely planted by Victorian gardeners. Water-borne weeds are solely aquatic plants, but some land weeds are carried to new homes this way, for example giant hogweed (*Heracleum mantegazzianum*) which is a particular problem on riverbanks. Other ornamentals that have absconded include slender speedwell (*Veronica filiformis*), mind-your-own-business (*Soleirolia soleirolii*, Plate 20), and Canadian pondweed (*Elodea canadensis*).

Even trampled areas have their weeds; annual meadow grass is adapted to exploit this environment. Weeds also need a way of spreading themselves and often hitch a ride with man. Better seed cleaning technology has largely eliminated contaminated seeds, but a modern example of man-made weed spread is the prevalence of mosses, liverworts, willow herb (*Epilobium* spp.) and hairy bittercress (*Cardamine hirsuta*) in container-grown nursery stock.

Contaminated tools, vehicles and manures are also involved in spreading weeds. Not only is the manure of farmed beasts liable to contain seeds, but birds and rabbits have been found to consume seeds and void viable ones later. Rabbits for example feed on redshank (*Persicaria maculosa*) seeds.

Useful weeds

Weeds are not all bad and some make tasty salad, purslane (*Portulaca* spp.) and dandelions (*Taraxacum officinale*) for example, while others such as fat hen (*Chenopodium bonus-henricus*) can be used as vegetables and ramsoms (*Allium* spp.) can be used to season food. Fashions change and what were once weeds, poppies (*Papaver* spp.) and corn cockle (*Agrostemma githago*), are now desirable garden plants for wild-flower gardens. Annual meadow grass makes passable lawns, as long as it is abundantly fed and watered and frequently and closely mown. Weeds act as green manures, scavenging nutrients from the soil after crops have been gathered. These nutrients are then safe from washing deep into the soil where plants cannot use them and from where they may pollute watercourses and wells. Recycled through the compost heap or dug into the soil, the decaying weeds can release nutrients to nourish subsequent crops.

Weeds provide cover for helpful predatory insects, especially in winter. Mixed populations of plants may be less likely to suffer devastating disease and insect attacks than populations entirely composed of one species or cultivar. The evidence for this is slim in gardens, although there is good evidence that diseases develop more slowly in mixtures of crops, for example wheat, than in monocultures. Those who subscribe to this view suggest that a few weeds in gardens may be helpful as long as they do not set seed.

You can see that gardeners need to consider three points in order to manage pests, diseases and weeds effectively and sensibly. They need to know *whether* control is necessary, then *when* to take action and *what* treatment to use. Science can help with many of these decisions, as described in the following sections.

DECIDING WHETHER AN ORGANISM IS A PEST

When is a pest a pest?

Animals only become pests when they cause damage that reduces either the yield or the quality of plants. If they are not going to do so, they may not need to be controlled, but how can gardeners tell when control is worthwhile?

Firstly consider the part of the plant that is affected. If the pest attacks the 'important' part, i.e. the edible parts of fruit and vegetables or the attractive parts of ornamentals, lower numbers usually warrant control than when other parts are attacked. Winter moth (*Operophtera brumata* and some other species) caterpillars feed on the leaves of apples so large numbers are necessary to cause serious damage, whereas codling moth (*Cydia pomonella*) larvae feed inside apples so relatively small infestations warrant control. Some pests can be more or less damaging on different crops. For example appreciably more cabbage white (*Pieris* spp.) caterpillars are needed to damage swedes significantly than to spoil cabbages. Conversely small numbers of cabbage root fly (*Delia radicum*) larvae, that feed on the roots, will disfigure swedes, but the same number does less harm to cabbages.

If the numbers of pests on crops are plotted against the amount of damage they cause the graph nearly always takes the same basic 'S' shape. Low numbers cause little damage and they can even improve growth, comparable to pinching out the tops of plants. As the numbers increase they reach the 'damage threshold level' at which significant damage starts to occur (Fig. 11.1). For a time the amount of damage done is almost proportional to the numbers of the pest, although the slope of the line varies. When a small increase in the numbers of the pest results in a lot more damage the slope is steep. This happens when the pest attacks the 'economically important' part of the plant. Conversely when a large increase in the numbers of the pest is necessary to increase damage significantly the slope is shallow. Eventually so much damage is done that it does not matter how many more pests there are.

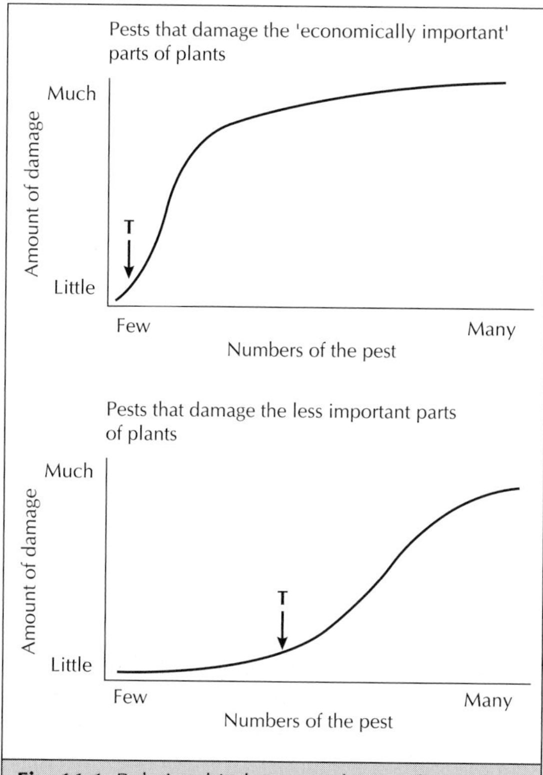

Fig. 11.1 Relationship between the numbers of a pest and the damage caused. T = threshold

Although plants are more susceptible to pests at some stages than at others, other factors also influence their tolerance to pests. Some or all of these should be considered when assessing the likelihood of damage or need for control.

Slugs will feed voraciously in the spring, but some plants are more susceptible to damage in the previous autumn. This is partly because the plants are smaller then, but it is also because the days are short so growth is slow and the plants cannot compensate for the damage. Flea beetles (*Phyllotreta* spp.) do more damage to brassica seedlings in hot dry summers than in cool wet ones, partly because the insects are more active, but also because the plants do not grow so well. Root-feeding nematodes similarly cause more damage in dry years than wet ones, particularly on light soils.

External factors may influence the reproduction rate and the behaviour of pests and their predators differently and this affects the success of biological control. Two-spotted spider mites (principally *Tetranychus urticae*) prefer hot, dry conditions, but the predatory mite used to control them (*Phytoseiulus persimilis*) prefers cooler, humid conditions. The pest therefore breeds more quickly than the predator in hot, dry weather, but this is compounded by the contrasting behaviour of the two mites in these conditions. When it is hot and dry the predators stay on the cooler and more humid lower leaves, but the spider mites congregate at the tops of plants where it is hot and dry, incidentally escaping their predators. Gardeners may not be able to alter the temperature or the day-length, certainly out doors, but in hot weather they can water and damp down, tipping the balance in favour of crops and predators, which dramatically improves control.

An understanding of the factors affecting the balance between plants, pests and any treatments enables the gardener to anticipate the degree of damage a pest may cause and to decide whether control measures are necessary and if so to apply the most appropriate treatment at the ideal time. Appreciating the importance of these interacting factors comes with experience, but having this 'feel' for plants is a logical process that can be learned or taught.

The importance of temperature

The development of most cold-blooded animals and plants (i.e. the rate at which they change from egg to adult, or from the vegetative to the flowering state) is governed mainly by temperature. Other factors affect the health, size, quality, fecundity and survival etc. of animals and plants, but it is the accumulated effect of the temperature over the season that is critical. Above a certain minimum temperature, called the 'base threshold', which varies with species, the amount an insect develops in a day is proportional to the average temperature that day. If the base threshold is 6°C, an insect develops three times as quickly at 9°C (three degrees above base threshold) as it will at 7°C (one degree above base threshold). Information about temperature is used by farmers and growers to predict when insects will emerge, so they can time the application of control measures accurately.

The emergence in the spring of cabbage root fly (*Delia radicum*) and carrot fly (*Psila rosae*) can be predicted by recording the daily temperature from February until April or May (although many carrot growers also put out sticky traps to catch the flies, to determine the size of the population). Temperature data can also be used to calculate when turnip moths (*Agrotis segetum*), the caterpillars of which are the familiar 'cutworms', are likely to start laying eggs and when these eggs will hatch.

Temperature is also responsible for synchronising the development of some insects and their host plants and this can be turned to gardeners' advantage. Cow parsley is an important source of the pollen which female cabbage root flies need before they can lay many eggs. Both have similar temperature requirements, so the flies always emerge as the plants begin to flower. This happens in different parts of the country and in early and late seasons alike. Noting when cow parsley starts to flower is a much easier way for gardeners to predict when cabbage root fly will emerge than recording the temperature continuously from February to April, but it is just as accurate.

Insect numbers increase geometrically from one generation to the next, so an insect laying 100 eggs has the potential to give rise to 100 in the first generation, 10 000 in the second and 1 000 000 in the third. The more generations the insect completes in a season, the larger the numbers, which is why pests that breed continuously, like aphids, spider mites and thrips are generally more numerous in hot seasons than in cool ones.

Insects like glasshouse whitefly (*Trialeurodes vaporariorum*) sometimes seem to appear overnight. The relatively low numbers present early in the season are easily missed, but if it is hot and reproduction is rapid, swarms of adults appear when the next generation hatches. This might seem like 'spontaneous generation' but is in reality a simple function of mathematics. This also illustrates why it is so important to check crops carefully to see what is happening to pests at an early stage.

When to control pests

Never underestimate the scientific worth of local knowledge and past experience when deciding whether and when pests need to be controlled. There are generally sound reasons for the local variations in the need or timing of pest control, even though these may not be fully understood. Pests are affected by local conditions that are unlikely to change overnight, although global warming appears to be having a widespread effect. A number of insects that live only in the south of England can now be found about 20 km further north than was the case ten years ago.

Think about the logic behind old sayings and traditions, temper them to suit changing situations and individual seasons if necessary, but do not simply dismiss them.

When is a disease a problem?

As with pests, the importance of a disease depends on many factors, notably the purpose for which the plant is grown, the time of disease attack in the plant's life cycle and the type of damage. Apple scab disfigures the fruit surface and minor attacks have no effect on edibility, but even the most minor cosmetic damage may reduce the value of a commercial crop to the extent that it is not worth selling, which can bankrupt the grower – an example of minor damage having a very serious effect. Conversely potato blight can attack the foliage late in the season, after the tubers are formed, and if they are harvested promptly before the fungus reaches them they will be healthy even though the foliage may be destroyed – in this case a serious disease has no commercial effect (though it is devastating if it does reach the tubers, or attacks early enough to prevent them forming).

Appearances of severity can be very deceptive:

cherry laurel is often very obviously attacked by powdery mildew (*Sphaerotheca pannosa*) which deforms the leaves, but has negligible effect on the growth of such a robust tree, whereas *Phytophthora* root rot can kill it with no visible presence above ground.

Identifying diseases

Symptoms can have more than one cause. For example, fungal infection of roots or conducting tissue can both cause wilting and so can insect damage to roots, or drought. The interpretation of symptoms is therefore not always straightforward. This is particularly the case with root disease, where the symptoms in the upper parts of the plants are seen by gardeners but the actual damage to the plant is done below ground and out of sight.

Yellowing leaves can be caused by virus or phytoplasma infection in the leaves, or a deficiency in the supply of nutrient arriving at the leaf. The deficiency might have a number of causes:

- a genuine lack of the vital nutrient in the soil
- an inappropriate soil chemistry that does not allow the nutrient to be taken up (for example, lime-induced iron deficiency)
- poor root development because soil factors such as a hard pan restrict root growth
- attack by nematodes or other soil animals
- root disease that damages normal root function
- stem disease that decreases nutrient flow to the leaves.

Because pathogens are microscopic they may not be detectable by direct observation, although some micro-fungi produce characteristic visible structures *en masse*: the dusty white spores of powdery mildews and the orange, brown or black pustules of rusts, for example. However, diagnosis often depends on indirect methods. The simplest is microscopic examination. In the field a hand lens is indispensable, but in a diagnostic laboratory, the first stop for a plant with a problem is usually the dissecting microscope. This low power microscope allows diagnosticians to identify some fungi directly. If a closer look is needed, pieces of suspect material can be picked off, teased out in a drop of stain and examined under the compound microscope at up to ×2000 magnification. With this power, fungi can usually be seen in enough detail to be identified if the right life stages are present. Bacteria may also be detected, but viruses cannot be seen with the light microscope.

Difficulties may arise even with fungi if they are not producing their characteristic spores – only very broad characterisation is possible from sterile fungal mycelium. In this case, it may be necessary to keep the specimen for a few days at high humidity, in the hope that the diagnostic spores will appear, or to culture the fungus onto a suitable culture medium, usually a simple jelly of agar with some nutrients and perhaps antibiotics to suppress bacteria. A pure culture obtained in this way can be subject to further tests, for example to induce spore production. Modern molecular methods of identification involve digesting chosen parts of the DNA with enzymes and running out the fragments in a gel matrix under an electric charge to generate characteristic banding patterns. These methods can usually only be used when a pure culture of the organism is available.

For bacteria, a culture is essential. The identification of bacteria is based on

- their colony characters on agar
- the chemistry of their cell walls
- the pigments they secrete into the medium and
- the enzymes they contain.

(Tests include the well-known cell wall staining reaction known as the Gram stain, which differentiates 'Gram positive' bacteria, whose cell walls retain a crystal violet-iodine stain, from 'Gram negative' ones which do not.) The enzyme tests involve placing some of the culture on a chosen substrate and observing changes that may occur. A species may be differentiated on its ability to reduce nitrate to nitrite, or whether it can digest the amino acid asparagine. Although the underlying chemistry may appear complex, the tests are often very simple; a positive result may simply mean the test medium becomes cloudy because of the presence of bacterial cells, or changes colour because their activity alters the acidity which changes the colour of a dye added to the substrate.

Diagnosis of virus infection is more difficult and time consuming. Viruses cannot be grown except in the cells of their host plants. If a virus is suspected on the basis of the symptoms in the plant, such as mottling, chlorosis or distortion, there are three ways that virologists will try initially to confirm its presence. Firstly some infected material will be ground up and examined directly with an electron microscope. In a light microscope, the maximum magnification is determined by the wavelength of visible light. An electron microscope uses a beam of electrons with a much shorter wavelength and therefore allows much greater magnification. The specimen must be placed in a vacuum and cannot be examined directly, the image being projected onto a screen. If the virus particles are numerous, they may be seen using this instrument, but the electron microscope can only be used to examine very small amounts of material and if the viruses are not numerous, or only in a few parts of the plants, they may be missed.

A second approach uses serology. Like the viruses which infect mammals, plant viruses cause mammal immune systems to react by the production of specific proteins called antibodies, which combine specifically with the protein components of the virus outer surface, and inactivate the virus by locking it up so that it cannot replicate. It is interesting to speculate why mammalian immune systems should be able to do this with organisms that are non-infective and apparently pose them no threat – in fact, mammals can produce antibodies to very many alien proteins, including those of plant pathogenic fungi. Thus, purified plant viruses injected into rabbits will cause the production of virus-specific antibodies, which can be extracted from a drop of the rabbit's blood; the blood product containing the antibodies is called an antiserum.

The antisera can be used in various ways, of which one of the most elegant is the DAS-ELISA test (double antibody sandwich enzyme-linked immunosorbent assay, Plate 21). A drop of antibody is first added to a small plastic well, where it adsorbs (sticks) to the plastic base. A separate batch is chemically joined to an enzyme, alkaline phosphatase, that reacts with a chosen chemical substrate to produce a distinctive colour. The ground-up plant material is then added to the well and the antibody traps the virus to the bottom. The well is then washed, but when the virus is present it sticks to the antibody and is retained. The enzyme-linked antibody is then added and sticks to the trapped virus. After another wash, the substrate is added. A colour change in the substrate indicates the presence of the enzyme, which is only present where the virus is too. Although chemically very complex to prepare, such tests are simple to carry out. The reaction is assessed by a light-sensitive plate reader which compares the intensity of the colour against a known standard.

Serology can only identify viruses to which antibodies have already been raised. If no antibodies are available, virologists may carry out plant inoculation tests. Some plants, such as *Chenopodium amaranticolor*, are easy to grow and react to virus infection in characteristic ways. Inoculations on these standard test plants may reveal symptoms that identify the virus, or at least confirm that a virus is present. The tests are particularly useful if virus levels in the infected material are low, and difficult to detect by the previous two methods, because a susceptible plant allows the virus to multiply to detectable levels.

Modern techniques of molecular biology have raised hopes of more rapid and specific tests for microorganisms. Some exist, but are not yet in widespread use. They are very valuable for precise identification, but to date they are not so useful for detection, because most rely on obtaining a DNA sample from the organism. To do this one usually needs a pure culture, which can be very time consuming to prepare. One of the main areas of research at present is to find techniques that can detect the pathogen DNA directly, among the diverse range of other microorganisms that occur in most infections.

If a microorganism is consistently associated with a disease, pathologists will want to prove its guilt. To do this, they must prove *Koch's Postulates*, put forward in the late nineteenth century by the German human pathologist Robert Koch. To satisfy his postulates, the pathologist must show that the microorganism is consistently associated with the problem, that it can be extracted and grown in a pure form, that it will cause the disease when reinoculated into the plant, and that it can then be reisolated. This can be a very time-consuming procedure and if the organism cannot be cultured, as is the case with viruses, it can be very difficult (though viruses can be purified). Diagnosticians do not confirm Koch's Postulates for routine cases, but they remain a very powerful tool to discriminate true pathogens from the many suspicious characters that are usually present on and around diseased parts of plants.

Weeds as pests

Competitive damage

Weeds restrict the yield of crops, by competing for moisture, nutrients and light. If uncontrolled they can restrict desirable plants' yield and development. Bigger weeds and numerous weeds clearly make greater demands upon the soil. However the bigger and better developed the crop, the more competitive it can be with its weeds, so vigorous crops can tolerate more weeds.

Research has shown that crops tolerate annual weeds to some extent, before yield is lost. Tolerance is often dependent on the growth stage of the crop. Seedling crops do not, initially, suffer much from seedling weeds, but vigorous weeds can 'overtake' the crop and damage it later. Ripening plants are not at much risk of weed damage at that late stage, but if left, the weeds are likely to shed seed, which will damage subsequent crops. In the middle period, there is a 'critical period' when the weeds have a very marked effect on the growth of crops. Removing weeds is essential at this time as the effort involved makes a big difference to the final results. Commercial growers can exploit knowledge of critical periods to save on weeding. Using pre-emergence weedkillers, they prevent weeds damaging crops at the vulnerable stages or even use selective weedkillers to eliminate weeds at the critical times. Weed seeds germinating in subsequent crops are of less concern in commercial crops, for they can always be controlled by weedkillers. However, weedkillers are not an option for gardeners in cropped areas, except for certain limited cases. Periods when weeds are not damaging cannot be exploited as the weed seed shed could cause later problems.

The number of weeds that causes a problem varies. Research with onions showed that 21 weeds per square metre caused no losses even if left for the entire length of the crop. More than 21 weeds reduced the yield, and the more weeds present the more quickly they adversely affected yield. Once more than 150 weeds per square metre were present, additional weeds caused little extra loss.

Weed control is often only undertaken when weed levels reach a point where the cost and disruption caused in removing them is less than the damage they will do if they remain. This use of a 'threshold level' is very common agricultural practice. Again this is less of an option for gardeners as they do not have many chemical means of removing weeds and non-chemical methods are often very damaging to the growing plants. Also the threshold depends on the type of weed and the length of time it has to develop before the crop is uprooted or mown. Even a very few weeds can cause serious loss in long term crops, if the weed has the capacity to get very large. Fat hen, for example, can grow in an *indeterminate* way; on the other hand groundsel has a limit to the size it can reach. This means that quite complex decisions would be required for each weed and crop combination. Where weedkillers are used and on a field scale this is practical, but it is excessively complicated on a garden scale.

Perennial plants typically suffer most from weed competition in the time following planting. At this time weeds can reduce the initial growth rates and the damage caused is not made up in subsequent years. Weed populations need not be high to inflict damage. Even a few weeds can be very harmful, where weeds are vigorous and the crop in question is unable to compete. Weed competition in the early years of perennial crops has subtle effects. By restricting nutrient and water supply overall growth is reduced, but plants can compensate by producing flower and fruit at the expense of foliage, so overall performance is not impaired. In fact, *cover crops* are sometimes used to produce a similar effect, as with grass grown on orchard floors for example. However, it is risky to allow weeds to substitute for this purpose as they can persist into subsequent crops or spread to adjacent areas. This suggests that eliminating every weed is sometimes more trouble than it is worth, although severe weed infestations should be dealt with.

Interference effects

Weeds interfere with cultivation. Annual meadow grass for example exerts little competitive effect, but removing it uproots seedlings, and annual nettles (*Urtica urens*) make weeding and harvesting unpleasant. Weeds also interfere with the harvesting and consumption of produce. An example of this is fragments of chickweed (*Stellaria media*) contaminating lettuce at harvest.

Weeds damage buildings, walls, paths and other man-made structures. They impair the function of garden equipment and features, blocking drains, clogging ponds and blunting mowers. Stinging, prickles, thistles, or even poison – giant hogweed (*Heracleum mantegazzianum*) sap, for example, sensitises the skin to sunlight – are other effects of weeds. Weeds may even poison other plants by secreting chemicals that inhibit garden plants. Happily few garden weeds parasitise other plants, although world wide, parasitic weeds are a menace to agriculture.

Weeds impair the visual quality of amenity areas. Lawns especially suffer from broad-leaved weeds, but grass weeds can occur. Annual meadow grass invades lawns where its pale colour, abundant seedheads and susceptibility to drought make it unwelcome. Gravel and paving are not enhanced by weeds colonising them. Borders of ornamental plants are rendered unattractive by an understorey of weeds. Even wild areas can be colonised by weeds which out-compete the desirable wild vegetation especially in fertile soil.

Not all weed damage is obvious. Weeds can act as hosts for harmful organisms. These can spread into adjacent susceptible crops, or the weeds act as 'green bridge' between susceptible crops, undoing much of the helpful effects of rotations. Cucumber-mosaic virus, a virus with a wide host range, but especially damaging to courgettes, persists in chickweed for example, while beet western yellows virus overwinters in many weeds, such as groundsel, ready to infect lettuces in the following season. Club-root disease caused by the fungus *Plasmodiophora brassicae* persists in cruciferous weeds such as shepherd's purse (*Capsella bursa-pastoris*). On the other hand, weeds are not usually important in the persistence of pests. Ornamental plants, especially vegetatively propagated ones, provide an alternative 'green bridge' for pests and diseases in many gardens. The importance of weed control to reduce 'carry over' problems is not likely to be great. In fact, crop residues such as carrots and potatoes left in the ground or not thoroughly composted, are likely to be more important than weeds as sources of pests and disease.

WHEN TO TAKE CONTROL MEASURES

Direct observation of pests

Monitor crops regularly, so that pest, disease and weed problems can be identified and controlled promptly. Learn to recognise the damage that pests cause, as this is often easier to see than the culprits themselves, which may be small or present on the undersides of leaves. Distorted leaves and growing points indicate the presence of aphids, leaf speckling is typical of spider mite and leafhopper damage, and black sooty moulds are often present in the honeydew produced by sucking insects, such as whitefly, aphids and scale insects.

Try to get a general impression of the health of the whole batch first before examining individual plants. The experienced eye will quickly spot an atypical plant, which can then be examined more closely (ideally with a $\times 10$ hand lens).

Pests prefer certain species and varieties of plants to others, so identify those that are particularly susceptible and concentrate on these 'indicator' plants when checking for pests. Even consider applying additional control measures (chemical or biological) to them.

Trapping and monitoring

Sticky traps are useful for monitoring some pests, particularly in glasshouses, because they catch insects before they reach levels that can realistically be detected just by looking. Yellow traps catch aphids, whitefly and thrips, but blue ones catch mainly thrips. Unfortunately yellow traps also catch beneficial insects, so use them sparingly in a glasshouse where biological control is being used (see below). Examine the traps regularly, although it is not necessary to make detailed counts. Instead note whether the numbers being caught are changing significantly. Change sticky traps regularly whilst it is still possible to assess numbers easily, generally every two to four weeks.

Traps are also useful for monitoring nocturnal pests. The numbers of slugs trapped under a flat stone or a tile baited with a teaspoonful of slug pellets can be used to decide if treatment is justified. The technique can underestimate the numbers present, because if it is too cold or dry slugs stay under ground so are not trapped. Catches therefore only reflect the numbers of slugs active on the surface of the soil, but as these are the only ones likely to be killed by slug pellets the exercise is still justified.

Pheromones

Pheromones are chemical 'messengers' that insects produce to communicate with each other in different ways. Some female insects produce volatile sex pheromones to attract males. These chemicals, although highly specific so that they only attract males of the same species, are easily synthesised artificially and are used in two ways to control insects.

Sticky traps, baited with artificial sex pheromone to attract and catch male moths, are used to time the application of conventional insecticides correctly to control pests such as codling moth (*Cydia pomonella*), carnation tortrix (*Cacoecimorpha pronubana*) and cutworms (Plate 22). Commercial growers use them extensively, but some are also available to gardeners. Sex pheromones, liberated in large quantities, have been used to confuse male moths to prevent mating. This technique is only practical where a crop is grown extensively in an area, as happens with cotton in the Sudan, where the technique has been used successfully to prevent caterpillar infestations.

Aphid parasites detect their hosts partly by scent and this pheromone can also be synthesised. Research is in progress to see if these chemicals can be applied to crops to attract parasites from hedges and surrounding areas earlier in the year, so they are present when the aphids arrive. If successful the technique has considerable scope for use in gardens.

Weather

The weather (past and present) influences many pests, both directly and indirectly, and it affects the amount of damage they cause. Average temperature and rainfall are the most important factors that affect pests, but wind strength and direction can be critical, as can extremes of temperature.

Pests are more active when it is warm, but slugs prefer it warm and wet, while others such as aphids, spider mites and thrips, prefer it warm and dry. Many soil-dwelling pests are largely unaffected by moisture themselves, although if they feed on roots, they may well do more damage in dry conditions.

Previous weather conditions may have a greater impact on pests than those that prevail when the damage is done. Leatherjackets, the larvae of crane flies, *Tipula* spp. (daddy long legs) cause most damage in the spring, but whether this occurs depends largely on the amount of rain in the previous autumn when the eggs are laid. The eggs and the young larvae do not survive well in dry soil, so leatherjacket damage is normally only serious after a wet autumn.

Two-spotted spider mite is mainly a pest of glasshouse plants, although in hot summers it damages outdoor crops such as strawberries, raspberries and runner beans. It overwinters as specially adapted adult mites, in cracks and crannies in glasshouses and in the stakes and bamboo canes used to support raspberries and runner beans. Vast numbers of mites may go into hibernation after a hot summer and if this happens expect early and serious attacks the following spring. A simple ploy to avoid damage to runner beans is to have two lots of canes and to use these in alternate years, because the mites will not survive without food for two seasons.

Most native pests survive just as well in cold winters as they do in mild ones, but aphids are an exception. Some species over-winter both as eggs and as wingless aphids in mild years, but only the eggs survive cold winters. Attacks of cabbage aphid (*Brevicoryne brassicae*) and the peach-potato aphid (*Myzus persicae*) in the spring following a mild winter are usually earlier and more severe than they are after a cold one.

Many small insects, such as aphids and thrips, are carried a long way by the wind. These insects will not take off from plants if the wind speed is more than 5–6 km per hour, but once airborne they have little control over where they are blown. Larger, more powerful insects can also be blown long distances. Serious attacks in mid-summer of caterpillars of the diamond-back (*Plutella xylostella*) and the silver-Y moth (*Autographa gamma*) generally occur only after a prolonged period of south easterly wind earlier in the year, which blows adults over from the Continent.

Diseases

The fungi and bacteria that cause disease in annual plants, or the deciduous parts of perennial plants, must survive when there is no suitable host material around, in the winter in temperate climates and in the dry seasons in the tropics. The host therefore imposes on its pathogens the need for a survival strategy. Just as plants have seeds, most fungi and bacteria have resting spores for this purpose (Plate 23). This reduction of the organism to a single, tough and microscopic speck in turn imposes on the pathogen a need to sense the right time to attack the host. The more specialised the pathogen, the more critical this is, and some wonderful examples of co-dependence exist. The camellia-petal-blight fungus, *Ciborinia camelliae*, newly arrived in UK, infects the flowers of camellias, grows in them and forms tough resting structures called sclerotia in the dead, fallen flowers. These stay dormant on the ground until next spring when they germinate at precisely the moment when the flowers are opening, release infective spores and reinfect. If the fungus missed the flower, only there for a few weeks every year, it could not survive.

The most specialised adaptations occur among those pathogens that have committed themselves to complete dependence on their hosts – the 'biotrophs' that feed on the tissues of the living host and have little or no ability to live apart from it. This is not so restrictive as it appears, because such sophisticated parasites, exemplified by the rusts and powdery mildews, must not damage the host so much that it dies, or they die with it. The most destructive pathogens, in fact, are often those that are able to live independently of the host, often on its decaying tissues, but also on other substrates. These include honey fungus (*Armillaria* spp.), the most destructive pathogen in gardens. Honey fungus invades the roots of almost any woody perennial, kills them and then feeds on the decaying wood. There is no advantage to keeping the host alive and this voracious life cycle makes the fungus particularly difficult to control, because there is no weak point to attack – it is never dormant, rarely reduced to a single spore survival stage and usually underground, out of sight.

Bacteria are simpler, but many of the same principles apply. Many bacteria form resting spores to survive when the host is not present, but they have a less complex cycle: resting spores germinate to form bacterial cells and these go back to resting spores when they run out of food. Those that do not form resting spores either survive at low levels in perennial host tissues, or can compete in the outside world, usually the soil, in their normal state.

Viruses cannot grow except in the nuclei of their host plants, so they survive outside the host only in their vectors, or passively in the environment. This appears to make them very amenable to control by attacking their vectors (usually insects or nematodes); nonetheless, viruses are very successful organisms. In particular, they are extremely infectious and this makes control by exterminating vectors much more difficult than it might appear. Even eliminating 99% of the vectors, which is difficult even with the most effective chemical insecticides, will not eliminate a virus which has only to survive in a tiny fraction of the vector population to break out again, relying on the unerring ability of its vector to find a host plant – which that vector must do, if it is to survive.

Given that pathogens have a typical life cycle of multiplication, reproduction, rest and multiplication again, much advantage can be gained in the disease war by attacking at the most vulnerable stage. Accurate knowledge of the life cycle is crucial and it is surprising that the life cycles of some pathogens have been very hard to unravel. In particular, where some temperate climate fungal pathogens go in winter is still not completely known: the peach-leaf-curl fungus *Taphrina deformans* is thought to spend the winter hidden in peach tree buds, waiting for the spring to invade the newly formed leaves, but pathologists are still not completely confident. In general, the ability of fungi to survive at low levels in their plant hosts is very important for the life cycle, even though this survival may be in a quiescent state that is not easily detectable. In recent years there has been renewed interest in this area with the demonstration that plants host a wide range of previously unsuspected 'endophytic' fungi, embedded in the tissues with all sorts of unexpected side effects. One that has been known for a long time is the fungal pathogen *Cryptostroma corticale* that causes sooty-bark disease of sycamore. Undetected while the tree is in good health, it breaks out during times of stress, especially very hot summer weather, to kill the bark and produce masses of black dispersal spores.

Correct diagnosis is obviously crucial to control, because it may then be possible to identify the weak point in a pathogen's life cycle at which it can be most easily attacked. This might be the survival stage, an attractive target because the pathogen is very low in numbers, but on the other hand the resting spores may be very tough and difficult to kill. If the pathogen infects via wounds, e.g. in the case of silver leaf (*Chondrostereum purpureum*), the gardener can protect wounds chemically to prevent the spores from colonising the wood, or prune during summer months when spores are not produced. If damage to the host depends on high levels of infection to a particular growth stage, good protection can be achieved by spraying to protect during this stage. A timely fungicide application to outdoor tomatoes will not prevent infection by blight (*Phytophthora infestans*) but it may allow a crop of tomatoes to be harvested from an infected plant – even a week's delay in disease development may be enough to save the crop.

Where disease control depends on fungicides, as it often does in commercial agriculture, timing of applications is crucial if the control is to be effective: too early and the chemical inhibition is diluted by the growth of the plant, too late and the pathogen has too strong a hold. Disease prediction to reduce pesticide wastage and maximise effectiveness is therefore attractive and much effort has gone into developing predictive 'models', mathematical equations combining the key data on climate and pathogen availability. The concept is easy enough to understand: in a crop such as a potato field or an apple orchard, spore traps continuously monitor airborne spores and climate sensors monitor temperature, rainfall, humidity and leaf wetness. The data are continuously analysed using a predetermined model that has shown that when an amount of fungal spores (x) is present during a period of (y) temperature and (z) leaf wetness, a dangerous level of disease will develop. Just before the danger level is reached, the farmers are warned to spray. Commercial systems do exist and some are effective, but they are not as widely used as could be wished, because local variations in climate (and imperfections in the models) make them expensive to implement and not always reliable enough. Gardeners can benefit from warnings of potato blight and apple scab, if they are in the right area for an accurate forecast, but of course, someone has to pay for this advice, which is expensive to produce. Also, gardeners may not have a useful range of control measures available to them. For example, they do not have access to the most effective chemical fungicides for potato blight that are available to professional growers, and control will therefore be more difficult where it has to depend on less effective 'amateur' products.

When to control weeds

Most gardeners regard weeds as unacceptable and try to keep their numbers very low. Partly this is because they are thought unsightly, but also because experience has shown that even low levels of weeds can quickly result in heavy infestations. This attitude can have a high price because the removal of weeds is often a very lengthy and labour intensive process. The disruption to soil by cultivating and pulling weeds, the compaction of soil by gardeners treading on the ground, and the effects of weedkillers all reduce the growth of the crop. Sometimes the damage caused by weeding is greater than that caused by the weeds, but gardeners often think this loss is worthwhile, if it reduces future infestation.

Weeds are highly adaptable and even low populations can soon lead to massive plants shedding huge numbers of seeds. Partial destruction, by hoeing in weather that allows them to re-root, for example, can be no more effective than leaving them alone. When time and resources are limited, weed control decisions have to be given different priorities.

Weed control tactics depend on forecasting the likely consequences of weed infestations. In practice it is very difficult to know what bare ground or apparently clear plantations have in store. Research into the seed content of soil has shown that it is very difficult to predict subsequent weed crops. Weeds seeds are hard to find and count. If found, they may be so dormant that they may not germinate for years. Trying to germinate weeds in a soil sample is no better, as weed seeds germinate in flushes through the year.

Producing many seeds allows even a few survivors to maintain populations. The vast number of seeds lets weeds 'search' out fresh habitats. Annual meadow grass may produce as many as 13 000 seeds per plant and groundsel as many as 38 000. The downside for weeds is that having many and highly mobile seeds means the seeds have to be small. A small seed has few reserves of nutrients to establish in adverse conditions and is unable to out-compete surrounding vegetation. To cope with this weeds have evolved mechanisms that allow seeds to 'sense' their surrounding and germinate when conditions are optimal for survival.

Light is required for many seeds to germinate. Annual nettles and annual meadow grass are garden examples; buried seeds don't germinate until they reach the surface. On the other hand, seeds germinating in light but in the presence of other plants are unlikely to survive. Where sunlight passes through vegetation red and blue light is absorbed as part of photosynthesis and green light is reflected. The light that is transmitted is rich in far-red light. Germination is inhibited by far-red light in weeds such as chickweed (*Stellaria media*) that cannot cope with shade. Temperature fluctuations are greatest near the soil surface. Weed seeds can also sense these, and germination may be triggered even in the absence of light.

Persistence of dormant seeds in the soil, as a 'seed bank' is a common weed attribute. Weed seeds remain viable in the soil for astonishing periods. Field speedwell (*Veronica agrestis*) has been recorded as surviving at least 30 years and common poppies (*Papaver rhoeo*) can last 26 years or more. On the other hand, some weed seeds are short lived. Groundsel for instance persists for about three years. Some weeds germinate soon after being shed while others may germinate at a low frequency over several years, when conditions favour rapid growth and development. These conditions are often required to occur in complex combinations. Disturbed soil and warmth triggers annual nettles; alternating temperatures, a preceding period of chilling at low temperatures, nitrates in the soil, moisture and light are required for annual meadow grass. Such fine tuning greatly enhances weeds' persistence.

Other mechanisms include different types of seed being produced. Fat hen produces many black and fewer brown seeds. Brown seeds are thin walled and germinate immediately after separating from the parent plant late in the season. Those few that survive the winter form large plants, shedding five times the normal numbers of seeds. The black seeds germinate over a long period beginning the following spring and are less likely to die out, but less efficient at exploiting a temporary opportunity.

Dormancy is conferred by other means. Hard seed coats, which require rotting or scratching away before germination, inhibitory substances that need to be leached out of the seed before germination can proceed, and slowly developing embryos are examples of these.

Seed bank control is an essential gardening practice. Leaving weed seed on the surface promotes early germination, leaving seedlings vulnerable to weedkillers and cultivation. Buried seed loses viability over time and seed 'turnover' can be rapid.

Perennial weeds have 'bud banks': these are the buds on perennating structures like rhizomes that perform a similar function to seeds and also need controlling. Buds are often less resistant to elimination measures than seeds, but exceptions to this include bulbous weeds, such as *Oxalis*, *Allium* and lesser celandine, for example which have robust small bulbs. Forecasting perennial weeds is often easier than annual weeds. They persist from year to year, often in 'refuges' where they are safe from attack, for example beneath or amongst buildings, paving, permanent plantings or deep in the soil.

HOW TO CONTROL

Integrated pest management

Pest control is often seen as a choice between natural, 'organic' means or the use of man-made pesticides. Both approaches have advantages and disadvantages and the best of both worlds can be achieved by using an integrated pest and disease management (IPM) approach in gardens, following from several decades of development in agricultural systems. Originally developed as a reaction against excessive and unnecessary routine use of chemical pesticides, IPM emphasises that all available techniques should be used to best advantage and chemical pesticides should only be used when needed, when pest monitoring techniques indicate genuinely threatening pest levels and when no effective alternative exists. More recently, the emphasis has shifted to integrated crop or farm management, to take account of the fact that pests and diseases do not exist in isolation from the crops and farm environment.

Application of IPM techniques has produced some startling reductions or even eliminations of pesticide use, but many farmers still spray routinely for many problems. Gardeners have probably been more ingenious practitioners than farmers, being forced to rely more on non-chemical means of control by the limited range of pesticides available to them. IPM will always make best use of plant resistance, cultural techniques and biological controls before resorting to chemical pesticides, but it is emphatically not a synonym for 'organic' gardening; growers who adopt IPM will use chemical pesticides as one weapon among many, not as the only one. The real key to successful IPM for farmers and gardeners is accurate and complete information about the target pest.

Controlling pests

Chemical controls

Chemical pesticides can give rapid control of pest problems. They must be appropriate for the problem and require thorough application at the right time. Most garden insecticides have a broad spectrum of activity, dealing with a variety of pests but also killing non-target insects, including bees, predators and parasites. Pests which survive treatment breed more prolifically once their natural enemies have been eliminated. Frequent use of the same type of chemical can result in pesticide-resistant pests developing, particularly glasshouse pests such as whitefly, red spider mite and some aphids.

Beneficial insects and mites are susceptible to most insecticides. Pirimicarb is an exception and is selective for aphids, leaving other pests and most beneficial insects unharmed (from 2001, this has been available to amateur gardeners only in combination with fungicides for a limited range of applications). Insecticidal soaps, rape oil, derris and pyrethrum have very short persistence and this limits harm to non-target insects. Use of these pesticides, where appropriate, helps to conserve predators and parasites of pests.

Predators and parasites

Predators and parasites sometimes keep pests at a low level and make insecticides unnecessary. Unfortunately, in gardens they are often not present sufficiently early in spring and early summer to prevent damaging infestations from developing. Although most pests have at least one natural enemy, many major pests lack predators or parasites that give effective control. The use of natural enemies or biological controls has had most success in greenhouses. Glasshouse whitefly, two-spotted spider mite and mealybugs are not native to Britain and good control has been achieved by using predators and parasites from the pests' countries of origin. Other predators and parasites are commercially available for controlling aphids, thrips, fungus gnats and some glasshouse scale insects. Pathogenic nematodes, which transmit bacterial diseases fatal to insects and slugs, are available for controlling pests such as slugs and vine weevil grubs. A bacterium, *Bacillus thuringiensis*, produces a highly specific toxin and when applied like a conventional spray it causes a fatal disease in moth or butterfly caterpillars that eat treated plants.

Plant breeding

Most popular garden plants are available in a range of cultivars and these may show variation in susceptibility to pests. Growing cultivars that are less attractive to pests or better able to tolerate damage is part of integrated pest control. Some plants have been bred for resistance to particular pests or diseases; they have a high degree of resistance making them immune to the problem. New techniques in plant breeding, such as genetic engineering, open up the possibility of artificially inserting genes for resistance, sometimes from unrelated organisms. The gene controlling production of the insecticidal crystal in *Bacillus thuringiensis* has been introduced into some plants, such as tobacco, maize and potato, to produce cultivars resistant to caterpillars. No genetically modified cultivars are currently available in Britain but this method of plant breeding will develop and may provide pest and/or disease-resistant plants for garden use.

Barrier methods

Rabbits can be excluded from gardens or flower beds with wire netting. Insects can also be kept away by covering plants with a finely woven material such as horticultural fleece (Plate 24). Light and rain can reach the plants but access is denied to all but the smallest of insects. This technique is mainly used in vegetable gardens to protect brassicas from cabbage root fly, cabbage butterflies and moth, or on carrots to exclude carrot fly. Crop rotation must be practised as many pests overwinter as pupae in the soil. Emerging adults will be trapped under the fleece with their host plants if the growing positions have not been changed.

Companion planting

Certain plant combinations are said to be beneficial partnerships because volatile chemicals produced by one plant makes the other less susceptible to attack by a pest or disease. Frequently quoted examples are onions and carrots to repel mutually onion fly and carrot fly; African marigold to repel glasshouse whitefly from tomatoes; garlic to protect plants against diseases such as peach-leaf-curl and rose blackspot. These claims are based on anecdotal evidence and scientific trials have failed to show worthwhile protection. There is, however, some evidence of how plants may interact. Many plants produce a chemical called methyl jasmonate, which mediates the biochemical mechanisms that lead to the synthesis of compounds conferring resistance to pests and diseases: for example proteinase inhibitors which reduce feeding by pests such as caterpillars and leaf beetles, and phenolic compounds which suppress fungal diseases such as rusts. Some plants, such as *Artemisia* spp., produce so much methyl jasmonate, which is volatile, that they may release it into the air and stimulate increased resistance in neighbouring plants. The level of control shown so far is inadequate to prevent damage but there may be some truth in the notion of companion planting.

Using biological controls

Using natural enemies to control pests is not new although most examples of successful application occur in the twentieth century. World-wide there have been some spectacular successes in the control of pests and weeds. These successes have mostly been against pests and weeds introduced accidentally into countries where they have not previously occurred. In these new countries there is often an absence of natural enemies capable of giving control, so the introduced species reproduce unchecked and become a major problem.

A well-established industry has developed since the 1960s for supplying predators, parasites and pathogenic organisms for controlling pests; these are mainly for glasshouse pests but some can also be used in outdoor situations (Plate 25). Glasshouse whitefly and glasshouse red spider mite had by the 1970s become resistant to a wide range of formerly effective pesticides. When used correctly biological controls give control that is as good as, or better than, chemicals at a competitive cost.

The biological controls given in Table 11.1 are available by mail order to amateur gardeners. The addresses of suppliers can be found in advertisements in gardening magazines, usually in the small-ads on the back pages.

To be effective, biological control agents must be used in an appropriate manner. They are pest specific so it is essential to identify the pest correctly. It must also be present when the biological control is introduced otherwise there is nothing for the predator to feed on or the parasite to breed in. They are not miracle workers and, unlike pesticides, cannot give an immediate reduction in

Table 11.1 Biological controls available by mail order to amateur gardeners.

Pest	Biological control
Aphids	– a fly larva predator, *Aphidoletes aphidimyza* – parasitic wasps, *Aphidius* or *Praon* species – lacewing larvae, *Chrysoperla carnea*
Two-spotted spider mite	– a predatory mite, *Phytoseiulus persimilis**
Glasshouse whitefly	– a parasitic wasp, *Encarsia formosa* – a ladybird predator, *Delphastus catalinae*
Mealybugs	– a ladybird predator, *Cryptolaemus montrouzieri*
Soft scale	– a parasitic wasp, *Metaphycus helvolae*
Fungus gnats/sciarid flies	– a predatory mite, *Hypoaspis miles*
Western flower thrips	– a predatory mite, *Amblyseius* species
Moth and butterfly caterpillars	– a bacterium, *Bacillus thuringiensis**
Slugs	– a pathogenic nematode, *Phasmarhabditis hermaphrodita**
Vine weevil grubs	– a pathogenic nematode, *Heterorhabditis megidis**

All can be used in greenhouses, conservatories or on houseplants but those marked* can also be used out of doors.

pest numbers. Introduce predators and parasites before heavy infestations develop in order to avoid damage occurring before the biological controls achieve control. All predators and parasites need warm sunny conditions if they are to be active and outbreed the pests. Their period of use is generally April to October when daytime temperatures are at or above 21°C (70°F). Most insecticides kill predators and parasites and permethrin and bifenthrin remain active for eight to ten weeks after their last use. If pesticides need to be used before introducing biological controls, short-persistence insecticidal soaps, rape oil, derris or pyrethrum are the preferred choices.

Pathogenic nematodes are watered into the soil to control slugs or vine weevil grubs. The nematodes enter the bodies of their host animal and release bacteria which cause septicaemia. Use the slug nematode when soil temperatures are above 5°C (41°F); it is best in the spring when seedlings and soft emergent shoots need protecting. Vine weevil nematode needs higher soil temperatures (14–21°C [57–70°F]), and is best applied in late summer before the grubs are large enough to cause serious root damage. The bacterium for caterpillars is sprayed directly on to the plant foliage when caterpillars are seen. Caterpillars ingest the bacterial spores and toxin as they feed and rapidly become fatally ill. All biological controls are harmless to humans and animals other than the intended pest targets.

Safe use of pesticides

Before any insecticide, fungicide or herbicide is approved for garden use it undergoes extensive testing and trialling to assess its effectiveness, toxicity to humans and wildlife, persistence and environmental effects. Careful attention to the manufacturers' instructions, especially the dilution rate and types of plants on which it can be used, will enable safe use to be made of these pesticides.

Regulations made under the Food and Environment Protection Act 1985 give legal force to the manufacturers' instructions, making it illegal to use the wrong dilution rate or to use a product for purposes other than those stated. Prosecution could follow if careless spraying of plants in flower results in a beekeeper losing his bees. Only products approved by the government can be used as pesticides; home-made pesticides brewed from rhubarb, cigarette butts or even washing-up liquid are now outside the law. Amateur gardeners cannot purchase, store or use chemicals marketed for professional growers. Chemicals must be stored in their original containers, so it is illegal to split a batch of pesticide among several gardeners.

Controlling diseases

Disease can seldom be eliminated and gardeners, like commercial growers, must learn to manage disease to keep levels within acceptable limits. Commercial growers manage disease to maximise profits and because public tolerance of defects in their fruit and vegetables is very low, growers aim for very low disease levels. Their reliance on chemical pesticides is often very heavy. Gardeners have different objectives and fewer pesticides, but more time, at their disposal.

A gardener's first line of defence against disease is quarantine. Much damage to plants is prevented on a world scale by national and international legislation to prevent the spread of plant disease: in Europe, the European Plant Protection Organisation (EPPO) ensures that national authorities are kept aware of the threat from exotic pathogens and publishes lists of those that are currently completely absent (A1 organisms), or of limited distribution (A2 organisms). Despite these international efforts, the spread of pests and diseases is inexorable and is increasing all the time with the massive increase in air travel and the abolition of trade restrictions (though plant quarantine can be an effective political weapon to protect national agricultural interests from unwelcome overseas competition).

Relatively recent arrivals in UK have included chrysanthemum white rust *Puccinia horiana* in 1963. (This is now established and attempts to eradicate it have been abandoned), the new and virulent pathogen *Ophiostoma novo-ulmi* causing Dutch-elm disease in the 1970s and Camellia petal blight (*Ciborinia camelliae*) in the 1990s. Still absent, but ever-threatening, are sweet chestnut blight (*Cryphonectria parasitica*) which is present in France, and oak wilt (*Ceratocystis fagacearum*) and its beetle vector, which have not yet crossed the Atlantic from North America.

Within countries, legislation may exist to limit spread of pathogens. For example fireblight (*Erwinia amylovora*) must be absent from commercial sources of susceptible plants in the UK. Some diseases must be compulsorily notified if found, such as wart disease of potatoes (*Synchytrium endobioticum*), a serious disease which is well controlled by a combination of legislation and resistant varieties.

Sometimes, diseases can be avoided completely by excluding them. The use of clean seed – and all reputable commercial seed sources are produced to very high standards of purity – will ensure almost complete freedom from such diseases as bacterial blight of beans (*Pseudomonas syringae* pv. *phaseolicola*). Attacking seed-borne diseases by treating seed with pesticides is also an extremely economical and efficient use of the chemical, with minimal hazard to the environment. Sometimes, it is impractical to eliminate disease completely from seed but good hygiene can still reduce levels effectively, as in the case of powdery scab on potatoes, which is controlled by ensuring that seed potatoes are infested only at low levels which are known not to result in significant disease in the following crop.

Most common garden diseases must survive, usually over winter, when the host is not present and good hygiene is therefore very important. By removing the sources of inoculum and ensuring that there is a clean start in the spring, onset of the disease can be delayed. This may even be true for wind-blown pathogens, which will inevitably reinvade eventually from over the neighbour's fence, because delaying the arrival of the pathogen by even a few weeks can make all the difference to the health of the crop. For example, blight on potatoes and tomatoes (*Phytophthora infestans*) can survive over winter in infected tubers or fruit left on the ground: careful removal will ensure a disease-free start. Spores will inevitably blow in later, but rapid application of a protectant fungicide can suppress and delay the infection, and a delay of one or two weeks may allow a crop to be harvested before the disease reaches tubers or fruit. Loss of leaves at this late stage is unimportant, but if the disease develops unchecked, the entire crop may be destroyed.

Most plants are immune to most diseases and susceptibility to a particular pathogen is the exception. When plants are attacked, wild relatives are usually less affected than garden cultivars: the inherent resistance of the wild ancestors has often been sacrificed (or simply allowed to dissipate unnoticed) during the relentless selection for yield, flower size or sweetness. Breeders are nevertheless very aware of the value of resistance, not least because when it is effective it offers disease control to the grower for only the price of the seed. Resistance is conferred by biochemical mechanisms and is under control of many genes, some of which confer high levels and some less. Good for gardeners, but hard for breeders to manipulate, is multigene resistance, where many genes each contribute a small amount to the overall resistance of the plant. This may not confer a very high overall resistance, but is more stable. Spectacular breakdowns of resistance can occur when breeders place too much emphasis on too few genes, or even only one, the so called 'major gene' resistance, for example in the case of rust of wheat and blight of potatoes. In these cases resistance is very good, because the breeders have selected one gene which has a major effect, but the pathogens rapidly develop genes in turn which overcome this host gene, and the host is then susceptible.

Resistant cultivars should always be considered by gardeners and are vital for those who do not wish to use chemical control (Plate 26). However, best-loved varieties of many plants are often disease-susceptible: for example the popular apple 'Cox's Orange Pippin' is very susceptible to apple scab (*Venturia inaequalis*). In some cases good levels of resistance do not exist; all varieties of strawberry are more or less prone to greymould (*Botrytis cinerea*) and research efforts are now underway to improve resistance by genetically engineering resistance genes into strawberries from other sources.

The science of biological control is much less advanced for diseases than insect pests. Some products have come onto the commercial market which rely on antagonistic organisms (bacteria or fungi) to control disease, but most are for commercial crops and they have yet to make as much impact as the biological control systems for insect pests in glasshouses. Product efficacy and stability have been difficult problems to overcome.

More promising for gardeners are systems of soil management that promote high levels of these naturally-occurring antagonists. Pathogen spores remain dormant in the soil when susceptible hosts are not available. The soil also contains bacteria, fungi and protozoa which live on soil organic matter, including pathogen spores. As a general principle, higher levels of microbial activity result in more rapid removal of pathogen spores. Even more generally, it makes sense that organic manure and compost will provide a better living for antagonists than inorganic fertilisers, and thus suppress diseases. However, gardeners who place heavy reliance on organic soil amendments should be aware that attempts at disease control using such materials have not always been successful and successes have not always been repeatable. Composted organic waste that has been heated properly will be free of most disease spores, but poorly prepared compost may simply add large amounts of disease from the previous season and generally, gardeners should avoid composting diseased plant material. The heating process sterilises compost and microorganisms recolonise it as it cools. This is an uncontrolled process and may explain why the effects of compost on disease suppression are unpredictable. Gardeners should also be aware that products sold for soil improvement may not be subject to such rigorous regulatory requirements for efficacy data as products sold as pesticides.

The range of chemical fungicides available to gardeners is very limited compared to the total range of 'professional' products. The list of amateur fungicides is growing steadily shorter as public pressure for ever-higher, and more costly, standards of safety renders the amateur pesticide market increasingly uneconomic. In addition to loss of active ingredients from the shelves, most products are now subject to precise and statutory labelling of disease targets and must therefore be used only for labelled purposes – see the above discussion of safe use of pesticides. Thus gardeners may not use carbendazim formulations for control of paeony wilt (*Botrytis paeoniae*) because none are labelled, even though this compound is effective and was formerly recommended. Nonetheless, where fungicides are available they can be very effective. Systemic compounds such as myclobutanil, carbendazim and penconazole can penetrate into the plant and give protection from within, protected from rain. They may also have some curative effects on disease already present. Protectants such as copper formulations and mancozeb do not penetrate the tissues and are not curative. They are prone to wash off in rain, but can still be very effective, particularly in the case of Bordeaux mixture which actually sticks very well to leaves if allowed to dry before rain arrives. Fungi are less likely to become resistant to the metallic protectants, whereas they have found it quite easy to adapt their biochemistry to coping with some of the systemics; carbendazim-resistant strains of many pathogens exist.

Where chemical control apparently fails, there are three likely reasons. The pathogen may have developed resistance and another pesticide should be tried (if a suitable one is available). The application may have been made too late, and gardeners should remember that disease often develops invisibly within a plant before symptoms appear. Finally, many garden sprayers are relatively crude machines and gardeners need to exert considerable skill in ensuring an even coverage of chemical, particularly underneath the leaves. Most manufacturers exhort users to spray to 'run-off', but it is easy to see that this results in less material being retained on the leaf than spraying to just BEFORE run-off. At the moment of run-off, much of the accumulated pesticide runs off onto the ground. The ideal application is a dense, even distribution of discrete droplets over all the parts that need protection.

How to control weeds

Weed research has to be interpreted with caution for garden use, because it has almost always been done with commercial growers in mind. Gardens are different, but the basic principles can be useful in outwitting weeds.

Weed control in cultivated soil

Cultivating ground has several uses; it uproots and buries weeds, gets rid of residues, mixes in lime, fertiliser and manure and breaks the soil into structures that favour crops. Weed vegetation is completely destroyed by cultivation, and must regenerate from survival organs, especially seeds.

The drawback is that cultivation will bury some weed seeds, sending them into dormant states so they will survive longer, at the same time bringing to the surface others that were previously buried and exposing them to light and other dormancy breaking factors. Also, where weeds are merely uprooted rather than buried, a proportion may survive long enough to set seed.

The majority of weed problems are already present in most gardens, lurking in seed or bud banks. Imported seeds usually represent a fraction of those already present in the ground. Seed banks are unlikely to be depleted by cultivation alone. Repeated cultivation ought in theory to expose all the weed seeds in the soil to dormancy breaking factors, leading to a reduction and perhaps elimination of soil weed seeds. Research has shown this to be broadly the case, but the process would take seven years to reduce the seed bank to less than 1% of the original population.

This suggests that digging and other cultivation methods are poor ways to control weeds, as well as being very hard work. Shallow cultivation has some potential, being easily accomplished and effective. Hoeing is the usual shallow cultivation method, preferably with tools such as a Dutch hoe that slice weeds off just below the surface, rather than uprooting them, as cultivators or ordinary hoes do. Hoeing works well between rows or widely spaced plants, but less well in rows. Here weeds can thrive and seed, unless laborious hand weeding is undertaken.

No-digging methods of gardening should avoid bringing weeds to the surface and replenishing the seed bank. Left on the surface, any weed seeds that fall are exposed to seed-eating wildlife, are not pushed into dormancy and should germinate quickly and be vulnerable to destruction. Some caution is needed in adopting this approach because under agricultural conditions no-tillage has sometimes resulted in a weed flora that thrives in the absence of ploughing. On a garden scale, annual meadow grass and other weeds that seed when young, are hard to spot and resist hoeing could become prevalent.

Research has shown that the seed bank in untilled ground declines at half the rate in tilled ground. Overall, whether digging or not, it would take up to four years to significantly erode the seed bank. The conclusion has to be that once a no-digging regime is started, the full weed control benefits will take several years to emerge.

When making a strategy to control the weeds present in a garden, prevention of seeding is the first priority. Imported weeds are unlikely to add significantly to the weed problem, unless a new weed species is inadvertently introduced. Depletion of the seed and bud bank by optimising rotation, cultivation and cultural techniques are important longer term aims. To hard-pressed gardeners this might sound unrealistic, but research found that a high standard of cultivation in vegetable cropping resulted in a fall from 40 000 viable seeds per square metre to 3800 seeds per square metre over four years. In a subsequent wet year, however, the seed bank temporarily tripled.

Heating weed seeds is an effective control of the seed bank, but is only applicable to sterilising potting media ingredients, especially loam. 'Solarisation' by covering the soil with a plastic sheet effectively traps sunlight. In warm conditions this can kill weed seeds. Solarisation has been successfully used in warm climates, but in cooler regions such as the UK it is unlikely to be useful. Chemical sterilisation of the soil is also possible; however the most effective chemicals are not available to amateur gardeners.

Chemical and heat sterilisation have important pest and disease control actions, and their expense is probably more easily justified on these grounds than weed control. Chemical control through selective weedkillers is not an option in cultivated garden soil. The necessity for matching the correct herbicide with crop species and growth stage is not worthwhile for garden-scale activities.

Total weed control by destroying weeds with a contact weedkiller before planting or sowing, can be effective in reducing later weed infestations as long as few seeds are brought to the surface in subsequent hoeing or other activities. An alternative to a chemical weedkiller is a flamegun.

Biological control of weeds in cultivated ground as an alternative to herbicides has been proposed. However, the specificity required to eliminate weeds without harming desirable plants means that a biological control agent would not have the broad range of activity that successful herbicides possess. These agents are usually introduced from the place where the weed originally evolved, to control it where it has become a weed because the biological agents which limit it in its home territory are absent. The most famous example is probably the control of prickly pear cacti (*Opuntia* spp.), in Australia and elsewhere, where they have become aggressive weeds, by the moth *Cactoblastis cactorum* and other insects which were collected in the original home of *Opuntia* in Latin America. The technique has been used particularly against weeds of low value crops, such as grazing land, where other controls are uneconomic, but can be applied to any introduced, exotic weed.

There is clearly a risk of introduced organisms running amok, but modern biological control programmes require very exhaustive host range testing of the proposed control agent before release, to ensure that non-target plants are unaffected. In the UK however, most weeds are native species and the technique has had very limited application. There have been extensive studies on the possibility of bracken (*Pteridium aquilinum*) control using leaf-feeding insects collected on southern Africa populations of this fern, but no releases have yet been made. More recently, there has been increasing interest in the biological control of two other introduced plants which have become weeds, Japanese knotweed (Fig. 11.2) and giant hogweed. See: http://www.cabi.org./BIOSCIENCE/japanese_knotweed_alliance.htm for more information on current proposals.

Birds, ants and other seed eaters may also eliminate useful amounts of seed. 'No-digging' gardens leave any weed seeds where they can be easily accessed by seed eaters. Cultural weed control methods include rotations, where it may be possible to leave ground fallow for a while to eliminate weeds more easily, close spacing to suppress weed growth and seed germination and use of smother crops where the thick canopy of leaves keeps weeds down. 'Rest periods' of perennial plants, (soft fruit for example) where mulches and herbicides can eliminate annual weed problems, are another cultural control.

Weed control in perennial situations

Amongst perennial plants such as fruit plantations and perennial borders, weed control by soil disturbance disrupts the root activities of the crop, as there is no pre- and post-cropping phase to allow unrestricted cultivation. Managing the weed population without cultivation improves growth and yields.

Weedkillers have potential here, as selective ones based on dichlobenil are available to

Fig. 11.2 Japanese Knotweed (*Fallopia japonica*) invading a garden boundary. Photo courtesy of R. Shaw (CABI Bioscience).

eliminate weeds from established woody plantings. It is often possible to use directed sprays of contact weedkillers (based on glyphosate, paraquat, diquat or glufosinate) to control weeds, but the weedkiller should not come into contact with young bark. Follow label recommendations in this respect. Herbicides are especially useful where perennial weeds are abundant, as translocated products such as dichlobenil or glyphosate are carried into the roots, eliminating the entire plant.

Weedkillers have many different modes of action; some interfere with cell division, others (such as glyphosate) prevent synthesis of certain amino acids, and hence proteins. Many gardeners are uneasy about using them, but there are alternatives such as flame weeders which sear off vegetation. Strawberries are a case in point, where the straw mulch can be burnt after cropping. Research suggests that burning straw will destroy many, but not all weed seeds.

Mulching is ideal for gardens where annual weeds are the main problem. Taking advantage of the inability of weed seeds to survive if they germinate below the surface, a covering of at least 50 mm (2 in) of weed-free organic material can prevent annual weeds from growing. Where perennial weeds occur, opaque plastic sheets make effective mulches.

Cover crops, usually mown grass, are often used to suppress weeds in permanent plantations. Other useful effects of cover crops are erosion control, boosting soil structure, improved access and being visually pleasing. However, experiment suggests that they may also act as weeds, depriving crops of nutrients and, especially, water. A compromise is to grow strips of cover crops in the alleys between crops, and leave the soil beneath crops bare or mulched. In this case there is less loss of yield, but most of the benefits of the cover crop are retained.

Weed control in special cases

Lawns, almost uniquely in gardens, can be weeded efficiently with selective weedkillers. These act on broad-leaved weeds, eliminating them and leaving the grass unharmed. Where weeds are resistant to these weedkillers selective cultural measures can sometimes work: liming to reduce woodrushes (*Luzula* spp.) or improving drainage to suppress mosses, for example. In extreme cases, where highly resistant weeds such as mind-your-own-business (*Soleirolia soleirolii*) are involved, removal of all vegetation by cultivation or herbicides and re-sowing can be required.

Chemical control of water weeds is not very satisfactory: algicides are only partially effective, and environmental considerations limit the use of other chemicals. Here cultural controls are often used, removing rampant weeds and endeavouring to suppress unwanted species by shading with highly desirable water lilies, for example.

Paved areas where there is no plant population are readily kept clear of weeds by killing established weeds with a translocated herbicide and preventing re-establishment by using a residual herbicide. Some residual herbicides travel in the soil reaching the roots of adjacent plants. Modern herbicides are only slightly soluble and are safe in this respect.

An alternative to weedkillers is to lay paving on weedproof membranes. These prevent perennial weeds coming up and growing in the paving. Flame weeders offer an alternative means of killing annual weeds.

CONCLUSION

The same scientific principles underlie pest and disease control for professional growers and gardeners alike. However, gardeners' needs and opportunities are different and this is reflected in their approach to managing their problems. Gardeners work on a smaller scale and have more time; they are driven by enthusiasm rather than economic necessity. They have fewer chemical pesticides available to them, less-effective application methods and in many cases, no enthusiasm to use them. All gardeners should make the best use they can of the scientific knowledge that is available, if for no other reason than a simple dependence on pesticides is often not their best option, and the integration of all available methods requires a more subtle and complete knowledge of the plants and the multitude of organisms that interact with them.

FURTHER READING

Agrios, G.N. (1997) *Plant Pathology*, 4th edn. Academic Press, San Diego, USA.

Aldrich R.J. & Kremer, R.J. (1997) *Principles in Weed Management*, 2nd edn, Iowa State University Press, Ames, USA.

Alford, D.V. (1984) *A Colour Atlas of Fruit Pests: their Recognition, Biology and Control*. Wolfe Publishing Ltd., London.

Alford, D.V. (1991) *A Colour Atlas of Pests of Ornamental Trees, Shrubs and Flowers*. Wolfe Publishing Ltd, London.

Alford, D.V. (1999) *A Textbook of Agricultural Entomology*. Blackwell Science, Oxford.

Buczacki, S. (2000) *Plant Problems*. David and Charles, Newton Abbot.

Buczacki, S. & Harris, K.M. (1998) *Collins Photoguide: Pests, Diseases and Disorders of Garden Plants*. HarperCollins, London.

Gratwick, M. (ed.) (1992) *Crop Pests in the UK*. Chapman and Hall, London.

Holm, L., Doll, J., Holm, E., Pancho, J. & Herberger, J. (1997) *Weeds: Natural History and Distribution*. John Wiley & Sons, New York.

Ingram, D.S. & Robertson, Noel F. (1999) *Plant Disease, a Natural History*. The New Naturalist. HarperCollins, London.

McGiffen, M.E. (1997) *Weed management in horticultural crops*. American Society for Horticultural Science & Weed Science Society of America Joint Workshop 6–7 February 1997, ASHS.

Salisbury, E. (1961) *Weeds and Aliens*. Collins, London.

Stephens, R.J. (1982) *Theory and Practice of Weed Control*. Macmillan, London.

Williams, J.B. & Morrison, J.R. (1987) *ADAS Colour Atlas of Weed Seedlings*. Wolfe Publishing Ltd, London.

12

Storage and Post-harvest

- Physiology of ripening and storage: vegetables, fruits and cut flowers
- Influences of the storage environment: refrigerated storage in air; controlled atmosphere storage; ethylene management
- Harvesting, handling and preparation for storage: time of harvest; how to handle harvested products; post-harvest treatments
- Pre-harvest influences on storage quality: diseases and disorders; climatic factors; orchard and field factors; varietal factors
- Future developments

INTRODUCTION

Garden produce is seasonal by nature: gardeners generally harvest their vegetables and fruits for immediate consumption after they have reached an acceptable stage of maturity or ripeness. In commercial horticulture the storage techniques employed provide virtually year-round availability of most fresh commodities, and this has led to the erosion of the idea of seasonal production. This chapter sets out to provide the amateur gardener with some understanding of the biology of fruit and vegetable storage, rather than practical advice on this subject, and students of horticulture with a theoretical basis for a most important aspect of commercial practice.

In the commercial sector, post-harvest handling and storage of fruit and vegetables provides consumers with fresh produce, irrespective of the distance between production areas and the point of sale or the time of the year. Continuing advances in our understanding of the underlying processes (physiological and biochemical) that occur in produce after it has been harvested are giving rise to improved techniques to maintain product quality and extend storage life.

The needs of individual consumers may vary, but we all want what we eat to taste good. Fruits and vegetables not only help to satisfy our hunger, but they also contribute to the quality of life in terms of health, happiness and pleasure. Both amateur gardeners and commercial growers are interested in extending the period of availability of fresh fruits and vegetables. Generally amateurs do not have access to the type of storage facilities that are required to prolong the life of fruit and vegetables after harvest, and so availability is extended primarily by choice of variety, delayed planting or repeat sowing. This chapter looks at the physiological changes that take place in fruit and vegetable crops at harvest and during storage, and at their response to changes in the post-harvest environment.

The storage of cut flowers is also discussed, as commercially these are economically important but often highly perishable. Customers demand high quality flowers that are fresh when purchased and last for a reasonable period: increasingly retailers guarantee the vase life of packaged cut flowers. Supporting this guarantee means that growers, hauliers and retailers have to monitor many factors that affect the longevity of the flowers.

ASPECTS OF PHYSIOLOGY

The physiology and storage of fruits and vegetables are often considered separately because it is only fruits that undergo marked and rapid changes during the process of ripening. These changes may become apparent in terms of appearance, particularly change in colour, smell, taste and texture. Visual, structural and chemical changes that occur in soft fruits such as strawberry (*Fragaria* × *ananassa*) and raspberry (*Rubus idaeus*), stone fruits such as cherry (*Prunus avium*) and plum (*Prunus domestica*) and pome fruits such as apple (*Malus domestica*) and pear (*Pyrus communis*) are associated with an orchestrated sequence of biochemical changes that are under genetic control. The extent to which these underlying changes have progressed determines the visual and eating quality of the product at the point of sale and consumption.

Once harvested, fruits and vegetables no longer have access to water or nutrients, and the only interchanges with the environment are loss of water and carbon dioxide and uptake of oxygen. Although anabolic (constructive) processes occur during the ripening of fruits, the balance of metabolism is catabolic (destructive) and ultimately the fruits become senescent. The *respiration rate* of a product (i.e. the amount of carbon dioxide produced or amount of oxygen consumed per unit of time) is an indicator of the rate of metabolism, and in general higher respiratory activity is associated with a shorter storage or shelf life.

The object of commercial storage is to delay senescence, preserving the desired characteristics of fresh horticultural produce and providing the market with a continuous supply of products with high nutritional value and acceptable sensory quality. To do this, technologies are used to reduce respiration and conserve moisture within the product. These will be described later below.

Vegetables

Vegetables are derived from virtually all types of plant tissue and include underground organs such as swollen roots (e.g. carrots, *Daucus carota*) and swollen stem tubers (e.g. potatoes, *Solanum tuberosum*), and above ground parts such as leaf blades (e.g. spinach, *Spinacia oleracea*), axillary buds (e.g. Brussels sprouts, *Brassica oleracea*), swollen inflorescences (e.g. broccoli, *Brassica oleracea* and seeds (e.g. sweet corn, *Zea mays*). Vegetables also include some fleshy fruits such as tomatoes (*Lycopersicon esculentum*) and cucumbers (*Cucumis sativus*) and immature fruits such as peas (*Pisum sativum*) and bean pulses (*Phaseolus* spp.).

The part of the plant that the vegetable develops from has important implications for its likely storage ability. The stage of development may also be critical in the maintenance of quality in the shorter or longer term. Vegetables such as lettuce (*Lactuca sativa*) and spinach, which are made up of young tissues, are respiring and transpiring actively and are likely to lose quality particularly rapidly; in commercial practice, immediate steps are taken to reduce the rate of metabolism and transpiration of the product. Practical measures for preserving the quality of vegetables (including those of a highly perishable nature) will be considered later.

Certain types of vegetable are more suited to storage, particularly those that are biennial in nature; normally some kind of storage organ is formed in the first season, followed by flowering and seed formation in the second year. Although biennial vegetables are extremely varied morphologically they all cease growth in the autumn. The *rate of metabolism* (respiration rate) of their storage organs declines naturally which makes them adapted for long-term storage. Storage techniques have been developed to extend the natural dormant period of biennial vegetables and increase the period of their availability. Commercial storage of biennial vegetables in the UK includes winter white cabbage (*Brassica oleracea*), onion bulbs (*Allium cepa*) and potatoes.

Fruits

The botanical definition of a fruit is a 'seed receptacle developed from an ovary' (see also Chapter 1); a consumer definition is more likely to concentrate on sensory qualities, the aromatic flavours and sweetness that distinguish a fruit from a vegetable. In nature the aromas and palatable nature of most fruits are important to ensure that seeds are consumed and dispersed by animals. The tissues (up to twelve types) that form various fruits are derived from different parts of the flower. Some examples from the temperate fruits that are common in gardens include strawberries (receptacle), apples (accessory tissue) and tomatoes (placental tissue).

Ripening is a term reserved for fruits and relates to changes in colour, texture, flavour and aroma. The state of ripeness at the point of harvest has particular implications for producers and consumers alike. Generally fruits that are allowed to ripen on the plant achieve a higher sensory quality, but are more susceptible to damage during post-harvest handling and may have a reduced storage life. The biochemical changes associated with fruit ripening and the controlling mechanisms continue to attract much research attention. Understanding the ripening processes may in future show how ripening can be controlled more effectively, and so increase product availability and improve the eating quality and nutritional value of products to the consumer.

Although fruit is important in a healthy diet, it is likely that many consumers eat fruit primarily for enjoyment. The perception of flavour is an important part of sensory quality, and it is vital that technologies to extend the period of availability of fruits do not result in an unacceptable loss of flavour quality. The flavour of fruits is derived from a complex mix of sugars, acids, phenolics and more specialised flavour compounds, including a wide range of volatile chemicals. Sugars (and acids) originate from photosynthesis and in some fruits such as apples and pears these sugars are converted to starch during fruit development. Fruits that accumulate carbohydrates such as starch are often picked at an immature stage and only achieve an acceptable flavour during subsequent ripening when starch is broken down (*hydrolysed*) to form sugars. The starch and sugar contents of apples and pears are often used as indicators of when to harvest for storage (see later). Fruits continue to accumulate sugars from the plant during maturation and ripening and for some types such as strawberry, this contributes significantly to flavour at harvest. Strawberries that are harvested too early will never develop sufficient flavour (sugars) for commercial acceptability.

The main organic acids of fruits are malate and citrate and these are significant components of taste. It is important to achieve the correct balance between the acid and sugar concentration in the fruit. Excessive acidity is often associated with harvesting fruits in an immature or unripe state. Acids are used in respiration, and consequently concentrations decline during ripening; these changes are complemented by corresponding increases in sugar concentration. *Phenolics* (e.g. tannins) are responsible for astringency, and generally result in an adverse reaction from consumers. Although most fruits lose their astringency during ripening, this remains a major quality problem in fruits such as persimmon (*Diospyros kaki*) because of the presence of water-soluble tannins.

Specialised volatile flavour compounds are produced in ripening fruit and these provide the unique sensory character associated with the different types of fruit. In apples and pears esters are the major flavour volatiles. In strawberries esters, alcohols, carbonyls and sulphur-containing compounds are important in flavour. In stone fruit, mostly of the genus *Prunus*, lactones have a role in flavour development, particularly in nectarines and peaches (*Prunus persica*).

Without going into the physiology of ripening in detail it is worth noting that an important distinction is made between fruit types on the basis of their respiratory behaviour. '*Climacteric*' fruits are characterised by a dramatic rise in respiration rate at the onset of ripening and this is accompanied by a burst in the production of ethylene, the 'ripening hormone' (see later in this Chapter). This upsurge in respiration rate provides the cells of the fruit with the chemical energy necessary for synthetic reactions that produce changes in colour, aroma and texture. Climacteric fruits common in commercial horticulture and in gardens in the UK include apples, pears, plums and tomatoes. '*Non-climacteric*' fruits such as strawberry and cherry generally show a decline in respiration rate during ripening and are generally slower to ripen and must be picked at a stage of ripeness suitable for consumption. As in the case of vegetables, high rates of respiration in harvested fruits are generally associated with a reduction in storage or shelf life.

Cut flowers

There is a large variation in the structure of flowers used commercially as cut flowers. Examples include cymes (spray carnation, *Dianthus carophyllus*), umbels (belladonna lily, *Hippeastrum* spp.) and spadix plus spathe (anthurium, *Anthurium andraeanum*). The functional life span of the flowers of different species varies from a few hours to several months, although for most commercial species the approximate storage period is a few days to a few weeks. Generally inflorescences have a low carbohydrate reserve and in this regard are similar to many leafy vegetables. They also have a large surface area in relation to their mass and this makes flowers very vulnerable to water stress. Flowers that have lost 10–15% of their original fresh weight are normally wilted. Ethylene has a major role in the senescence of some cut flowers such as carnation, but some other species are insensitive to ethylene. These include chrysanthemum (*Chrysanthemum morifolium*), aster (Michaelmas daisy, *Aster novi-belgii*), sunflower (*Helianthus annuus*), nerine (*Nerine bowdenii*), liatris (*Liatris spicata*), zinnia (*Zinnia elegans*) and rudbeckia (*Rudbeckia hirta*).

Senescence in cut carnation flowers is characterised by a rise in respiration rate and ethylene production and in this respect the carnation behaves much like a climacteric fruit. In carnation and some other species of flowers, pollination stimulates ethylene production and causes senescence of the petals. The functional significance of pollination-induced senescence may be to save energy and sugar reserves that would otherwise be used in maintaining elaborate flower structures, and to prevent further visits by pollinators such as bees, conserving pollen and economising on the energy requirements of the pollinators themselves. The processes that cause the deterioration of cut flowers have many similarities with those that operate in fruits and vegetables (see above). This is not surprising in view of the fact that some vegetables are derived from flower buds (e.g. artichokes, *Cynara scolymus*) or swollen inflorescences (e.g. broccoli). Florist greens (decorative foliage) are likely to behave in a similar way to leafy vegetables such as spinach.

There are many causes of deterioration in cut flowers.

- Depletion of 'food' reserves, mainly carbohydrates, may reduce the amount of energy available to maintain cell structure and function.
- As with all fresh horticultural crops there is a risk of attack by bacteria and fungi.
- Excessive loss of moisture by transpiration may cause wilting of the flowers and foliage.
- Bruising and crushing during handling may stimulate respiration and hasten the process of senescence.
- Exposure to warm temperatures is likely to increase respiration and hasten senescence.
- Conversely, low temperatures may induce an abnormal metabolism in the flowers, resulting in chilling injury.
- Colour changes or fading can effect flower quality and acceptability.
- Accumulation of ethylene in the storage environment can have serious consequences by accelerating the ageing process and increasing floret shattering (abscission).

Other causes of deterioration include the use of poor quality water and adverse cultural practices.

INFLUENCES OF THE STORAGE ENVIRONMENT

Refrigerated storage in air

From the last section it will be apparent that deterioration may be expected to occur in fruit, vegetables and flowers under conditions that accelerate their rate of metabolism (respiration rate) and water loss. The consequences of subjecting products to high temperatures and low humidities will vary according to the extent to which a particular type of fruit or vegetable is adapted to withstand such stressful conditions.

Temperature is one of the most important factors affecting the keeping quality of horticultural products. Lowering the temperature reduces the rate of respiration and other metabolic processes and thereby reduces the rate of senescence and of ripening in fruits. The highest freezing point for fruits varies between −3.0°C (pomegranates, *Punica granatum*) and −0.3°C (avocados, *Persea americana*) and for vegetables between −2.2°C (Jerusalem artichokes, *Helianthus tuberosus*) and −0.1°C (endive and escarole, *Cichorium endivia*). The highest recorded freezing point for blooms varies from −0.5°C for Easter lilies (*Lilium longiflorum*) and roses (*Rosa* spp.) to −0.7°C for carnations.

It might be assumed that the longest storage or shelf life of fruits, vegetables and cut flowers is likely to be achieved by maintaining temperatures at slightly above their freezing points. However, a number of plants, particularly those originating in tropical or sub-tropical regions, are injured when stored at temperatures well above freezing. For these plants the full potential benefit of refrigeration cannot be realised. Chilling stress leads to an altered metabolism and to the development of a range of injury symptoms. Among vegetables affected by chilling injury probably the most economically significant are squash (*Cucurbita* spp.), cucumber, water melon (*Citrullus lanatus*), aubergine (*Solanum melongena*) tomato, snap beans (*Phaseolus vulgaris*), sweet pepper (*Capsicum annuum*) and potato; these have lower temperature limits ranging from 4 to 13°C.

The storage of potatoes is of particular interest because of their economic importance in the UK and because the changes that take place have a marked effect on their acceptability for different markets. Storage at temperatures below about 6°C results in the hydrolysis of starch to form sugars. This 'low-temperature sweetening' imparts an unacceptable dark colour to potato crisps and chips.

The recommended temperature for the storage

of most types of vegetables of temperate origin is 0°C. This applies to many of the field vegetables grown in the UK such as roots and onions, brassicas and legumes. Even at 0°C the storage life of vegetables varies from five to eight days (sweet corn) up to eight months (onion). Chilling-sensitive fruits that are most familiar to UK consumers include avocados, bananas (*Musa* spp.), grapefruits (*Citrus* × *paradisi*), lemons (*Citrus limon*), mangos (*Mangifera indica*), melons (*Cucumis melo*) and pineapples (*Ananas comosus*). Of the fruit types important in UK production (apples, pears, plums, cherries, strawberries, raspberries, and blackcurrants, *Ribes nigrum*) only apple is regarded as susceptible to chilling injury and this is highly dependent on the cultivar. This is reflected in the temperatures that are recommended for storage. Cut flowers are commonly kept at about 4°C at the wholesale level and during transport but there are some types that require much higher temperatures in order to avoid chilling injury. These include some orchids (Orchidaceae) such as *Cattleya* (7–10°C), and *Anthurium* (13°C). Chilling injury may result in failure of flowers to open properly or discoloration of sepals and petals.

Refrigeration is of prime importance in slowing respiration and extending the storage and shelf life of fresh horticultural crops. It is particularly important in the distribution of highly perishable vegetables such as asparagus (*Asparagus officinalis*) and lettuce and soft fruits such as strawberries and raspberries. Refrigeration cannot significantly extend the season for these types of crops. This can only be achieved by choice of cultivar, repeated sowing, cultural techniques and importation from different sources.

Refrigerated storage is expensive and it should be used judiciously. Root crops such as carrots and parsnips (*Pastinaca sativa*) are generally kept in the ground; they are naturally dormant during the winter, when lower soil temperatures reduce respiration rate and developmental changes and high soil moisture prevents desiccation. Ground storage of carrots has become quite sophisticated. Polythene sheeting is used to cover the crop followed by a layer (30–60 cm) of straw. This provides protection against rain and frost and delays the warming up of the soil in the spring with the result that ground storage is possible through to May. It may be necessary to harvest and store other types of biennial vegetables such as onion and white winter cabbage to avoid a loss of quality, and to extend the period of availability past that of the natural break of dormancy under field conditions.

For perennial fruit trees that produce one crop a year, refrigeration may be used to extend the marketing period. Fruits such as plums and apples that are left on the tree will ripen and eventually fall to the ground. Decay of fruits on the orchard floor will occur rapidly.

Refrigeration is the major way of slowing deterioration and extending the life of cut flowers, and is used extensively during transport and distribution. It is perhaps more appropriate to talk about refrigerated transport than storage of cut flowers. For most species of cut flowers maximum storage life can be achieved without water in moisture retentive containers at −0.5 to 0.5°C. This 'dry-pack' method of storage has given important extensions in the storage life of many cut flower species such as carnations and roses.

Various types of mechanical refrigeration systems are used for the commercial storage of fruits and vegetables. The construction and refrigeration systems required for any particular application are dictated by the characteristics of the product, and in particular by the rate of cooling required to maintain quality after harvest. The conventional type of building used for the storage of apples in the UK is illustrated in Figs 12.1 and 12.2. Suggested cooling times vary from three hours for highly perishable products such as soft fruit, sweet corn, asparagus, calabrese (*Brassica oleracea*) and spinach to up to six weeks for potatoes and onions. Quick methods of cooling that remove the field heat from perishable crops rapidly (such as forced air-cooling, vacuum cooling and hydro-cooling) are described elsewhere. Forced-air cooling has been used increasingly to pre-cool cut flowers prior to shipment under continuous refrigeration. Forced-air systems are available that maintain a high humidity (95–98% RH) while cooling to the required temperature within one hour.

Precise control of the temperature is a major factor limiting the ability of the amateur to store fruits and vegetables effectively. In the UK there is a strong tradition of storing apples and pears

Fig. 12.1 The interior of a typical UK apple store showing the 'stack' cooler that houses the cooling pipes (base) and air circulatory fan (top).

from the garden and advice is available to the amateur as a 'spin-off' from the research done over the years for the commercial fruit grower. It is also a tradition for the amateur to store dry bulb onions and to construct clamps for red beet (*Beta vulgaris*) and potatoes.

Controlled atmosphere storage

Modifying the oxygen and carbon dioxide concentration in the atmosphere surrounding fruits and vegetables may have important additional benefits in maintaining quality and extending the duration of storage and shelf life. During the respiration of plant tissues, sugars, acids and lipids are broken down in the presence of oxygen (aerobic respiration) to provide the energy required for maintaining the integrity of cells. The other products of respiration are carbon dioxide and water. Removing heat from fruits and vegetables by refrigeration reduces the energy available for the numerous chemical reactions involved in respiration. However, the effectiveness of refrigeration is limited by the susceptibility to chilling injury of each type of fruit or vegetable. The discovery that increased carbon dioxide and reduced oxygen concentration in the atmosphere around fruits and vegetables would further reduce their respiration had important implications in extending storage life. Franklin Kidd and Cyril West, who worked initially in the Botany School at Cambridge and subsequently at the Ditton Laboratory in Kent, are recognised as the pioneers of *controlled atmosphere storage* (CA). However, the concept of controlling ripening by modifying the atmosphere had originated before their systematic investigations.

CA storage was shown to have major advantages over cold (air) storage for the preservation of many types of fruits and vegetables, and the technique is widely practised in many of the world's production areas for horticultural crops. The technology for the CA storage of apples in the UK and elsewhere has become highly sophisticated. Oxygen concentrations recommended for the storage of apples are typically 1–2% (air is nominally 21% oxygen) and carbon dioxide may be as high as 10% (air contains approximately 0.03% carbon dioxide). Recommended concentrations of oxygen and carbon dioxide are available for apples, pears, Nashi (Asian pear), fruits other than pome fruits and vegetables. Several thousand tonnes of UK-grown onions are stored in CA in order to extend the period of availability. White winter cabbage is also kept for long periods in CA storage to provide continuity of supply to the fresh market and for processing into coleslaw. Although CA

Fig. 12.2 A part-loaded apple store showing typical stacking arrangement of bulk bins of fruit.

was traditionally practised in purpose-built stores the technology has progressed to include shipping containers and packaging of produce in semi-permeable films to generate CA conditions within the package.

CA technologies are generally unavailable to the amateur grower although storing apples and pears in polythene bags has been recommended as a means of reducing water loss and preventing shrivelled or leathery fruit. The bags should not be tightly sealed, or carbon dioxide will accumulate and oxygen will deplete to damaging concentrations; this will also impart off-flavours to the fruit because of the accumulation of fermentation products such as ethyl alcohol and ethyl acetate. Although it cannot be claimed that the use of perforated or unsealed polythene bags prolongs the storage life of the fruit (apart from preventing shrivelling) there is the possibility of carbon dioxide accumulation and some positive effects on storage life and quality retention.

CA storage is not generally recommended for cut flowers for three main reasons:

- the small margin of safety between effectiveness and toxicity
- the small volumes of any one cultivar
- the high cost of treatment.

Ethylene

Ethylene (C_2H_4) is a colourless gas produced by most plants. It has particular significance in the ripening of fruits, particularly those of the climacteric type, and in the senescence of cut flowers. In climacteric fruits the production of ethylene will not normally increase until the fruits are almost fully developed. From an evolutionary viewpoint this could be a safeguard against the fruit ripening prematurely, at a stage when the seeds are immature and non-viable, and when the attractiveness of the fruit – designed to encourage animals to consume it and disperse the seeds – is less than optimal. In fruits such as apple, pear, plum and tomato, a dramatic increase in ethylene concentration is required to trigger and synchronise ripening processes such as softening and aroma production. Clearly ethylene has important commercial implications for the storage and marketing of fruit, vegetables and cut flowers.

The ability of ethylene to initiate ripening is exploited in the marketing of several types of fruit such as avocado, banana and tomato. The commercial ripening of bananas is a routine operation carried out in importing countries to provide fruit to the consumers at a specified colour stage. Bananas are harvested in an unripe (green) condition and kept at cool temperatures (13–14°C) during their transport to distant markets; subsequent ripening with ethylene ensures uniformity in ripening and product quality.

Most of the uses of ethylene to stimulate ripening relate to climacteric fruit but there are some instances where ethylene treatment is useful for non-climacteric fruit. One important use is in the degreening of citrus where certain cultivars become edible before the green colour of the peel has disappeared. In such cases ethylene gas is administered to encourage the degradation of the green pigment (chlorophyll) in the peel.

Knowing how this ethylene treatment can help ripen immature climacteric fruits may be useful to the amateur grower. For example, it is possible to stimulate the ripening of immature tomatoes at the end of the growing season by enclosing the green fruits with ripening (red) fruits or other ripe climacteric fruits such as apples or bananas. Greatest stimulation of ripening will be achieved in warm temperatures.

Although ethylene gas can be used to improve the quality of certain types of fruit, its presence in the atmosphere can also have detrimental effects on the quality of fresh produce. The build-up of ethylene in storage chambers can cause premature ripening in bananas and softening in kiwi fruits (*Actinidia chinensis*). The presence of ethylene in greenhouses, storage rooms or anywhere in the distribution chain can have serious detrimental effects on the keeping quality of flowers. These include *epinasty* (downward bending of leaves), withering, ageing, yellowing and abscission of flower parts and leaves.

Various types of technology exist to reduce ethylene concentrations around stored fruit and vegetables: these include ventilation with fresh air, its removal (by oxidation) using chemical agents such as potassium permanganate and ozone, or physical/chemical methods such as heated catalyst systems. This method is preferred for removing ethylene from storage rooms, and commercial ethylene converters are commonly used in the storage of kiwi fruits. High conversion efficiency (ethylene to carbon dioxide) is achieved by heating the catalyst (usually platinum) to 200°C and calculating the maximum flow of store atmosphere through the catalyst bed. Removing ethylene from refrigerated CA stores can have important benefits in delaying ripening and so improving fruit quality or extending storage life. Although important benefits of ethylene removal have been established for apples, there has been no commercial development of the technique in the UK.

A build-up of ethylene is generally detrimental to the quality of vegetables, too, although there is a range in sensitivity. Vegetables that display high sensitivity to ethylene include broccoli, Brussels sprouts, Chinese cabbage (*Brassica chinensis*), cauliflower (*Brassica oleracea*), cabbages, cucumber, endive, sweet corn, lettuce and spinach. Adverse effects caused by ethylene include loss of visual quality due to accelerated degreening, senescence and abscission and impaired eating quality due to toughening (e.g. asparagus) and the development of bitterness (e.g. carrots).

During the distribution and retailing process vegetables may experience high concentrations of ethylene, chiefly because of the presence of ripening fruit. Several measures have been suggested to avoid ethylene damage, including not mixing vegetables with fruit unnecessarily. When storing ethylene-sensitive types of vegetable in a domestic refrigerator, the consumer should avoid keeping apples, pears or stone fruits, which produce large amounts of ethylene, alongside them. However, the cool temperatures in a domestic refrigerator will reduce ethylene production and slow the overall rates of deterioration.

With cut flowers it is particularly important that the source of ethylene pollution is identified so that a logistical approach can be taken to growing, storage and marketing. It is clear that ethylene-sensitive flowers should not be stored with climacteric fruits such as apples and pears, that produce vast quantities of ethylene during ripening, but for many species the greatest risk of ethylene exposure is during transportation, from the emissions from motor vehicles.

Good ventilation can be particularly effective in reducing the concentration of ethylene around cut flowers; also important are chemical inhibitors of ethylene action such as silver thiosulphate (STS) that are used as a single treatment after harvest before dry packing and transport. Silver ions inhibit the action of ethylene and delay senescence. STS treatments have been mandatory for many cut-flower species sold in auctions in the Netherlands, although recently there have been growing environmental concerns about the use of heavy metals such as silver. There is now increasing interest in more environmentally friendly alternatives such as 1-methylcyclopropene (1-MCP), a gaseous compound that inhibits a range of plant responses to ethylene including senescence of carnation flowers.

HARVESTING, HANDLING AND PREPARATION FOR STORAGE

Time of harvest

The physiological maturity of a vegetable does not always coincide with its commercial maturity. Most vegetables are picked at a stage of maturity determined by the demands of consumers or processors rather than at a stage that might provide a more sustained storage or shelf life. *Physiological maturity* refers to progress through a series of developmental stages that typically include the growth, maturation, ripening and senescence of an organ or organism. *Commercial maturity* reflects the stage at which the market requires the plant organ. For example, asparagus, celery (*Apium graveolens*), lettuce and cabbage are harvested when the stems and leaves are sufficiently developed and artichoke, broccoli and cauliflower are harvested when the inflorescence is sufficiently developed. Cucumbers, green beans (*Phaseolus vulgaris*), okra (*Abelmoschus esculentus*) and sweet corn are marketed as partially developed fruits.

The exceptions to this are the biennial types of vegetable: their time of harvesting is determined by their progress to a dormant state. Harvesting is undertaken when growth ceases in the autumn, when the diminished rate of metabolism (respiration) makes them suited for long-term storage.

There may be other factors that determine the time of harvest; these include

- the development of pest and disease problems in vegetables left in the ground
- the likelihood of adverse weather conditions that might affect the ability to harvest
- the time of harvest may be critical in order to prevent loss of quality in the harvested product. For example, late harvesting of onions leads to a greater proportion of split and shed skins and to the staining of onions left in the field.

Most of the ripening changes in non-climacteric fruits such as strawberries and cherries take place while the fruit is still attached to the plant, so these are normally harvested when they are fully ripe, at 'optimum consumer quality'. Although climacteric fruits such as plums, apples and pears will also ripen on the tree (except avocados, which only ripen when detached from the plant) they are commonly harvested in an under-ripe condition but at a stage where optimum consumer quality will be achieved after storage. Of course, where storage is not a requirement, apples can be left on the tree to ripen – which is likely to provide superior eating quality.

For the long-term storage of apples, fruit must be picked before the onset of the climacteric rise in respiration rate. This applies to storage by both the amateur and the commercial grower. Growers do not usually have access to the specialist laboratory skills and equipment that can be used to determine the stage of ripeness of fruit from respiration or ethylene measurements. However, simpler methods are available that can be related to physiological or commercial maturity. Commercial judgements of the correct time to pick apples for long-term storage are based primarily on starch-iodine staining, firmness and sugar (soluble solids) concentration.

Although the main objective with cut flowers is the longest possible vase life, the flowers have to open fully – reach their full 'floral development'. Some types such as carnations and daffodils (*Narcissus pseudonarcissus*) can be picked at the bud stage, but others such as orchids should be fully developed before they are cut.

Handling

Fruits and vegetables are susceptible to decay caused by a wide range of bacteria and fungi. They can be infected during their growth and development, at harvest and during storage. In many cases it is only possible to control post-harvest diseases by eradicating the causal organism, or protecting the crop in the field or orchard before harvest. Examples of this include such fungal diseases as grey mould (*Botrytis cinerea*) of strawberry, bull's-eye-rot (*Gloeosporium* spp.) of apple and neck rot (*Botrytis allii*) of dry bulb onions. Great care is needed in both the harvesting and post-harvest handling of fruits and vegetables to avoid allowing access to pathogens.

The ideal method for long storage life is to harvest all commodities by hand, but this may only be an option for the amateur. It cannot be over-emphasised that only perfect, unblemished apples should be selected for storage. Commercial fruit crops destined for the fresh market are picked by hand but those for processing (such as blackcurrants) may be harvested mechanically. Harvesting field vegetables by hand may not be feasible, particularly with root crops, but mechanical harvesting has to be carried out very carefully to avoid damage. This could result in

- direct loss of visual quality
- internal damage (such as bruising in potatoes)
- and access to pathogens.

Cut flowers should be considered as highly perishable products and should receive prompt careful handling. Any physical impacts will damage and bruise the blooms and may directly affect their visual quality and reduce vase life by stimulating ethylene production. Many cultivars have specific handling requirements. Sanitation is particularly important: dead or decaying plant material remaining in containers used for fresh cut flowers is likely to be a source of ethylene and of microorganisms that may cause decay. Some of these, particularly bacteria, are implicated as a cause of stem blockage in freshly harvested flowers. This reduces water uptake and hastens the onset of wilting. Cutting the base of stems at intervals will ensure that stems continue to take up water.

Post-harvest treatments

There are various types of treatment that can be applied to harvested horticultural crops before they go into storage; all are aimed at maintaining quality or minimising decay. For example, some crops such as dry bulb onions require fast and efficient drying until the necks of the bulbs are tight and dry (*cured*). This can be achieved by blowing heated air through the stored onions until they have lost about 3–5% of their original weight. In countries where fine weather is normally experienced after harvest, onions can be cured adequately in the field over a period of two to four weeks. Amateur growers generally use field drying, although further drying can be achieved by moving the onions to open sheds. In subsequent storage, a comparatively low humidity (65–70% RH) is needed to prevent re-rooting and shoot growth and to minimise the development of neck rot.

Potato tubers require curing before storage; this stimulates the production of a corky outer skin and so reduces moisture loss during storage. It also increases resistance to infection by *Fusarium* spp. and other rot-forming bacteria and fungi. The corky tissue that forms around wounds inflicted during harvesting effectively heals cuts and bruises and thereby forms a physical barrier to protect the tuber from infection by pathogens. Potatoes are cured at intermediate to high temperatures (10–15.5°C) and at 95% relative humidity for up to two weeks. Washing may be important to remove surface deposits (debris, dirt and sap) from certain crops such as carrots but in others contact with water may increase the spread of disease. Where washing is required, clean water is essential and disinfectant may be added to the rinse water to kill bacteria and fungal spores.

Other post-harvest treatments include waxing, which enhances the appearance and reduces water loss in commodities such as apples, citrus and sweet potatoes (*Ipomoea batatas*). Chemicals such as tecnazene and maleic hydrazide (MH) are applied to suppress sprouting in potatoes and onions respectively. Sprout inhibitors such as tecnazene are normally applied to potatoes after they have been loaded into the store; MH is applied to onions as a pre-harvest spray when 50–80% of the foliage is down.

Insect disinfestation treatments such as methyl bromide are important in the international trade of a wide range of fruit and vegetable crops. Other post-harvest chemical treatments include the use of specific fungicides, antioxidants and calcium. Fungicides such as carbendazim are applied to apples and pears as a pre-storage drench when a significant risk of rotting in store is predicted for particular consignments. Rot risk in stored apples is estimated on the basis of the weather during fruit development and on orchard management practices. Ozone injection into CA stores of Vidalia onions is used in some countries to control surface pathogens. Chemical antioxidants such as diphenylamine and ethoxyquin are applied to cultivars of apples and pears that are susceptible to a form of skin browning known as *superficial scald*. This physiological disorder affects many apple cultivars grown in hot dry conditions but in the UK it is potentially a serious problem only on the culinary cultivar 'Bramley's Seedling'. Scald occurs following the death of cells in the epidermal and hypodermal layers of the fruit; this is due to the oxidation of alpha-farnesene, a compound that forms naturally in the cuticle of the fruit as it ripens. Chemical antioxidants prevent the conversion of alpha-farnesene to harmful compounds.

Post-harvest treatment with calcium is restricted mainly to apples, and in particular to cultivars such as 'Cox's Orange Pippin' and 'Egremont Russet'. These are regarded as susceptible to a range of physiological disorders that occur when insufficient calcium accumulates in the fruit during its development. The most common of these is *bitter pit*, which appears as small pockets of brown, dry, tissue; these may extend throughout the edible flesh (*cortex*) but are usually at the calyx end of the fruit. Although affected fruits are edible, the pitted areas are bitter in taste and severely affected fruits may be unpleasant to eat.

For cultivars susceptible to bitter pit orchard sprays containing calcium can be applied routinely from June until harvest. Calcium chloride is the preferred form of calcium for bitter pit control. Amateurs can improve the storage potential of their apples by spraying their trees with 0.8% (w/v) flake calcium chloride (contains 78% actual calcium chloride) at intervals of 10–14 days from late June to harvest. Damage to the foliage can be avoided by spraying at temperatures below 21°C and preferably in the evening. Post-harvest application of calcium compounds such as calcium chloride provides an additional means of supplementing calcium in the fruit.

The use of preservative solutions throughout the marketing chain is particularly important for flower quality and vase life. Several commercial preservative materials are available. These contain sugar (usually sucrose) to provide the energy that the flower requires to maintain its normal cell functions and complete its development (where this is not already completed at the time of harvest). *Biocides* such as chlorine and quaternary ammonium compounds are included to kill microorganisms. An acidifying agent such as citric acid is included to lower the pH. This has the effect of improving water uptake, although the mechanism is unknown.

PRE-HARVEST INFLUENCES

Diseases and disorders

Fresh fruits and vegetables are highly perishable and spoilage during storage may occur as a result of attack by fungi, bacteria and viruses or by the development of functional or physiological disorders (see Chapter 11). Soft rots caused by bacteria such as *Erwinia* and *Pseudomonas* can affect practically all vegetables and especially carrot, celery and potato. Fungi cause most of the spoilage that occurs in fresh fruit and vegetables during storage. Genera such as *Penicillium*, *Botrytis* and *Sclerotinia* commonly affect most commodities including apples and pears and are likely to affect produce stored by the amateur. Some of the physiological disorders of apple such as superficial scald (a skin browning disorder) and bitter pit, an internal calcium-deficiency disorder characterised by brown 'corky' lesions, are also likely to be seen by the amateur (see above).

Stems and leaves of cut flowers and florist greens are usually contaminated with bacteria and quickly contaminate clean vase water. Bacteria isolated from cut-flower containers include species of *Achromobacter*, *Bacillus*, *Micrococcus* and *Pseudomonas*. There is little information on the loss of quality or level of wastage in cut flowers that can be attributed to pathogens. Good sanitation practice combined with the use of biocides is usually sufficient to prevent significant decay of cut flowers.

Climatic factors

The potential of fruits and vegetables to develop particular diseases or disorders is influenced by climatic and field or orchard factors. The extent to which diseases can develop in stored crops is determined by the number of pathogenic micro-organisms that are present and their viability, i.e. their ability to infect. The susceptibility to infection of particular consignments of fruits or vegetables is also a major factor in determining the amount of rotting that takes place during storage. As the *inoculum* of many of the important fungal diseases of apple is dispersed by rain, it is not surprising that wet growing seasons are associated with a higher incidence of storage rots. High rainfall in August and September is associated with higher rotting in stored 'Cox's Orange Pippin' apples due to *Gloeosporium* spp. Infection of apples by *Phytophthora syringae* occurs during periods of heavy rain just prior to or during harvest. Spores are spread from the soil to the fruit either by direct contact or by rain-splash.

Weather conditions during their development and at the time of harvesting affect the susceptibility of vegetable crops to disease after harvesting. Phytophthora rot (*Phytophthora porri*) of winter white cabbage is most likely to develop when cabbages are harvested in very wet field conditions. Similarly harvesting root crops in wet weather can result in a considerable amount of soil and disease inoculum being taken into store with the produce.

Climatic conditions during the development of the fruit affect the susceptibility of stored apples to physiological disorders. Cool, wet summers are known to reduce the tolerance of apples for low storage temperatures and in some years it may be necessary to raise the storage temperature slightly to avoid a deterioration of the fruit flesh, referred to as *low temperature breakdown*. Dull, sunless summers may increase the risk of *core flush*, which as the name implies is a physiological disorder that results in a pink discoloration in the core of the fruit. Warm, dry summers increase the risk of superficial scald, bitter pit and water core. In such years it is particularly important to apply chemical antioxidants to the harvested fruit to prevent scald development later in the storage period. To reduce the susceptibility of apples to bitter pit and water core, calcium sprays should be applied in the orchard to supplement the natural uptake of calcium into the developing fruit (see below).

It is particularly important for the commercial grower to be able to evaluate the storage potential of each orchard. In this way consignments with the least storage potential will be marketed early in order to avoid losses. More recently, mathematical models have been developed in order to predict the risk of disorders in different consignments of 'Cox's Orange Pippin' apples on the basis of temperature and rainfall measurements made in the orchard during fruit development.

As most cut flowers are produced as protected crops under glass, it may be less relevant to consider pre-harvest climatic influences because these are largely under the control of the grower. If flowers are exposed to high light conditions before cutting, sugars accumulate and storage life is improved. Any subsequent exposure to light may be beneficial to cut flowers since they retain their capacity to photosynthesise and produce carbohydrates.

Orchard and field factors

A great deal of research has been carried out on the effects of orchard factors on the storage quality of apples and pears. There have been numerous investigations into the effects of tree factors such as rootstock, age, cropping level and pruning and of soil and nutritional factors and the use of orchard sprays (fungicides, insecticides and growth regulators). For some cultivars of apple, nutritional factors have a major influence on the keeping quality of the fruit. The importance of achieving adequate calcium nutrition in fruit and vegetables is recognised universally. Calcium deficiency impairs membrane permeability in plant cells and membrane leakage is followed by a major disintegration of membrane structure and reduction in growth of meristematic tissues. An under-supply of calcium results in necrosis in apple (bitter pit), tomatoes (blossom-end rot), celery (blackheart) and many other fruits and vegetables.

Insufficient calcium in apples at harvest increases their rate of respiration, promotes ion leakage from cells, accelerates their rate of senescence in store and increases their susceptibility to a range of physiological disorders of which the most common is bitter pit. The characteristic brown 'corky' lesions in the flesh of the fruit are familiar to amateurs and commercial growers alike. The most effective remedy is the repeated application of orchard sprays containing calcium chloride or calcium nitrate.

Calcium deficiency is a major cause of losses in vegetable crops. In the USA it is estimated that in the period 1951–60 annual losses caused by calcium-dependent disorders such as blossom-end rot in tomato, blackheart in celery and tipburn in lettuce amounted to 4.5 million US$. Although these disorders develop before harvest, internal disorders may not be apparent at that stage and other problems may develop during storage, such as *tipburn* in white winter cabbage. The importance of nutrition during an apple's development is so great in relation to its storage quality that mineral analysis standards are available for growers, to help them to decide the most suitable storage and marketing strategy for any particular consignment of fruit.

Varietal factors

The storage potential of apples varies widely with the rate of ripening and senescence of each particular variety. Amateurs should not attempt to store early varieties of apple and it is not worth taking too much trouble with mid-season varieties. Apples picked after the middle of September are likely to keep for periods varying from several weeks to several months. In commercial practice, the strong genetic component to the storage potential of different apple cultivars is recognised and despite the application of modern storage techniques, maximum storage varies from a few weeks for cultivars such as 'Discovery' to up to ten months for cultivars such as 'Bramley's Seedling'.

Of the major commercial cultivars currently in production in the UK, the most suitable for storage by amateurs would include

- 'Cox's Orange Pippin'
- 'Red Pippin' ('Fiesta')
- 'Bramley's Seedling'
- 'Gala'
- 'Jonagold'.

Storage recommendations provided to the UK fruit industry take into account that there are marked differences in the storage potential of pear cultivars. Pears are an extremely difficult crop for the amateur to store, regardless of variety, as the fruit is very easily bruised and hard to ripen to a satisfactory quality.

There is relatively little information on how cultural practices affect the post-harvest performance of cut flowers, although excessive applications of fertiliser are thought to encourage more rapid deterioration of flowers.

FUTURE DEVELOPMENTS

Over the past decades there have been major changes in storage requirements for horticultural crops. Storage initially provided consumers with fresh produce during the winter months, and the major objective was to provide producers with the means to store various crops without undue wastage. In due course, advances in storage technology for permanent structures and for transportation contributed to international trade and eventually the seasonal aspect of fruit and vegetable production was eroded. In a highly competitive global market it was increasingly important to provide fresh produce of the highest visual and eating quality. The development of storage recommendations reflected the changing need for high quality standards, in addition to the control of diseases and disorders.

In the future it is likely that consumer demands for quality will change. In the more health-conscious society of today, storage techniques will need to be developed that maintain or improve components of food that contribute to a healthy diet. Research will be directed towards maximising sensory attributes such as taste, aroma and texture, essential nutritive compounds such as carbohydrates, proteins, vitamins and minerals, and bioactive substances such as polyphenols, carotenoids, phyto-oestrogens and dietary fibres. There is also likely to be increased consumer awareness of undesirable attributes such as *mycotoxins* and pesticide residues.

Changes in consumer awareness of the health, nutritional and ecological aspects of food quality will lead to continual changes in the market place; in the future changes in pre- and post-harvest management practices will have to reflect these changing needs. To provide fruit and vegetables in the freshest condition possible there will be an increased use of cool-chain marketing, chilled display cabinets in retail shops and perhaps an increased use of polymeric film packaging. The demand for ready-to-use fruit and vegetables is likely to grow, with consumers willing to pay for both quality and convenience.

CONCLUSIONS

Although this chapter was not intended to provide practical guidance on the storage of garden products, it may be useful to summarise key aspects of post-harvest biology and storage that can help the amateur gardener to manage their crops effectively after harvest.

Soft fruits such as strawberries and raspberries are highly perishable and should be picked mature but not over mature. Eating quality will not improve after harvest. Fruits should be picked during the coolest part of the day and stored in a refrigerator until required. It is important not to overload the refrigerator. The heat produced by the fruit may result in temperatures that exceed those recommended for other foods such as meat, fish and dairy products that may also be present in the refrigerator.

Stone fruits will also benefit from refrigerated storage, but it is unlikely that a domestic refrigerator has sufficient capacity to store significant quantities. Cherries should be picked when they have achieved the desired eating quality but plums should be harvested slightly under ripe (background colour green/yellow as opposed to yellow) if they are to be stored. Only keep perfect fruit as any physical damage or lesions caused by insects or disease are potential sites for infection by rot-forming organisms.

Pome fruits, especially apples, are most amenable to storage, particularly those that mature late in the season, i.e. late September onwards. Pick apples for storage when the fruit can be detached easily from the tree and before the fruit achieves good eating quality in terms of sugar content and aromatic properties. Store sound apples in a cool (4°C) frost-free cellar or outbuilding. Storage in perforated polythene bags will help to prevent shrivel. The bags must not be sealed completely otherwise the oxygen in the bag will be depleted and alcoholic off-flavours will develop. As the apples are in contact there is the potential for the spread of disease within the bags. Regular inspection is necessary in order to remove any rotted fruits.

Pears do not ripen properly when left on the tree. It is difficult for the amateur to store pears since they will ripen unless kept below 0°C. Keep sound pears as cool as possible and inspect regularly and squeeze gently to gauge the extent of softening. Softer pears should be placed at room temperature to complete ripening. Although polythene bags can be used for pears it is particularly important to provide sufficient ventilation, otherwise a build-up of carbon dioxide may damage the fruit.

The traditional method of storing vegetable root and tuber crops is to make a *clamp* or 'pie'. This was the original method of storing potatoes following their introduction into the UK during the sixteenth century. Garden clamps are used mostly for potatoes, but also for other vegetables such as beets, carrots, swedes (*Brassica napus*) and turnips (*Brassica campestris*). Clamps are piles of roots or tubers on a straw base, often triangular in vertical cross-section, that are covered with a layer of straw and a layer of soil to provide protection from frost and rain. Clamps should be constructed on well-drained land and have a trench dug out around the structure to provide drainage for rain running off the sides. Ventilation holes plugged with loose straw are required in the ridge of the clamp to prevent a build-up of carbon dioxide and the depletion of oxygen through respiration. Only healthy roots or tubers should be placed in the clamp.

Onions are particularly suitable for storage by the amateur. They should be harvested when the foliage is brown and brittle, pulled or dug from the ground on a fine sunny day and left on the surface of the soil to dry. When the onions are dry, remove any soil, dead roots and loose, dry skins. Store the onions in trays in a cool dry place or rope the crop if preferred. Check the condition of the bulbs at intervals and use before sprouting signals the break of dormancy. Commercial harvesting of onions is carried out when the foliage is down but still green. Harvesting at this stage provides better storage and avoids staining and the splitting and shedding of skins that is associated with later harvesting. Early harvesting is only possible where there are facilities for fast and efficient drying.

The amateur gardener is not likely to need to store cut flowers, but more general methods of extending the vase life of flowers from the garden or from retail stores may be relevant. Cutting flowers at the bud stage may be appropriate for flowers such as roses and gladioli. Flowers should be handled carefully and placed in fresh clean water, preferably distilled, with leaves removed from stem sections below the water level. There is no advantage in a depth of water greater than 10–15 cm. Commercial preservative treatments should be used where available and excessively warm dry atmospheres should be avoided.

Wherever possible practical advice has been provided on the storage of fruits and vegetables that might be grown by the amateur gardener. In addition there are many other procedures used by commercial growers and also described in this chapter that may provide the more adventurous amateur with a basis for experimentation. One of the joys of gardening is to keep experimenting, learning from one's failures and rejoicing in one's successes.

FURTHER READING

Blanpied, G.D., Bartsch, J.A. & Hicks, J.R. (eds) (1993) *Proceedings from the Sixth International Controlled Atmosphere Research Conference*. Cornell University, Ithaca, New York.

Fidler, J.C., Wilkinson, B.G., Edney, K.L. & Sharples, R.O. (1973) *The Biology of Apple and Pear Storage*. Research Review No. 3, Commonwealth Agricultural Bureau, Slough.

Hardenburg, R.E., Watada, A.E. & Wang, C.Y. (1986) *The Commercial Storage of Fruits, Vegetables, and Florist and Nursery Stocks*. USDA, Agriculture Handbook No. 66.

Johnson, D.S. (1999) Controlled atmosphere storage of apples in the UK. Proceedings of the International Symposium on Effect of Preharvest and Postharvest Factors on Storage of Fruit. *Acta Horticulturae*, **485**, 187–93.

Kays, S.J. (1991) *Postharvest Physiology of Perishable Plant Products*. Chapman & Hall, London.

MAFF (1979) *Refrigerated storage of fruit and vegetables*. Reference Book 324, The Stationery Office, London.

Roberts, J.A. & Tucker, G.A. (eds) (1985) *Ethylene and Plant Development*. Butterworths, London.

Seymour, G. B., Taylor, J. E. & Tucker, G. A. (eds) (1993) *Biochemistry of Fruit Ripening*. Chapman & Hall, London.

Shewfelt, R.L. & Bruckner, B. (eds) (2000) *Fruit and Vegetable Quality: An Integrated View*. Technomic Publishing Company Inc, Pennsylvania.

Wills, R., McGlasson, B., Graham, D. & Joyce, D. (1998) *Postharvest: An Introduction to the Physiology & Handling of Fruit, Vegetables & Ornamentals*, 4th edn. CAB International, Oxford.

Glossary

Abscisic acid (ABA) A plant hormone that is particularly involved in managing the water economy of plants. ABA plays a central role in the control of stomatal closure.
See also **stoma**.

Abscission The separation or falling away of leaves, flowers, fruits or other plant organs. See also **abscission zone**.

Abscission layer See **abscission zone**.

Abscission zone A zone at the base of a leaf, flower, fruit or other plant organ in which a layer of cells (the **abscission layer**) disintegrates to aid the falling away of that organ, as in leaf fall from **deciduous** trees.

Accessory pigment A pigment such as **chlorophyll b** and **carotenoids**, that absorbs light energy in different parts of the spectrum from **chlorophyll a** and then passes it to **chlorophyll a** so that it may be utilised in photosynthesis.

Acclimation Biochemical and/or physiological changes made by plants in response to their environment which improve their survival capacity.

Achene A small, **indehiscent** fruit, usually with a single seed, surrounded by a thin, dry fruitwall (the **pericarp**), as in members of the family Ranunculaceae (e.g. *Clematis* spp.).

ADP: Adenosine diphosphate A chemical molecule formed when **ATP** is broken down to release its energy.

Adventitious Of buds, shoots or roots that grow from an unusual place on the plant, as when roots arise direct from a stem in a cutting.

Aerenchyma A tissue consisting of long files of gas-filled spaces that allow oxygen to diffuse through the plant. This tissue is important for plants that grow in waterlogged conditions or in water.

After-ripening A period of air-dry storage that some seeds must undergo before they will germinate.

Algae (sing. **a**). A term used loosely to describe simple, unicellular or multicellular photosynthetic plants that lack a true vascular system and are not differentiated into roots, stems and leaves. They are found in wet places or, more usually, freshwater or marine habitats. See also **phytoplankton**.

Allele Usually two, sometimes more, forms of a **gene** occurring at the same relative position (**locus**) on each of a pair of **homologous chromosomes**. Alleles may be **dominant** or **recessive**. The effects of a **dominant allele** may be seen in the form or growth of a plant (the **phenotype**), whether the allele is paired with an identical dominant allele or with a recessive form of the allele. The effects of a **recessive allele** are only observed in the phenotype if it is paired with another identical, recessive allele or if the tissue is **haploid**.

Allelic variation The difference in a particular characteristic (trait) of a plant **phenotype** controlled by two or more **alleles**.

Anion A negatively charged **ion** such as nitrate (NO_3^-) or sulphate (SO_4^{2-}).

Anion exchange capacity (AEC) A measure of the ability of a soil to adsorb (bind) **anions**. To adsorb anions, the surface must be positively charged. In soils of temperate regions, few such surfaces generally occur, but in tropical soils the clay mineral kaolinite and oxides of iron and aluminium are widespread and can have an AEC.

Antheridium (pl. **ia**). Male sex organ (**gametangium**), within which **gametes** are produced in fungi and lower plants such as **algae**, **liverworts**, **mosses** and **ferns**.

Anthocyanins Coloured pigments belonging to a group of chemicals called **flavonoids**. They accumulate in **vacuoles** and are particularly important in determining the colours of flowers, although they also occur in stems, leaves and fruits.

Antiflorigen See **florigen**.

Apical dominance The term used for the process in which the apical bud prevents the growth of the lateral buds below it.

Apical meristem See **meristem**.

Apomixis The production of a seed by a non-sexual process.

Apoplast The space, permeable to water and dissolved substances, comprising the non-living, unthickened **cell walls** and **intercellular spaces** of a plant tissue.

Arbuscular mycorrhiza See **vesicular-arbuscular mycorrhiza**.

ATP: Adenosine triphosphate A chemical molecule of fundamental importance in plant cells as a carrier of energy. See also **ADP**.

Auxins The first category of plant hormones to be recognised. The major naturally occurring auxin is β, **indolyl-acetic acid (IAA)**, but several synthetic auxins are used in horticulture. They have several effects in plants including the promotion of root initiation.

Available Water Capacity (AWC) The portion of water in a soil that can be readily absorbed by plant roots. It is the water held between field capacity (the water content at which a soil ceases draining) and permanent wilting point (the water content at which a plant will wilt and not recover).

Axil The upper angle between a leaf **petiole** or small branch and a stem.

Axillary bud A bud occurring in an **axil**.

Bacterium (pl. **ia**) A very simple, usually unicellular **microorganism** that reproduces by dividing or by forming spores. Bacteria do not have a clearly defined nucleus and the genetic material usually takes the form of a single, circular chromosome. They may occasionally be **photosynthetic** (see **Cyanobacteria**) but most gain their nutrients by breaking down dead organic matter or by developing a **symbiotic** association with a plant or animal host.

Betanin The pigment which gives the colour to red beets.

Bilateral symmetry Exhibited by a flower that is symmetrical in one plane only and therefore may appear flattened (e.g. the flowers of sweet peas,*Lathyrus* spp.).

Binomial A system of naming plants, formalised by Carl von Linné (Linnaeus), in which two names are given in Latin: the **genus** (with a capital initial letter), followed by the **species**, with a lower case initial letter (e.g. *Bellis perennis*, the common daisy). Both names are written in italics.

Biocide A substance that kills living organisms.

Bioregulant A substance that, when applied to a plant, changes or controls its growth. See also **plant growth regulant**.

Biotroph A plant **pathogen** that obtains its nutrients from the living cells of its host and usually has little ability to live apart from it.

Bitter pit A condition of apple fruit, caused by calcium deficiency, in which the skin develops sunken brown spots less than 1 mm in diameter and the flesh exhibits numerous pale brown spots. Affected apples may have a bitter taste.

Blue-greens See **Cyanobacteria**.

Botanical key A device for identifying plants in which a series of paired, mutually exclusive, statements each leads to a further pair of statements and so on, and eventually to the identity of the plant.

Boundary layer A layer of air between the leaf surface and the surrounding air. This normally contains more water vapour than the surrounding air.

Bract A reduced or modified leaf with a flower in its **axil**. Bracts may function as petals, as in *Euphorbia* spp. (spurges).

Bud An immature shoot or flower protected by a casing of **scale leaves**.

Buffering The ability of a soil to maintain the concentration of an **ion** in solution. Because soils contain charged surfaces, exchange of adsorbed (bound) **cations** from these surfaces with the soil solution can maintain the concentration despite, for example, uptake by plant roots. Generally, clayey soils are better buffered than sandy soils.

Bulb A storage organ consisting of fleshy modified leaves (often called **scales**) that are attached to a base plate, which is a compressed stem from which the roots grow.

Bundle sheath A single layer of cells ensheathing a vascular bundle.

C-3 plants Plants in which the first products of photosynthesis contain three carbon atoms.

C-4 plants Plants in which the first product of photosynthesis contains four carbon atoms.

C:N ratio The ratio of the weight of carbon to that of nitrogen in plant material or soil. Soils typically have a ratio in the range 10–15:1, but in plant materials the ratio varies considerably (from 10–100:1) depending on the age and nature of the material.

Calcicole A plant that prefers chalky or calcareous soils.

Calcifuge Plants that are adapted to growing on acid soils and do not grow well on chalky soils, often showing yellowing of the leaves.

Callus tissue The undifferentiated mass of cells produced at the site of a wound.

CAM/crassulacean acid metabolism A type of **photosynthesis** in which plants open their **stomata** at night and temporarily store carbon dioxide in the form of malic acid. During daylight, the stomata are closed and the malic acid is broken down to release carbon dioxide, which is then re-fixed into the usual products of photosynthesis, namely sugars.

Cambium A layer of cells, between the **xylem** and **phloem** of stems and roots, capable of dividing to produce secondary xylem and phloem, thereby increasing the girth of the organ. A specialised cambium, the **cork cambium (phellogen)**, arises in the outer layers of older stems or roots to produce a protective, corky layer of cells.

Carotenes **Carotenoid** pigments consisting of only carbon and hydrogen atoms. They are mainly orange in colour.

Carotenoids A group of pigments that are orange or yellow in colour. Some are **accessory pigments** in photosynthesis, while others are responsible for some of the yellow and orange colours in flowers and fruits.

Carpel A female reproductive organ comprising an **ovary**, enclosing the **ovules**, usually with a terminal **style** tipped by a **stigma**. The ovules mature to become the seeds and the carpel wall thickens and matures to form the fruit.

Cation A positively charged **ion** such as potassium (K^+) or calcium (Ca^{2+}).

Cation exchange capacity (CEC) A measure of the ability of a soil to adsorb (bind) **cations**. To adsorb cations, the surface must be negatively charged. There are many such negatively charged surfaces in soils, including various clay minerals, living roots and dead organic matter.

Catkin A pendulous **spike** in which the individual flowers are usually of one sex only.

Cell The basic unit of the body of a living organism. Organisms may be unicellular (e.g. some **algae**) or multicellular, as in most plants. Plant cells have a **cell wall**, **plasmalemma**, **cytoplasm**, **nucleus**, **organelles** and **vacuoles**.

Cell wall The rigid wall, comprised principally of **cellulose microfibrils**, proteins and sometimes **lignin** or **suberin** embedded in a **matrix** of **hemicelluloses** and **pectins,** that surrounds the plant cell.

Cellulose A **polysaccharide** molecule composed of chains (**polymers**) of glucose molecules that is the principal structural material of the plant **cell wall**.

Cellulose microfibril An aggregation, usually crystalline, of **cellulose** molecules found in plant **cell walls**.

Centromere A structure which joins the two halves of a **chromosome** (the **chromatids**). The centromere becomes attached to the nuclear spindle during nuclear division.

Chelate An organic compound (a compound containing carbon) containing a metal ion.

Chemical dwarf A plant that has been dwarfed by the application of a chemical known as a growth retardant.

Chimera A plant made up of genetically different cells.

Chipping A method used to propagate **bulbs** by dividing them vertically into small pieces (chips). Chipping can also refer to the practice of removing a small part of a seed coat in order to facilitate water uptake and germination of hard seeds.

Chlorophyll a The green pigment in plants capable of absorbing the radiant energy of sunlight (mainly from the red and blue part of the spectrum) during the first stages of **photosynthesis** Chlorophyll normally occurs in the **chloroplasts**.

Chlorophyll b An **accessory pigment** that

absorbs light energy and passes it to **chlorophyll a** for use in photosynthesis.

Chloroplasts Microscopic, membrane-bound **organelles** in the cells of plants that contain a complex system of membranes (**thylakoids**)and **chlorophyll** pigments. Chloroplasts are the sites of **photosynthesis**.

Chromatids The two daughter strands of a **chromosome** that has undergone division. As a **nucleus** divides, the chromatids are pulled apart, so that one ends up in each of the two daughter nuclei.

Chromosome A thread of **DNA** and protein that carries genetic information in the form of **genes** arranged in a linear manner. Chromosomes are located in the **nuclei** of all plant cells. They occur in **homologous** pairs and each species has a characteristic number of chromosomes.

Chromatophore A specialised type of cell containing **carotenoid** pigments.

Chromophoric group The light-absorbing region of a pigment that is responsible for its colour.

Circadian clock An internal clock that keeps time by going through a cycle that returns to the same point at approximately 24-hour intervals. The term circadian means 'about a day' and is used because the cycles not exactly 24 hours long.

Cladistics An approach to taxonomic analysis in which organisms are classified according to their evolutionary relationships. See also **taxonomy**.

Cladode (phylloclade) A specialised, photosynthetic stem that resembles a leaf.

Clamp A traditional structure for storing vegetables (such as potatoes, beet, turnips, swedes and carrots). A clamp consists of a pile of the vegetables to be stored, encased with layers of straw and soil to provide protection from frost and rain, and with ventilation holes plugged with loose straw to prevent the build up of carbon dioxide and the depletion of oxygen.

Clone A group of genetically identical genes, cells or individual plants derived from a single, common ancestor by non-sexual means.

Colour wheel A wheel of colours that begins with yellow and circles through orange, red, purple, blue, green and back to yellow. Contrasting colours lie opposite to each other, while harmonious colours lie next to each other.

Codon A group of three **nucleotides** within a molecule of messenger **RNA** that acts as a unit specifying the code for making a specific amino acid during protein synthesis.

Coleoptile A sheath surrounding the apical meristem of a grass seedling that provides protection as the shoot grows towards the surface of the soil following germination.

Companion cells Specialised, metabolically active cells, each linked to a **sieve cell** in **phloem** tissue.

Composite flower A structure resembling a true flower that is in fact made up of a large number of smaller flowers, tightly pressed together on the surface of a **receptacle**.

Compost Organic residues (often with soil materials) that have been piled in a moist condition and allowed to decompose. The term is also used to describe a medium used to grow plants in pots and containers.

Conidium (pl. **ia**) An asexual, non-motile, usually wind dispersed spore produced by a fungus.

Controlled atmosphere storage A form of storage used mainly for fruit in which the atmosphere is controlled by increasing the concentration of carbon dioxide and, in some cases, also reducing the concentration of oxygen.

Convergent evolution The evolution of similar characteristics by unrelated organisms adapted to grow in a similar environment.

Core flush A physiological disorder of stored apples, resulting in a pink discolouration of the core, especially in fruit exposed to dull, sunless summer weather before harvest.

Cork A tissue made up of **cells**, impregnated with the fatty substance **suberin**, that have a protective function, especially in woody plants.

Cork cambium (phellogen) See **cambium**.

Corm/cormel A corm is a storage organ consisting of a compressed stem covered with dry scales. Cormels are small corms.

Corpus The cells comprising the central region of an **apical meristem**.

Cortex The layer of tissue in a plant stem or root, bounded on the outside by the **epidermis** and enclosing the **stele**.

Cotyledon A seed leaf, attached to the **embryo**. Cotyledons usually have a storage function in the seed and in some species act as the first photosynthetic organs following germination. In the Monocotyledoneae only one cotyledon is normally present, whilst in the Dicotyledoneae there are normally two and sometimes more.

Corymb An **inflorescence** in which the stalked flowers arise from a single axis, with the stalks of the flowers being progressively shorter from the base to the tip of the axis.

Cover crop A crop, usually mown grass, grown primarily to suppress weeds in permanent plantations.

Critical day-length This is the daily duration of light above which short-day plants do not flower or are delayed, or below which long-day plants do not flower or are delayed. The critical day-length differs between plants but is a constant for any one **species** or **cultivar**.

Cultivar The name used for a plant variety that has been raised in cultivation, in order to distinguish it from a wild species.

Cuticle A thin, non-cellular, waterproof, waxy layer containing **cutin** that covers the outer surfaces of the leaves and stems of higher plants.

Cutin A polymer of fatty acids that is the major component of the **cuticle** and gives it its waterproofing properties.

Cyclic lighting A schedule in which flowering is controlled by giving light intermittently (rather than continuously) during a night-break.

Cyanobacteria (sing. **um**). A group of **bacteria** which possess **chlorophyll a** and carry out **photosynthesis** Cyanobacteria are sometimes called **blue-greens**.

Cytokinins A group of plant hormones mainly originating in the roots. They have many effects, including the promotion of cell division and the inhibition of **senescence**.

Cytoplasm The living part of a **cell** enclosed by the **plasmalemma**, but excluding the **nucleus**.

Cytoplasmic Male Sterility A condition in which functional pollen is not produced, determined by genetic information contained within **organelles** (usually **mitochondria**) in the **cytoplasm**.

Dark-dominant Plants in which flowering depends on whether or not a critical duration of darkness is exceeded Most are short-day plants.

Day-neutral plants Plants in which flowering is not controlled by **day-length**.

Day-length The duration of light in a 24-hour period.

Deciduous Used to describe parts of a plant, usually leaves, that are shed seasonally, for example at the onset of winter or of a dry season. Also used to describe a plant that sheds its leaves seasonally.

De-etiolation Changes that occur in dark-grown plants when they are exposed to light, such as a reduction in stem elongation and the expansion of leaves.

Dehiscent Of a structure, usually a fruit, that bursts or splits open as it matures to release its contents, usually seeds.

Dehydrins Proteins that accumulate in plants in response to the experimental application of **abscisic** acid and any environmental influence that has a dehydration component, such as freezing, salinity and drought.

De-vernalisation The process by which the vernalising effect of a previous exposure to cold is reversed by subsequent exposure to high temperatures. See also **vernalisation**.

Differentially permeable membrane A membrane that allows the passage of some molecules, such as those of water, but not of others, such as those of dissolved minerals (see also **semipermeable membrane**).

Diffusion A process in which different substances become mixed as a consequence of the random movement of their component atoms, ions or molecules.

Diffusion pathway The path followed in the diffusion of a gas or a solute from a region of relatively high concentration to a region of lower concentration.

Diploid Of a tissue or organism in which the majority of cells contain two complementary sets of chromosomes (the normal state). See also **haploid**.

Discontinuous variation Clear differences in a character that can be observed in a population of organisms. Such differences are brought about by **isolating mechanisms** that for plants may be: reproductive (plants cannot hybridise because they flower at different times); ecological (plants cannot hybridise because they have become adapted to separate environments); and distributional (plants cannot hybridise because they have become stranded on different mountain tops or in different valleys).

DNA: Deoxyribonucleic acid The genetic material of most living organisms, comprising a double helix of **nucleotides** in which the principle sugar is deoxyribose, linked by the hydrogen bonds between complementary pairs of bases (cytosine and guanine and thymine and adenine). The sequence of the bases constitutes the genetic code.

Dominant allele See **allele**.

Dormancy Seeds or buds that fail to grow when the environmental conditions are favourable for growth are termed 'dormant'.

Dormancy: double A type of **dormancy** in which seeds have first to be exposed to a warm temperature followed by a chilling treatment, a second exposure to a warm temperature and finally a further exposure to cold.

Dormancy: epicotyl A type of seed dormancy in which the root emerges in warm temperatures but the shoot apex is dormant and fails to grow until after the seed plus root system is exposed to cold.

Dormancy: enforced Seeds or buds that fail to grow because the environmental conditions are unfavourable. This is not considered to be true dormancy.

Dormancy: induced This occurs when seeds that are capable of germination are exposed to conditions that lead to the dormant state.

Dormancy: innate A type of dormancy that is under genetic control but can be influenced by the environmental conditions during maturation.

Ectomycorrhiza A **mutualistic symbiosis** involving a fungus and the roots of a plant, in which the fungus is restricted to a sheath around the outside of the root and the spaces between the surface cells.

Embryo The rudimentary diploid plant which develops after fertilisation and is contained within the seed; or a rudimentary plant which develops from a plant cell grown in **tissue culture**.

Endodermis The innermost layer of cells of the root **cortex,** surrounding the **stele** The cell walls are partially or completely thickened with **suberin**, restricting the free diffusion of water and other solutes into and out of the **stele**, thereby allowing the cells of the endodermis to regulate the passage of such substances.

Endomycorrhiza A **mutualistic symbiosis** involving a **fungus** and the roots of a plant in which the fungus grows within the tissues of the root and may penetrate the cells. See **vesicular-arbuscular mycorrhiza**.

Endosperm A nutritive tissue, derived from nuclei in the **embryo** sac, within the developing **seed**.

Entropy A measure of the disorder in a system: the higher the entropy, the greater the disorder.

Enzyme A protein that acts in minute amounts in biological systems to promote chemical changes without being changed itself.

Epidermal hair (trichome) A hair-like structure that is an outgrowth of an epidermal **cell** on a leaf or other plant organ. See **epidermis**.

Epidermis The outermost layer of **cells** in a plant.

Epinasty The rolling-under of leaves usually caused by **ethylene**.

Ethylene A gaseous hormone that is important in **senescence** and fruit ripening. It also causes **epinasty** in plants growing in water-logged soils.

Etiolation Plants growing in the dark do not produce **chlorophyll** and are said to be etiolated. Dark-grown seedlings are also elongated, have few fibres and reduced leaf expansion.

Evocation A term used for changes at the shoot tip that result in flowering.

Explant A small piece of tissue cut from a plant for propagation or establishment of a **tissue culture**.

F1 The first (filial) generation of plants following cross pollination of two parental lines. See also **F1 seeds**.

F1 seeds The first offspring of a cross between two distinct cultivars. Plants produced from F1 seeds do not breed true. See also **F1**.

F2 The second (filial) generation of plants, produced by self-pollinating or inter-pollinating individuals of the **F1** generation.

False fruit A fruit in which the **receptacle**, rather than the ripened ovary and its contents, constitutes the most conspicuous part, as in a strawberry.

Fan and pad cooling An evaporative cooling system used by commercial growers in which the outside air is drawn over wet pads into the greenhouse.

Fasicular cambium A **cambium** involved in the production of secondary **xylem** and **phloem** that occurs within the vascular bundles, between the primary xylem and phloem.

Feeder roots Delicate, short-lived, branch roots that are actively involved in the uptake of minerals and water.

Ferns (Filicinophyta). The most advanced and numerous **pteridophytes**, in which there is in the life cycle an alternation between a large, leafy **sporophyte** generation (the 'fern'), which reproduces by means of asexual, **haploid spores** and a very small, simple, free living **gametophyte** generation that bears the sexual structures and, after fertilisation, gives rise to a new **sporophyte**.

Fertiliser Any organic or inorganic material of natural or synthetic origin that is added to the soil/plant system to supply nutrients essential for plant growth. Most fertilisers are applied to soil, but some can be applied direct to foliage.

Fertiliser: compound A **fertiliser** containing more than one nutrient; for example 'Growmore'.

Fertiliser: controlled release A **fertiliser** that is treated so that it does not dissolve rapidly and the nutrients are released in a controlled manner over a prolonged period.

Fertiliser: straight A **fertiliser** containing a single compound; for example, muriate of potash.

Field heat (or **sensible heat**) The heat held by a crop at the time of harvest.

Flavonoids A group of water-soluble pigments responsible for the colours of many flowers and some fruits. They include anthocyanins, betalains, flavones and flavonols.

Flora All the plants in a given area or ecosystem, or a book or database designed to list and assist in the identification of individual plants from a specific area, region or country.

Floral stimulus An unknown chemical stimulus that is exported from the leaves and causes flowering at the stem apex and lateral buds.

Florigen/antiflorigen Florigen is the name that has been historically given to the unknown chemical substance that is exported from the leaves and causes flowering. In contrast, antiflorigen is the name that has been given to an unknown chemical substance that is exported from the leaves and prevents flowering. See also **floral stimulus**.

Fogging The injection of water-vapour into the air in order to maintain a high humidity around plants in a greenhouse.

Fruits: climacteric Fruits that produce a burst of **ethylene** gas (the climacteric) during ripening.

Fruits: non-climacteric Fruits that do not produce a burst of **ethylene** gas during ripening.

Fungus (pl. **i**). Organisms (often **microorganisms**) that lack chlorophyll, usually grow as filaments (**hyphae**) and reproduce by means of **spores**. Fungi obtain their nutrients by breaking down dead organic matter or by forming a **symbiotic** relationship with a plant or animal host.

Gametangium (pl. **ia**). A structure in which **gametes** are formed, as in mosses, ferns and fungi.

Gamete A specialised, **haploid** sex cell, which fuses, during fertilisation, with another gamete of the opposite sex or mating type to form a **diploid zygote**.

Gametophyte The **haploid** stage in the life cycle of a plant, during which **gametes** are produced.

Gemmae (sing. **a**). Asexual (i.e. non-sexual) reproductive structures, usually comprising small balls of cells, produced by some **thallose liverworts** in gemmae cups.

Genetic code The series of triplets of bases of **DNA** that controls the processes leading to the synthesis of a specific protein or proteins.

Genetic engineering See **genetic modification**.

Genetic locus The position of a specific gene on a chromosome (see **allele**).

Genetic modification (GM) A process whereby **genes** specifying a particular characteristic are identified and extracted from the cells of a plant species and transferred in the laboratory to the cells of another individual (which may be of a different species) where they are incorporated into the genetic material and expressed.

Genome All the genes carried by a single set of **chromosomes** of an individual.

Genome sequencing The determination by research of the entire **genetic code** of an organism.

Genotype The genetic information contained

within an organism, as opposed to its physical appearance (**phenotype**).

Genus (pl. **era**). The rank in the taxonomic hierarchy in plants, occurring between family and species. In a **binomial** plant name the genus occurs first, has a capital initial letter and is italicised (e.g. *Bellis* in *Bellis perennis*, the common daisy).

Germ cell A cell that produces a **gamete**.

Germplasm The hereditary material of an individual that is transmitted to the **germ cells** of offspring during sexual reproduction. Often used loosely to describe seeds or plants held in a collection such as 'gene bank'.

Gibberellins A group of plant hormones that is widespread throughout the plant kingdom. A major function is the control of stem elongation but they also have many other effects including the modification of flowering and the promotion of germination.

Glycoprotein A complex chemical substance made up of a carbohydrate linked to a protein.

Graft A union between two different plants.

Granum (pl. **a**). Part of the internal contents of a chloroplast, consisting of a stack of membranous discs (**thylakoids**). There may be up to 40 to 80 grana in a single chloroplast.

Growth retardant A generic term for a group of synthetic chemicals that reduce stem length.

Guard cell A specialised leaf or stem **epidermal** cell, pairs of which surround each **stomatal pore**. Guard cells have differentially thickened walls and changes in the **turgor** of the cells cause the opening or closing of the **stomata**. See also **epidermis**.

Haploid (n) Used of a **nucleus** or **cell** containing only a single set of chromosomes, formed as a result of meiosis of a diploid nucleus or cell, as in the production of **gametes**.

Hard seeds Hard seed coats inhibit germination by preventing the uptake of water into the dry seed. Germination cannot occur until the seed coat is rendered permeable by abrading, chipping or natural breakdown in the soil.

Hartig net The network of fungal hyphae that permeates the outer layers of root cells in an **ectomycorrhiza**.

Heartwood The wood occupying the central region of a tree trunk or branch, consisting mainly of **xylem** vessels and fibres, impregnated with oils, gums and resins that may give it a dark appearance. The xylem of the heartwood provides support but is not normally involved in the transport of water.

Hemicelluloses A miscellaneous group of chains (**polymers**) of various sugars, mainly glucose and xylose, that occur in the **matrix** of the plant **cell wall**.

Herbarium (pl. **ia**). A collection of dried or otherwise preserved plants, together with detailed information concerning their collection from the field, used in the study of plant classification and evolution.

Heterozygous Having different genetic information at a particular point (**locus**) on the complementary chromosomes of a **diploid cell**.

Hexaploidy A form of **polyploidy** in which an individual possesses three times ($6n$) the **chromosome** complement of a normal **diploid** ($2n$) individual.

Homologous chromosomes In diploid organisms, a pair of chromosomes having the same pattern of genes along their length, one chromosome being derived from the female parent and the other from the male.

Homozygous Having identical genetic information at a particular point (**locus**) on the complementary chromosomes of a **diploid** cell.

Hormone See **plant hormone**.

Horsetails Primitive, spore forming land plants belonging to the Phylum Sphenophyta, a group that together with the ferns (Filicinophyta) is often placed in the informal group **Pteridophyta**.

Hybrid vigour The increase in vigour resulting from a cross (hybridisation) between two, often inbred, lines of a species or cultivar. It is always accompanied by increased heterozygosity.

Hybridisation To produce seed by crossing two genetically different individuals. A hybrid is a plant resulting from such a cross.

Hydrolysis A chemical reaction between a compound and water, resulting in the splitting of the compound which is then said to have been hydrolysed.

Hypertrophy Excess swelling (usually of shoots at the base of the stem) as a result of treatment with hormone weedkillers or in response to water-logged soils. In both cases, the gaseous hormone, **ethylene**, is probably involved.

Hypha (pl. **ae**). The tube like filament that is the basic body form of most **fungi**.

Hypocotyl In an embryo or developing plant, the region of the axis between the cotyledons and the root.

IAA (β, indoly-acetic acid) The major naturally occurring **auxin** in plants.

Illuminance A measurement of light in terms of the sensitivity of the human eye.

Imbibition A physical process through which water can enter cells. The first stage of water entry into dry seeds is by imbibition.

Immobilisation The conversion of an element (commonly nitrogen, phosphorus or sulphur) from an inorganic to an organic form by microorganisms, so that the element is not readily available to plants or other organisms.

Inbreeding depression The loss of vigour, fertility and yield often associated with inbreeding (fertilisation of the flowers of a plant with its own pollen, resulting in an increase in the number of homozygous recessive genes).

Incompatibility The inability of one plant to successfully fertilise another; or of a particular scion to form a successful **graft** with a rootstock; or of a pathogen to infect a particular host. See also **incompatibility (graft)**.

Incompatibility (graft) When two different plants are unable to produce a stable graft union.

Indehiscent Not **dehiscent**.

Indeterminate Describing the growth of a flower spike in which extension continues unchecked as the lower flowers open; or of any other plant or fungal structure in which growth continues unchecked.

Inflorescence An arrangement of flowers on a stem of a plant.

Inoculum Spores, cells or particles of a pathogen (e.g. **fungus**, **bacterium** or **virus**) with the potential to infect a plant or a batch of soil.

Integrated pest and disease management The integrated use of all methods available for the control of pests and diseases that minimises the use of chemicals.

Integuments The outer coats (usually two) of the **ovule** which, after fertilisation, develop into the **testa** (seed coat).

Intercellular spaces The spaces between cells, occupied by (for example) water vapour, air, carbon dioxide and oxygen, depending on circumstances.

Interfasicular cambium A **cambium** involved in the production of secondary **xylem** and **phloem** that arises between the vascular bundles in the unthickened stem.

Intermediate-day plant A plant that will only flower, or will flower most rapidly, when the day-length is neither too long nor too short.

Intermittent mist system A system used in propagation that employs an artificial leaf to switch on spray jets of water each time its surface dries out.

Interstock A piece of another plant inserted between two incompatible plants during grafting. In some cases, the interstock overcomes the incompatibility and enables a successful **graft** to be established.

Ion An atom or group of atoms with an electric charge. See also **anion** and **cation**.

Irradiance The amount of light energy received at any one time.

Isolating mechanisms (distributional, ecological and reproductive) See **discontinuous variation**.

Juvenile phase The period during which plants exhibit **juvenility**.

Juvenility The name given to an early phase of growth (often lasting for many years in trees) during which flowering cannot be induced by any treatment. Plants sometimes have a characteristic morphology during their juvenile phase of growth, as in ivy (*Hedera helix*) and many *Eucalyptus* spp.

Koch's postulates A series of requirements that must be fulfilled if a specific pathogen is to be unequivocally identified as the cause of a particular disease.

Lamina The flat blade of a leaf.

Leaf primordium (pl. **ia**). The group of **cells** at the stem apex or in a **bud** that is the precursor of a leaf.

Leafy liverworts Members of the primitive Phylum of spore-producing land plants called the Hepatophyta in which the **gametophyte** is a small plantlet with rows of thin flattened leaves on either side of and below a delicate stem.

Lenticel A pore, filled with spongy cells, in the bark of a woody plant that allows air to reach the underlying tissues.

Lichen A composite organism formed by the

symbiotic association of a **fungus** with a photosynthetic **alga** or **cyanobacterium** (blue-green). See also **symbiosis**.

Light integral The total amount of light received during a period of time.

Light-dominant Plants in which flowering depends not only on the duration of the dark period but also on the kind of light during the day. Most are **long-day plants**.

Light saturation point The rate of photosynthesis increases with increasing light levels until, above a certain value, there is no further increase. This value is known as the light saturation point.

Lignin A tough, aromatic polymer deployed as a strengthening material in plant cell walls, particularly in the **xylem** The major component, therefore, of wood.

Liverwort The common name for the primitive, flowerless land plant belonging to the group Hepatophyta. See **leafy liverwort** and **thallose liverwort**.

Locus See **genetic locus**.

Long-day plant A plant that will only flower, or will flower more rapidly, when the **day-length** is longer than a particular value, known as the critical day-length.

Low temperature breakdown Deterioration of apple fruits during low temperature storage following a cool, wet summer.

Lux A unit of light (illuminance) that is related to the sensitivity of the human eye.

Macronutrient A chemical element that is essential for plant growth and generally needed in large quantities if the plant is to grow well. The macronutrients are: nitrogen, potassium, phosphorus, sulphur, calcium and magnesium.

Manure The excreta of animals, often mixed with either bedding materials or litter or both, in varying stages of decomposition.

Marker-assisted breeding Plant breeding in which a **phenotype** that may be difficult to identify by eye (e.g. disease resistance) is identified by the presence of a particular DNA sequence always associated with the presence of that **phenotype**.

Matrix The amorphous mass of **hemicellulose** and **pectic polymers** in which the **cellulose fibrils** of the **cell wall** are embedded.

Maturity (commercial) The stage of maturity at which plant structures are at their optimal stage of development for commercial use.

Maturity (physiological) This refers to progress through a series of developmental stages that typically include the growth, maturation, ripening and **senescence** of a plant or plant organ.

Medullary rays See **rays**.

Meiosis (reduction division) Two nuclear divisions, one after the other, resulting in the reduction of the chromosome complement of the **nucleus** from **diploid** to **haploid**, as in the production of **gametes**. In meiosis, one diploid nucleus gives rise to four haploid nuclei, each in a new daughter cell.

Membranes Sheet-like, living structures that cover, line or occur in cells. Plant membranes consist of two layers (a bilayer) of lipid (fatty molecules) and associated proteins. See also **plasmalemma**, **tonoplast** and **thylakoid**.

Meristem A plant tissue composed of **cells** capable of dividing and thereby giving rise to new growth. Such tissues occur principally at the stem and root **apices**, the tissue between the **xylem** and **phloem** (**cambium**), and in developing leaves.

Meristem culture A type of **micropropagation** involving the growth and proliferation of excised stem **apical meristems**, on a culture medium in a laboratory to produce large numbers of new, **clonal** plants. Meristem culture is used frequently to rid vegetatively-propagated plants of viruses.

Methyl jasmonate A gaseous chemical messenger that may be involved in the triggering of resistance to pathogens in plants, including those adjacent to an infected plant.

Micronutrient A chemical element that is essential for plant growth but generally needed in only small quantities. The micronutrients are: iron, zinc, chlorine, copper, manganese, boron, molybdenum and nickel.

Microorganisms Organisms such as **bacteria**, **fungi** and **viruses** that are so small that they can only be observed with the aid of a microscope.

Micropyle A pore in the developing **ovule** through which the pollen tube passes during fertilisation; and later, when the seed matures and begins to germinate, through which water enters.

Micro-cuttings Shoot tissues that are removed from **tissue cultures** and rooted in a normal rooting medium.

Micropropagation **Vegative propagation** of plants by induction of the proliferation in culture of new plantlets from small **explants** of parental tissues. See **meristem culture**.

Middle lamella The gluey material, mainly **pectins**, between adjacent plant **cell walls** that serves to cement them together.

Midrib The main, central vein of a leaf or leaflet.

Mineralisation The conversion of an element from an organic (i.e. containing carbon) form to an inorganic (i.e. not containing carbon) form as a result of decomposition by microorganisms. The elements nitrogen, phosphorus and sulphur are commonly made available to plants by this process.

Mitochondrion (pl. **ia**). A membrane-bound, subcellular **organelle** in which respiration occurs in living cells.

Mitosis Nuclear division, normally followed by cell division, in non-sexual cells of living organisms, resulting in two identical daughter nuclei, each containing the same number of chromosomes, identical to those of the mother nucleus.

Monopodial The pattern of growth in which the apical growing point does not die back at the end of the season. See also **sympodial**.

Morphogenesis Changes in the shape and form of the plant during development or in response to changes in the environment.

Mosses The common name for flowerless, primitive land plants belonging to the Phylum Bryophyta. Mosses have an alternation of a leafy **gametophyte** generation and smaller, non-leafy **sporophyte** generation attached to the gametophyte.

Mutation Any change in the **genetic code**, often leading to a changed **phenotype**, that may occur by chance or may be induced by certain chemicals, radiation, **transposons** or the activity of **viruses**. Natural mutations are known to gardeners as **'sports'**.

Mutations: germ cell **Mutations** occurring in the **cells** that give rise to **gametes**. They may thus be transmitted to the next sexual generation.

Mutations: somatic **Mutations** occurring in cells that comprise the body (leaves, stems, roots, etc.) of the plant. They are not transmitted from one sexual generation to another, but may be perpetuated by **cloning**.

Mutualistic symbiosis A **symbiosis** in which both partners benefit.

Mycorrhiza A **mutualistic symbiosis** in which a fungus associates with the roots of a plant.

Mycotoxin A toxin produced by a **fungus**.

Night-break lighting A term used to describe schedules in which the onset of flowering is controlled by giving a relatively short exposure to light (a night-break) at a particular time in the night.

Nitrogen fixation The process whereby some **microorganisms** incorporate nitrogen gas (N_2) from the atmosphere into water soluble nitrogenous compounds. Nitrogen-fixing microorganisms include free-living soil **bacteria** and **blue-greens (Cyanobacteria)**, **symbiotic bacteria** (e.g. *Rhizobium*) that form nodules on the roots of plants in the family Papilionaceae (e.g. peas and beans) and symbiotic bacterium-like Actinomycetes that form nodules on the roots of alders (*Alnus* spp.) and *Elaeagnus* spp.

Nitrogen-fixing bacteria See **nitrogen fixation**.

Node The point on a stem or branch at which one or more leaves, shoots, flowers or branches are attached.

Nucellus The tissue within the **ovule** of the plant that contains the **embryo** sac.

Nucleotide A unit structure of a nucleic acid (**DNA** or **RNA**), made up from a nucleoside (a molecule composed of ribose or deoxyribose sugar bound to a base) bound to a phosphate group.

Nucleus The membrane-bound **organelle** that in **cells** contains the genetic material.

Ooganium (pl. **ia**). The female sexual organ (**gametangium**) of some **algae** and **fungi** in the group Oomycota (e.g. *Phytophthora* spp.). Each ooganium contains one or more egg cells which, after fertilisation, form resting **spores** called oospores.

Organelles Structures within a cell that have specialised functions and are usually enclosed by a membrane (e.g. **chloroplasts**, **mitochondria**).

Organogenesis The organisation of an undifferentiated mass of cells into new **meristems** that can produce shoots and roots.

Osmosis The flow of water (a solvent) across a semi-permeable or differentially permeable membrane from a region of a relatively low concentration of dissolved substances (solutes), and therefore of high concentration of water (high water potential), to a region of higher concentration of solutes, and therefore of a lower concentration of water (low water potential).

Ovary Of a flower, the female reproductive structures.

Ovule A structure within an **ovary** that, after fertilisation, develops into a seed.

Oxaloacetic acid The first product of photosynthesis in **C-4 plants**.

Palisade mesophyll A tissue, usually on the upper side of the leaf, made up of columnar cells containing many chloroplasts. In most plants it is the principal tissue in which photosynthesis occurs.

Panicle A complex, branched **inflorescence**.

Parasitic A **symbiosis** in which one organism (the parasite) invades another (the host) and derives nutrients from it, often causing disease and debilitation. See also **pathogen**.

Parenchyma A tissue made up of large, thin-walled, relatively unspecialised **cells**.

Pathogen A parasite such as a **bacterium**, **fungus** or **virus** that causes disease in its host. See also **parasitic**.

Pattern genes The genes that control the variegation patterns in some plants. In this case, the plants will breed true from seed.

Pectic polymer (pectin) A molecule made up of long chains of simple sugars. Pectic polymers usually have a gel-like consistency and are major components of the **matrix** of the plant cell wall and of the **middle lamella**.

Periclinal chimeras Plants in which an outer layer of mutated cells overlays an inner core of the original genotype. See also **chimera**.

Pericarp The wall, sometimes three layered, of a ripe ovary or fruit.

Periderm An outer, protective layer of woody stems and roots, consisting principally of corky cells. A component of bark.

Perisperm A tissue derived from the remains of the **nucellus** that provides the main store of nutrients in some seeds.

Petals Leaf-like organs, usually brightly coloured, borne in a tight spiral or whorl in a flower.

Petiole The leaf stalk.

pH Negative logarithm of the hydrogen **ion** activity of a solution or cell content. See also **soil: pH**.

Phellogen See **cambium**.

Phenetics An approach to taxonomic analysis that brings together a wide range of characteristics (e.g. leaf length) of each plant in a broad comparison of similarity. See also **taxonomy**.

Phenolics A group of aromatic, often bitter compounds in plants that in the presence of air may be oxidised to dark compounds, as when a cut apple browns.

Phenotype The visible manifestation of a **genotype** (i.e. the actual appearance of a plant) resulting from the interaction between its genes (its **genotype**) and the environment in which it grows.

Pheromones Chemicals that are the sexual attractants for many insects.

Phloem The tissue, consisting mainly of **sieve cells**, **companion cells** and **fibres**, which transports organic substances (principally sugars) in vascular plants.

Photometric units Units that measure light in terms of the sensitivity of the human eye.

Photon A quantum of energy in the visible spectrum between 400 and 700 nm wavelength.

Photon flux density The number of **photons** falling on a unit area at any one time.

Photoperiodic induction The detection of **day-length** and the subsequent changes that take place in the leaf.

Photoperiodism A response to the length of day that enables an organism to adapt to seasonal changes in the environment.

Photorespiration A light-induced form of respiration in the chloroplasts of **C-3 plants** that may prevent the build up of toxic superoxide during periods of high photosythetic activity.

Photosynthesis The process whereby green plants capture the physical energy of sunlight (light reaction) and use this in the chemical reactions (dark reaction) in which carbon dioxide and water are used to make organic compounds (i.e. those containing carbon), notably sugars and other carbohydrates. Photosynthesis is dependent upon the green primary pigment **chlorophyll a**, which absorbs light energy in the blue and red regions of the spectrum, reflecting green light. In plants it occurs in the **chloroplasts**.

Photosynthetically active radiation The number of quanta per unit area in the visible spectrum between 400 and 700 nm. This is the most useful measurement of the amount of light available for **photosynthesis**.

Photosystems (I and II). The systems of photosynthetic reactions involved in the capture of light energy by chlorophyll.

Phototrophy The synthesis of foods using light energy (see **photosynthesis**).

Phototropism A process in which plants grow towards the direction of the incoming light.

Phylloclade See **cladode**.

Phyllomorphs These are found in some members of the family Gesneriaceae and consist of the leaf blade plus its petiole.

Phyllotaxy The arrangement of leaves in pairs, spirals or whorls on a stem.

Phytochrome A light-sensitive pigment that controls many developmental responses of plants to light. These include germination, stem elongation and flowering.

Phytoplankton The photosynthetic component of the **plankton**, comprising minute, mainly microscopic **algae**; they are confined to the surface layers of water, where light is available, and are responsible for 40–50% of the earth's photosynthetic productivity.

Phytotoxicity Of a substance that is toxic to plants.

Pith The tissue, usually composed of large, rounded **cells**, occupying the centre of a stem.

Pits Regions in a plant cell wall through which strands of cytoplasm pass (the **plasmodesmata**), linking the cell to adjacent cells.

Placenta The fused margins of the **carpel** or carpels on which the **ovules** are borne.

Plankton Minute, mainly microscopic organisms that float and drift with the currents in the surface layers of lakes or the sea. There are many millions of such organisms per cubic metre of water, and a significant proportion, the **phytoplankton**, is photosynthetic.

Plant growth regulator A synthetic chemical that regulates growth in a similar way to a naturally occurring **hormone**.

Plant hormone An organic compound that is synthesised by the plant and causes a physiological response at very low concentrations. In many but not all cases, the hormone is translocated and causes a response in a target region remote from the site of synthesis.

Plasmodesmata See **pits**.

Plasmalemma The outer membrane of a plant **cell** It is a living structure consisting of two lipid (fatty) layers (a bilayer) and associated proteins.

Ploidy The number of sets of **chromosomes** contained within a cell (see **haploid**, **diploid**, **triploid, tetraploid** and **hexaploid**.).

Plumule An embryonic shoot.

Polar nuclei Two **haploid nuclei** contained within the **embryo** sac of a flowering plant which fuse with another haploid nucleus (a sperm nucleus) to form the triploid (i.e. having three sets of chromosomes) **endosperm**.

Polarity Plant tissues have a polarity, which means that they can detect which way is up, even if they are removed from the plant.

Polyembryony An unusual situation in which there are many embryos in the same seed, which may have arisen spontaneously from body cells of a fruit (as in *Citrus* spp.) or by repeated division of a sexually-derived embryo (as in many orchids).

Polymer A chemical substance made up of chains of repeated molules.

Polyploidisation To render **polyploid** (i.e. having more than the normal two sets of chromosomes), either as a result of hybridisation or using a chemical (e.g. colchicine) that disrupts **mitosis**.

Polysaccharide A substance made up of chains (**polymers**) of sugar molecules.

Polytunnel A hooped supporting frame covered with a flexible plastic film inside which plants can be grown.

Procambial strand A strand of cells in a bud that will ultimately give rise to the **xylem** and **phloem** tissues of the stem.

Propagule The name given to any part of a plant that is used to start a new plant.

Prothallus (pl. i) A small, flattened structure that is the **gametophyte** (sexual) generation in the life cycle of a **fern** or **horsetail**.

Protoplast The living contents of a plant **cell**, bounded by the **plasmalemma**.

Protoplast fusion A laboratory process whereby naked **protoplasts** of plants are released from cells (either mechanically or using enzymes), induced to fuse in pairs and then induced to regenerate a new plant. The process may be used to produce hybrids of plants that cannot be crossed by normal pollination (e.g. in making interspecific hybrids).

Pulvinus A swollen region at the base of the **petiole** of some species capable of causing leaf movement.

Pteridophytes (Pteridophyta) A collective term used informally to refer to **ferns**, **horsetails** and their relatives.

Quantum (pl. **a**) A discreet unit of electromagnetic energy. Quanta within the wavelengths of the visible spectrum are known as **photons**.

Raceme An **inflorescence** in which flowers with stalks of equal length arise from a single axis.

Radial symmetry Exhibited by a flower that is circular in outline (e.g. the flowers of *Primula vulgaris*, primrose) and in which the flower parts radiate from a central point.

Radicle An embryonic root.

Rate of metabolism The rate at which a cell or tissue is respiring and carrying out metabolic functions (the chemical reactions of a living cell).

Rays Sheets of living cells extending radially across the secondary **xylem** and **phloem**. Those rays that link the **pith** with the **cortex** are called medullary rays, whilst those confined to the xylem and phloem are called vascular rays.

Receptacle The enlarged, elongated, flat, concave or convex end of a stem that bears the flower or flowers and later the fruit.

Recessive allele See **allele**.

Recombination The process during sexual reproduction whereby **genes** in an offspring are rearranged in combinations that are different from those in either parent.

Relative humidity The amount of water vapour present in the air expressed as a percentage of that which would be present if the air were saturated.

Replication: semi conservative A form of replication of **DNA** that uses an existing string of bases to create another double strand.

Respiration The process, which normally takes place in the **mitochondria**, whereby organic molecules (those containing carbon), usually sugars, are oxidised (combined with oxygen) to release energy for metabolism (the chemical processes of a living cell). Carbon dioxide is a byproduct of respiration.

Respiration rate The rate at which **respiration** is taking place.

Rhizoid In, for example, **mosses** and **liverworts**, a thread-like multicellular or unicellular structure that serves to anchor the plant and to absorb water and minerals.

Rhizome A specialised underground stem, which also serves as a storage organ.

Rhizosheath The soil attached to plant roots.

Rhizosphere The soil surrounding and influenced by plant roots.

Rhodopsin See **visual purple**.

Ribosomes Minute structures that are the sites of protein synthesis in a **cell**.

RNA: ribonucleic acid A nucleic acid characterised by the presence of the simple sugar ribose and the base uracil. It may occur in three forms, messenger-RNA, ribosomal-RNA and transfer-RNA, all of which are involved in protein synthesis.

Root cap The tissue, consisting mainly of large **cells**, many of which secrete mucilage, that covers the tip of the root, protecting it and facilitating its progress through the soil as the root grows.

Rootstock The plant upon which is grafted a shoot or bud (the **scion**) from another specimen, thereby providing the root system of the new combined plant.

Ruderal A plant with a short life cycle that is able to grow, flower and set seed rapidly if the ground is disturbed and competition is thus reduced. Many weeds are ruderals.

Sapwood The outer wood of a tree trunk or branch, consisting principally of xylem tissues which both transport water and provide support. See also **heartwood**.

Scale A modified leaf or leaf base that performs a storage function in a **bulb**.

Scale leaf A reduced leaf.

Scale/scaling/twin scaling Scales are the fleshy leaves of **bulbs** and can be used to propagate them by methods known as scaling and twin scaling.

Scion A term used in grafting for a piece of shoot placed in contact with a root system from another plant (the **stock**).

Scooping A method of propagating **bulbs** in which the basal plate (compressed stem) and growing point of the **bulb** are scooped out from below, leaving the outer rim of the basal plate intact.

Scoring A method of propagating **bulbs** in which deep cuts are made in the basal plate (compressed stem) and wedges of tissue are removed.

Seed: orthodox Seeds that can be dried to low moisture contents and stored at –18°C for long periods.

Seed: recalcitrant Seeds that are killed if their moisture content is reduced below some relatively high level (12–31%).

Seed vigour A measure of seed quality which relates to the ability of the seed to germinate and establish under a wide range of environmental conditions.

Self fertilisation The fusion of male and female gametes from the same individual. See also **selfed**.

Self incompatibility (SI) A genetically determined system to prevent or deter self fertilisation.

Selfed Pollinated with pollen from the same individual, leading to **self fertilisation**.

Semi-permeable membrane A membrane whose structure allows the passage only of solvent molecules (water in the case of plants) and prevents the passage of dissolved substances. See also **differentially permeable membrane**.

Senescence The deteriorative processes leading to death of a plant or a plant organ.

Sensible heat See **field heat**.

Sepals Leaf-like organs, usually green but sometimes coloured, that form the outer tight spiral or whorl of structures in a flower.

Shade avoider A term used for plants that elongate in response to shady conditions. This response enables them to outgrow their neighbours and reach better light conditions. Many arable weeds are shade avoiders.

Shade tolerator A plant, often from woodland habitats, that does not elongate in response to shade. Such plants have mechanisms that improve their ability to collect light under shady conditions.

Sheath See **bundle sheath** and **ectomycorrhiza**.

Short-day plant A plant which only flowers, or flowers earlier when the **day-length** is shorter than a particular duration known as the **critical day-length**.

Sieve cells Long cells that lack a nucleus and have perforated end walls (the **sieve plates**) and are joined together end to end to form **sieve tubes**. They are a major component of the **phloem** and transport sugars and other organic molecules such as **hormones** about the plant.

Sieve plates See **sieve cells**.

Sieve tubes See **sieve cells**.

Sink A tissue or organ in which a transported substance (e.g. sugars) or a diffused substance (e.g. CO_2) is consumed or sequestered, thus reducing its concentration at that place.

Soil: aggregate A group of soil particles cohering in such a way that they appear and behave for many purposes as a single unit.

Soil: horizons Horizontal layers in a soil that are usually distinguishable from each other by changes in soil colour or consistency.

Soil: organic matter The organic materials (i.e. those containing carbon) in soils which are a mix of live and dead animal, microbial and plant materials.

Soil: pH The negative logarithm of the hydrogen **ion** activity of a soil. A neutral soil has a pH of 7 while acidic soils have values less than this and alkaline soils have greater values. The typical range of soil pH is from 4 to 9.

Soil: pores The space in a soil not occupied by solids.

Soil: structure A description of the results of the ways in which the individual clay, silt and sand particles interact with each other and are bound together by other materials such as organic matter and various oxides into larger units. The arrangement of these secondary, structural, units and of the gaps between them allows a description of structure.

Soil: texture A description of the proportions of clay, silt and sand sized particles in a soil.

Solarisation Light-dependent inhibition of **photosynthesis** followed by bleaching of leaves.

This can occur when plants are transferred from shady conditions to bright sunlight.

Solutes **Ions** present in the soil solution or cell contents.

Somatic hybridisation See **protoplast fusion**.

Spadix A fleshy spike of flowers, often enclosed by an ensheathing leaf/petal-like structure (strictly a **bract**) called the **spathe**, as in *Arum* spp.

Spathe See **spadix**.

Species A taxonomic category or rank occurring below the level of the **genus** and referring to a group of related, morphologically similar, usually interfertile individuals. May be subdivided into subspecies, varieties and forma. It is a basic unit of nomenclature, being written after the genus, in italics, with a lower case initial letter, in a **binomial** (e.g. *perennis* in *Bellis perennis*, the common daisy).

Spike An **inflorescence** in which flowers without stalks arise from a single axis.

Spiral phyllotaxy The arrangement of the leaves in a spiral around the stem.

Spongy mesophyll A tissue located below the **palisade** in a leaf and consisting of large, irregular cells with wide air spaces between them.

Spore A minute propagule of a **fungus** or **bacterium**, functioning as a seed but without a preformed **embryo**.

Sporophyte The **diploid**, vegetative, often **spore**-bearing stage in the life cycle of plants (as opposed to the **gametophyte** or sexual stage). In the **fern** life cycle, the large, familiar leafy 'fern' is the **sporophyte** stage. In flowering plants the **sporophyte** and **gametophyte** generations are combined.

Sport See **mutation**.

Standard A plant specimen chosen by horticultural **taxonomists** to serve as the basis for naming and describing a new **cultivar**. The **standard** exhibits the precise characteristics of that **cultivar**.

Stele The vascular tissues of a root or stem, comprising the **xylem**, **phloem**, **pericycle**, **pith** and, in a secondarily thickened stem, the **rays**.

Stigma The surface on which pollen grains germinate in the female part of a flower.

Stipule A leafy structure, usually paired with another, at the base of the **petiole** in some species.

Stock See **rootstock**.

Stolon A specialised stem that may arise above or below ground and grows outwards to colonise new ground.

Stoma (pl. **ata**) A pore whose size may be varied (see **guard cell**) in the **epidermis** of a leaf or stem, allowing the controlled exchange of gases (usually carbon dioxide, oxygen and water vapour) between the plant and the atmosphere.

Stomatal pore See **stoma**.

Stooling The repeated cutting back of parent plants required for hardwood cuttings, in order to maintain their juvenility.

Stratification Layering seeds in moist soil or sand and maintaining them at low temperatures in order to break **dormancy**.

Stroma The matrix of a **chloroplast**, in which the membranes comprising the **grana** are embedded.

Style An extension of a **carpel** that supports the **stigma**.

Suberin A fatty, hydrophobic, waterproofing substance deposited in the cell walls of the **endodermis** and of **cork** tissue.

Substrate Either, a substance acted on by an enzyme; or, the soil or other medium to which an organism is attached and on which it grows.

Superficial scald A physiological disorder of apples in which the death of surface cells leads to skin browning during storage following growth in hot dry conditions. In the UK it is a serious problem only on 'Bramley's Seedling'.

Symbiosis Two organisms living together, either in a **mutualistic** or **parasitic** relationship.

Symplast The interconnected, living part of a plant. The symplast is a unit because protoplasts of adjoining cells are connected by the **plasmodesmata** passing through the **pits**.

Sympodial A type of branching in which the main axis is formed by a series of lateral branches, each having arisen from the one before. It occurs because the apical bud withers at the end of each growing season.

Tap root A large central root that is the main axis of a root system.

Taxon (pl. **a**) A named group of organisms of any taxonomic rank, such as a **genus** or **species**. See also **taxonomy**.

Taxonomist One who studies the scientific classification of organisms.

Taxonomy The scientific classification of organisms.

Teliospore A thick-walled, diploid resting **spore** in the life cycle of a rust fungus (Uredinales).

Telomere The end of a **chromosome**, consisting of repeated sequences of **DNA** that ensure that during nuclear division, each cycle of DNA replication has been completed.

Temperature inversion The term is used to describe the situation when frost occurs inside a **polytunnel** although the outside temperature remains above freezing.

Terminator gene technology A type of **genetic modification** that prevents germination of the seeds produced by the modified plant.

Testa The seed coat, which develops from the **integuments**.

Tetraploid A form of **polyploidy** in which a plant possesses twice the **diploid** number of chromosomes ($4n$).

Thallose liverwort A **liverwort** (Hepatophyta) in which the **sporophyte** consists of a flattened, lobed structure (the thallus) growing flat on the ground.

Thigmo-seismic effect The physiological responses of plants to shaking or stroking.

Thylakoid One of the membranous discs that makes up a **granum** in a **chloroplast**.

Tipburn Browning and death of leaf tips of lettuce and stored cabbage, resulting from calcium deficiency.

Tissue culture Plant tissue grown in the laboratory on a sterile, synthetic medium containing sugars and **hormones**.

Tonoplast The membrane enclosing a **vacuole** in a plant **cell**.

Topophysis A term used to describe cuttings taken from horizontally growing branches that retain a prostrate habit of growth.

Totipotent A term used to describe a plant **cell** that can be induced to undergo division and produce a complete new individual.

Tracheid An elongated, spindle-shaped dead cell with its walls thickened with bands of **lignin**.

Transcription factors Enzymes involved in **transcription**; i.e. the copying of **DNA** to **RNA** prior to protein synthesis.

Transpiration The movement of water vapour from a plant to the atmosphere, mainly through the **stomata** of the leaves and stems.

Transport proteins Proteins involved in the transport of substances across **membranes** in the **cell**.

Transposon Naturally occurring fragments of **DNA** that move around the **genome** by cutting themselves out of one region of the **DNA** and splicing themselves in, at random, elsewhere. In so doing they may disrupt the function of a **gene** or genes, as in *Rosa mundi* where the red spots and stripes on the petals are caused by the disruption of pigment formation by transposon activity.

Trichome See **epidermal hair.**

Triploid A form of **polyploidy** in which a plant possesses three sets of chromosomes ($3n$).

True breeding Term used to describe a plant which, when self-pollinated, produces progeny similar to the parent.

True fruit A fruit in which the **pericarp** (fruit wall) is derived from the wall of the **ovary**.

Tuber Root tubers consist of swollen portions of roots, while stem tubers consist of swollen portions of stem. Both have a storage function.

Tunica The layer of cells overlaying the **corpus** in the **apical meristem** of flowering plants.

Tunicate Enclosed by layered coats, the outermost being the tunic, as in an onion.

Turgid Term used to describe a **cell** so full of water that further uptake is prevented by the hydrostatic pressure (wall pressure) exerted by the cell wall. See also **turgor pressure.**

Turgor pressure The pressure exerted by the **cell** contents on the **cell wall** When water is lost from the leaves at a greater rate than it can be supplied from the roots, turgor is lost and the plant wilts.

Umbel An **inflorescence** in which the stalked flowers all arise from the tip of the axis.

Vacuole A **membrane**-bound sac, filled with liquid, contained within the **cytoplasm** of a plant cell.

Vapour pressure The absolute amount of water vapour present in the air. Water vapour moves along a concentration gradient from a region of higher vapour pressure to a region of lower vapour pressure.

Vascular bundle A long strand largely composed of **xylem** and **phloem** and therefore responsible for the transport of water, sugars and other substances in the plant.

Vascular rays See **rays**.

Vector Carrier (often an insect or other arthropod, but sometimes a fungus or human) of a virus from one host to another.

Vegetative propagation The propagation of a plant by non-sexual means such as division or by taking cuttings.

Vein In a plant, a visible strand of conducting tissue (**xylem** and **phloem**) in a leaf, petal or other organ.

Velamen A tissue comprising layers of dead cells that surrounds the aerial roots of certain tropical orchids that is involved in the absorption of water vapour.

Vernalisation A term used to describe the increase in flowering that occurs in response to a cold treatment given to seeds or young plants.

Vesicular-arbuscular mycorrhiza (VA mycorrhiza) A form of **endomycorrhiza** in which the fungus may form specialised feeding branches called arbuscules and swollen storage hyphae called vesicles. Now often simply called **arbuscular mycorrhiza**.

Vessel See **xylem**.

Virus A **microorganism**, capable of replication (reproduction), but only in association with another, more complex organism, and having no cellular structure, consisting only of **nucleic acid** (**RNA** or **DNA**) with a protein coat.

Virus infection An infection by a **virus** causing a disease.

Visual purple The light-sensitive pigment, also called **rhodopsin**, in the rods cells of the eye that enables people to see.

Water potential A description of the potential of water to do work. By definition, the potential of pure water is zero, so the water potential in unsaturated soils or plant cells is usually negative.

Wall pressure The inwardly directed pressure exerted by the **cell wall** that balances the **turgor pressure** in a **turgid** cell.

Water stress An insufficient supply of water to the roots or a too rapid loss from the leaves leads to water stress. Many **cell** functions are depressed when plants are subjected to water stress.

Xanthophylls A group of **carotenoid** pigments that have an oxygen atom in addition to carbon and hydrogen. They are responsible for the yellow colours of flowers and fruits.

Xylem A plant tissue involved in the transport of water and in support, in which most of the cells have walls thickened with **lignin**. These cells may be elongate, dead, without end walls and joined end to end, forming **vessels** for the transport of water; or they may be spindle-shaped, dead **tracheids**, involved in the transport of water; or they may be dead fibres, involved in support only.

Zeatin/Zeatin riboside Naturally occurring **cytokinins**.

Zygote The cell containing a **diploid nucleus**, resulting from the fusion of two **haploid gametes** in sexual reproduction, from which the embryo develops.

Index